RANDOM FIELDS ESTIMATION

RANDOM FIELDS ESTIMATION

Alexander G. Ramm
Kansas State University, USA

World Scientific

NEW JERSEY · LONDON · SINGAPORE · BEIJING · SHANGHAI · HONG KONG · TAIPEI · CHENNAI

Published by

World Scientific Publishing Co. Pte. Ltd.
5 Toh Tuck Link, Singapore 596224
USA office: 27 Warren Street, Suite 401-402, Hackensack, NJ 07601
UK office: 57 Shelton Street, Covent Garden, London WC2H 9HE

British Library Cataloguing-in-Publication Data
A catalogue record for this book is available from the British Library.

RANDOM FIELDS ESTIMATION

ISBN 981-256-536-1

Printed in Singapore by World Scientific Printers (S) Pte Ltd

To the memory of my parents

Preface

This book presents analytic theory of random fields estimation optimal by the criterion of minimum of the variance of the error of the estimate. This theory is a generalization of the classical Wiener theory. Wiener's theory has been developed for optimal estimation of stationary random processes, that is, random functions of one variable. Random fields are random functions of several variables. Wiener's theory was based on the analytical solution of the basic integral equation of estimation theory. This equation for estimation of stationary random processes was Wiener-Hopf-type of equation, originally on a positive semiaxis. About 25 years later the theory of such equations has been developed for the case of finite intervals. The assumption of stationarity of the processes was vital for the theory. Analytical formulas for optimal estimates (filters) have been obtained under the assumption that the spectral density of the stationary process is a positive rational function. We generalize Wiener's theory in several directions. First, estimation theory of random fields and not only random processes is developed. Secondly, the stationarity assumption is dropped. Thirdly, the assumption about rational spectral density is generalized in this book: we consider kernels of positive rational functions of arbitrary elliptic self-adjoint operators on the whole space. The domain of observation of the signal does not enter into the definition of the kernel. These kernels are correlation functions of random fields and therefore the class of such kernels defines the class of random fields for which analytical estimation theory is developed. In the appendix we consider even more general class of kernels, namely kernels $R(x, y)$, which solve the equation $QR = P\delta(x - y)$. Here P and Q are elliptic operators, and $\delta(x - y)$ is the delta-function. We study singular perturbation problem for the basic integral equation of estimation theory $Rh = f$. The solution to this equation, which is of interest in estimation theory, is a distribution, in general. The perturbed equation, $\epsilon h_\epsilon + Rh_\epsilon = f$ has the unique solution in $L^2(D)$. The singular perturbation problem consists of the study of the asymptotics of h_ϵ as $\epsilon \to 0$. This theory is not only of mathematical interest, but also a basis for the numerical solution of the basic integral equation in distributions. We discuss the relation between estimation theory and quantum-mechanical non-relativistic scattering theory. Applications of the estimation theory are also discussed. The presentation in this book is based partly on the author's earlier monographs [Ramm (1990)] and [Ramm (1996)], but also contains recent results [Ramm (2002)], [Ramm (2003)], [Kozhevnikov and Ramm (2005)], and [Ramm and Shifrin (2005)].

The book is intended for researchers in probability and statistics, analysis, numerical analysis, signal estimation and image processing, theoretically inclined electrical engineers, geophysicists, and graduate students in these areas. Parts of the book can be used in graduate courses in probabilty and statistics. The analytical tools that the author uses are not usual for statistics and probability. These tools include spectral theory of elliptic operators, pseudodifferential operators, and operator theory. The presentation in this book is essentially self-contained. Auxiliary material which we use is collected in Chapter 8.

Contents

Chapter 1

Introduction

This work deals with just one topic: analytic theory of random fields estimation within the framework of covariance theory. No assumptions about distribution laws are made: the fields are not necessarily Gaussian or Markovian. The only information used is the covariance functions. Specifically, we assume that the random field is of the form

$$\mathcal{U}(x) = s(x) + n(x), \quad x \in R^r, \tag{1.1}$$

where $s(x)$ is the useful signal and $n(x)$ is noise. Without loss of generality assume that

$$\overline{s(x)} = \overline{n(x)} = 0, \tag{1.2}$$

where the bar denotes the mean value. If these mean values are not zeros then one either assumes that they are known and considers the fields $s(x) - \overline{s(x)}$ and $n(x) - \overline{n(x)}$ with zero mean values, or one estimates the mean values and then subtracts them from the corresponding fields. We also assume that the covariance functions

$$\overline{\mathcal{U}^*(x)\mathcal{U}(y)} := R(x,y), \quad \overline{\mathcal{U}^*(x)s(y)} := f(x,y) \tag{1.3}$$

are known. The star stands for complex conjugate. This information is necessary for any development within the framework of covariance theory. We will show that, under some assumptions about functions (1.3), one can develop an analytic theory of random fields estimation. If the functions (1.3) are not known then one has to estimate them from statistical data or from some theory. In many applications the exact analytical expression for the covariance functions is not very important, but rather some general features of R or f are of practical interest. These features include, for example, the correlation radius.

The estimation problem of interest is the following one. The signal $u(x)$ of the form (1.1) is observed in a domain $D \subset R^r$ with the boundary Γ. Assuming (1.2) and (1.3) one needs to linearly estimate $As(x_0)$ where A is a given operator and $x_0 \in R^r$ is a given point. The linear estimate is to be best possible by the criterion of minimum of variance, i.e. the estimate is by the least squares method. The most general form of a linear estimate of u observed in the domain D is

$$L\mathcal{U} := \int_D h(x,y)\mathcal{U}(y)dy \tag{1.4}$$

where $h(x,y)$ is a distribution. Therefore the optimal linear estimate solves the variational problem

$$\epsilon := \overline{(L\mathcal{U} - As)^2} = \min \tag{1.5}$$

where Lu and As are computed at the point x_0. A necessary condition on $h(x,y)$ for (1.5) (with $A = I$) to hold is (see equations (2.11) and (8.423))

$$Rh := \int_D R(x,y)h(z,y)dy = f(x,z), \quad x,z \in \overline{D} := D \cup \Gamma. \tag{1.6}$$

The basic topic of this work is the study of a class of equations (1.6) for which the analytical properties of the solution h can be obtained, a numerical procedure for computing h can be given, and properties of the operator R in (1.6) can be studied. Since z enters (1.6) as a parameter, one can study the basic equation of estimation theory

$$Rh := \int_D R(x,y)h(y)dy = f(x), \quad x \in \overline{D}. \tag{1.7}$$

A typical one-dimensional example of the equation (1.7) in estimation theory is

$$\int_{-1}^{1} \exp(-|x-y|)h(y)dy = f(x), \quad -1 \le x \le 1. \tag{1.8}$$

Its solution of minimal order of singularity is

$$h(x) = (-f''+f)/2 + \delta(x+1)[-f'(-1)+f(-1)]/2 + \delta(x-1)[f'(1)+f(1)]/2. \tag{1.9}$$

One can see that the solution is a distribution with singular support at the boundary of the domain $\mathcal{D}$. By sing supp h we mean the set having no open neigborhood to which the restriction of h can be identified with a locally integrable function.

In the case of equation (1.8) this domain $\mathcal{D} = (-1, 1)$. Even if $f \in C^\infty(D)$ the solution to equations (1.7), (1.8) are, in general, not in $L^2(D)$.

The problem is: in what functional space should one look for the solution? Is the solution unique? Does the solution to (1.7) provide the solution to the estimation problem (1.5)? Does the solution depend continuously on the data, e.g. on f and on $R(x, y)$? How does one compute the solution analytically and numerically? What are the properties of the solution, for example, what is the order of singularity of the solution? What is the singular support of the solution? What are the properties of the operator R as an operator in $L^2(D)$?

These questions are answered in Chapters 2-4.

The answers are given for the class of random fields whose covariance functions $R(x, y)$ are kernels of positive rational functions of selfadjoint elliptic operators in $L^2(R^r)$. The class $\mathcal{R}$ of such kernels consists of kernels

$$R(x, y) = \int_\Lambda P(\lambda) Q^{-1}(\lambda) \Phi(x, y, \lambda) d\rho(\lambda) \tag{1.10}$$

where Λ, $d\rho$, $\Phi(x, y, \lambda)$ are respectively the spectrum , spectral measure and spectral kernel of an elliptic selfadjoint operator $\mathcal{L}$ in $L^2(R^r)$ of order s, and $P(\lambda)$ and $Q(\lambda)$ are positive polynomials of degrees p and q respectively. The notions of spectral measure and spectral kernel are discussed in Section 8.2. If $p > q$ then the operator in $L^2(D)$ with kernel (1.10) is an elliptic integro-differential operator R; if $p = q$ then $R = cI + \mathcal{K}$, where $c = \text{const} > 0$, I is the identity operator, and $\mathcal{K}$ is a compact selfadjoint operator in $L^2(D)$; if $p < q$, which is the most interesting case, then R is a compact selfadjoint operator in $L^2(D)$. In this case the noise $n(x)$ is called colored. If $\phi(\lambda)$ is a measurable function then the kernel of the operator $\phi(\mathcal{L})$ is defined by the formula

$$\phi(\mathcal{L})(x, y) = \int_\Gamma \phi(\lambda) \Phi(x, y, \lambda) d\rho(\lambda). \tag{1.11}$$

The domain of definition of the operator $\phi(\mathcal{L})$ consists of all functions $f \in L^2(R^r)$ such that

$$\int_\Gamma |\phi(\lambda)|^2 d(E_\lambda f, f) < \infty \tag{1.12}$$

where E_λ is the resolution of the identity for $\mathcal{L}$. It is a projection operator with the kernel

$$E_\lambda(x, y) = \int_{-\infty}^{\lambda} \Phi(x, y, \lambda) d\rho(\lambda). \tag{1.13}$$

In particular, since $E_{+\infty} = I$, one has

$$\delta(x - y) = \int_{-\infty}^{\infty} \Phi(x, y, \lambda) d\rho(\lambda). \tag{1.14}$$

In (1.13) and (1.14) the integration is actually taken over $(-\infty, \lambda) \cap \Lambda$ and $(-\infty, \infty) \cap \Lambda$ respectively, since $d\rho = 0$ outside Λ. The kernel in (1.8) corresponds to the simple case $r = 1$, $\mathcal{L} = -i\frac{d}{dx}$, $\Lambda = (-\infty, \infty)$, $d\rho = d\lambda$, $\Phi(x, y, \lambda) = (2\pi)^{-1} \exp\{i\lambda(x - y)\}$, $P(\lambda) = 1$, $Q(\lambda) = (\lambda^2 + 1)/2$, $e^{-|x|} = (2\pi)^{-1} \int_{-\infty}^{\infty} \left(\frac{\lambda^2+1}{2}\right)^{-1} \exp(i\lambda x) d\lambda$, and formula (1.9) is a very particular case of the general formulas given in Chapter 2. Let $R(x, y) \in \mathcal{R}$, $\alpha := \frac{1}{2}s(q - p)$, $H^\ell(D)$ be the Sobolev spaces and $\dot{H}^{-\ell}(D)$ be its dual with respect to $H^0(D) = L^2(D)$. Then the answers to the questions, formulated above, are as follows.

The solution to equation (1.7) solves the estimation problem if and only if $h \in \dot{H}^{-\alpha}(D)$. The operator $R : \dot{H}^{-\alpha}(D) \to H^\alpha(D)$ is an isomorphism. The singular support of the solution $h \in \dot{H}^{-\alpha}(D)$ of equation (1.7) is $\Gamma = \partial D$. The analytic formula for h is of the form $h = Q(\mathcal{L})G$, where G is a solution to some interface elliptic boundary value problem and the differentiation is taken in the sense of distributions.

Exact description of this analytic formula is given in Chapter 2. The spectral properties of the operator $R : L^2(D) \to L^2(D)$ with the kernel $R(x, y) \in \mathcal{R}$ are given also in Chapter 2. These properties include asymptotics as $n \to \infty$ of the eigenvalues λ_n of R, dependence λ_n on D, and asymptotics of $\lambda_1(D)$ as $D \to R^r$, that is D growing uniformly in directions. Numerical methods for solving equation (1.7) in the space $\dot{H}^{-\alpha}(D)$ of distributions are given in Chapter 3. These methods depend heavily on the analytical results given in Chapter 2.

The necessary background material on Sobolev spaces and spectral theory is given in Chapter 8 so that the reader does not have to consult the literature in order to understand the contents of this work.

No attempts were made by the author to present all aspects of the theory of random fields. There are several books [Adler (1981)], [Yadrenko (1983)], [Vanmarcke (1983)], [Rosanov (1982)] and [Preston (1967)], and many papers on various aspects of the theory of random fields. They have practically no intersection with this work which can be viewed as an extension of Wiener's filtering theory. The statement of the problem is the same as in Wiener's theory, but we study random functions of several variables, that is random fields, while Wiener (and many researchers after him) studied filtering and extrapolation of stationary random processes, that is random functions of one variable. Wiener's basic assumptions were:

1) the random process $u(t) = s(t) + n(t)$ is stationary,
2) it is observed on the interval $(-\infty, T)$,
3) it has a rational spectral density (this assumption can be relaxed, but for effective solution of the estimation problems it is quite useful).

The first assumption means that $R(t, \tau) = R(t - \tau)$, where R is the covariance function (1.3). The second one means that $D = (-\infty, T)$. The third one means $\tilde{R}(\lambda) = P(\lambda)Q^{-1}(\lambda)$, where $P(\lambda)$ and $Q(\lambda)$ are polynomials, $\tilde{R}(\lambda) \geq 0$ for $-\infty < \lambda < \infty$, $\tilde{R}(\lambda) := \int_{-\infty}^{\infty} R(t)\exp(-i\lambda t)dt$. The analytical theory used by Wiener is the theory of Wiener-Hopf equations. Later the Wiener theory was extended to the case $D = [T_1, T]$ of a finite interval of observation, while assumptions 1) and 3) remained valid. A review of this theory with many references is [Kailath (1974)].

Although the literature on filtering and estimation theory is large (dozens of books and hundreds of papers are mentioned in [Kailath (1974)]), the analytic theory presented in this work and developed in the works of the author cited in the references has not been available in book form in its present form, although a good part of it appeared in [Kato (1995), Ch. 1]. Most of the previously known analytical results on Wiener-Hopf equations with rational $\tilde{R}(\lambda)$ are immediate and simple consequences of our general theory. Engineers can use the theory presented here in many applications. These include signal and image processing in TV, underwater acoustics, geophysics, optics, etc. In particular, the following question of long standing is answered by the theory given here. Suppose a random field (1.1) is observed in a ball B and one wants to estimate $s(x_0)$, where x_0 is the center of B. What is the optimal size of the radius of B? If the radius is too small then the estimate is not accurate. If it is too large then the estimate is not better than the one obtained from the observations in a ball of smaller radius, so that the efforts are wasted. This problem is of practical importance in many applications.

We will briefly discuss some other applications of the estimation theory, for example, discrimination of hypotheses, resolution ability of linear systems, estimation of derivatives of random functions, etc. However, the emphasis is on the theory, and the author hopes that other scientists will pursue further possible applications.

Numerical solution of the basic integral equation of estimation theory was widely discussed in the literature [Kailath (1974)] in the case of random processes ($r = 1$), mostly stationary that is when $R(x, y) = R(x - y)$, and mostly in the case when the noise is white, so that the integral equation for the optimal filter is

$$(I + R)h := h(t) + \int_0^T R_s(t - \tau)h(\tau)d\tau = f(t), \quad 0 \le t \le T, \tag{1.15}$$

where R_s is the covariance function of the useful signal $s(t)$. Note that the integral operator R in (1.10) is selfadjoint and nonnegative in $L^2[0, T]$. Therefore $(I + T)^{-1}$ exists and is bounded, and the numerical solution of (1.15) is not difficult. Many methods are available for solving one-dimensional second order Fredholm integral equation (1.15) with positive-definite operator $I + R$. Iterative methods, projection methods, colloqation and many other methods are available for solving (1.15), convergence of these methods has been proved and effective error estimates of the numerical methods are known [Kantorovich and Akilov (1980)]. Much effort was spent on effective numerical inversion of Toeplitr matrices which one obtains if one discretizes (1.15) using equidistant colloqation points [Kailath (1974)]. However, if the noise is colored, the basic equation becomes

$$\int_0^T R(t - \tau)h(\tau)d\tau = f(t), \quad 0 \le t \le T. \tag{1.16}$$

This is a Fredholm equation of the first kind. A typical example is equation (1.8). As we have seen in (1.9), the solution to (1.8) is a distribution, in general. The theory for the numerical treatment of such equation was given by the author [Ramm (1985)] and is presented in Chapter 3 of this book. In particular, the following question of singular perturbation theory is of interest. Suppose that equation

$$\epsilon h_\epsilon + Rh_\epsilon = f, \quad \epsilon > 0, \tag{1.17}$$

is given. This equation corresponds to the case when the intensity of the white-noise component of the noise is ϵ. *What is the behavior of h_ϵ when $\epsilon \to +0$?* We will answer this question in Chapter 3.

This book is intended for a broad audience: for mathematicians, engineers interested in signal and image processing, geophysicists, etc. Therefore the author separated formulation of the results, their discussion and examples from proofs. In order to understand the proofs, one should be familiar with some facts and ideas of functional analysis. Since the author wants to give a relatively self-contained presentation, the necessary facts from functional analysis are presented in Chapter 8.

The book presents the theory developed by the author. Many aspects of estimation theory are not discussed in this book. The book has practically no intersection with works of other authors on random fields estimation theory.

Chapter 2

Formulation of Basic Results

2.1 Statement of the problem

Let $D \subset R^r$ be a bounded domain with a sufficiently smooth boundary Γ. The requirement that D is bounded could be omitted, it is imposed for simplicity. The reader will see that if D is not bounded then the general line of the arguments remains the same. The additional difficulties, which appear in the case when D is unbounded, are of technical nature: one needs to establish existence and uniqueness of the solution to a certain transmission problem with transmission conditions on Γ. Also the requirement of smoothness of Γ is of technical nature: the needed smoothness should guarantee existence and uniqueness of the solution to the above transmission problem.

Let $\mathcal{L}$ be an elliptic selfadjoint in $H = L^2(R^r)$ operator of order s. Let Λ, $\Phi(x, y, \lambda)$, $d\rho(\lambda)$ be the spectrum, spectral kernel and spectral measure of $\mathcal{L}$, respectively. A function $F(\mathcal{L})$ is defined as an operator on H with the kernel

$$F(\mathcal{L})(x, y) = \int_\Lambda F(\lambda)\Phi(x, y, \lambda)d\rho(\lambda) \tag{2.1}$$

and $\text{dom}F(\mathcal{L}) = \{f : f \in H, \int_{-\infty}^{\infty} |F(\lambda)|^2 d(E_\lambda f, f) < \infty\}$, where

$$(E_\lambda f, f) = \int_{-\infty}^{\lambda} \left\{ \iint \Phi(x, y, \mu) f(y)\overline{f(y)}dxdy \right\} d\rho(\mu), \quad \int := \int_{R^r}. \tag{2.2}$$

Definition 2.1 *Let $\mathcal{R}$ denote the class of kernels of positive rational functions of $\mathcal{L}$, where $\mathcal{L}$ runs through the set of all selfadjoint elliptic operators* *in* $H = L^2(R^r)$. *In other words,* $R(x,y) \in \mathcal{R}$ *if and only if*

$$R(x,y) = \int_\Lambda P(\lambda)Q^{-1}(\lambda)\Phi(x,y,\lambda)d\rho(\lambda), \tag{2.3}$$

where $P(\lambda) > 0$ *and* $Q(\lambda) > 0$, $\forall \lambda \in \Lambda$, *and* Λ, Φ, $d\rho$ *correspond to an elliptic selfadjoint operator* $\mathcal{L}$ *in* $H = L^2(R^r)$.

Let

$$p = \deg P(\lambda), \quad q = \deg Q(\lambda), \quad s = \operatorname{ord}\mathcal{L}, \tag{2.4}$$

where $\deg P(\lambda)$ stands for the degree of the polynomial $P(\lambda)$, and $\operatorname{ord}\mathcal{L}$ stands for the order of the differential operator $\mathcal{L}$.

An operator given by the differential expression

$$\mathcal{L}u := \sum_{|j|\le s} a_j(x)\partial^j u, \tag{2.5}$$

where $j = (j_1, j_2 \dots j_r)$ is a multiindex, $\partial^j u = \partial_{x_1}^{j_1}\partial_{x_2}^{j_2}\dots\partial_{x_r}^{j_r}u$ $|j| = j_1 + j_2 + \dots j_r$, $j_m \ge 0$, are integers. The expression (2.5) is called elliptic if, for any real vector $t \in R^r$, the equation

$$\sum_{|j|=s} a_j(x)t^j = 0$$

implies that $t = 0$. The expression

$$\mathcal{L}^+u := \sum_{|j|\le s} (-1)^{|j|}\partial^j(a_j^*(x)u) \tag{2.6}$$

is called the formal adjoint with respect to $\mathcal{L}$. The star in (2.6) stands for complex conjugate. One says that $\mathcal{L}$ is formally selfadjoint if $\mathcal{L} = \mathcal{L}^+$. If $\mathcal{L}$ is formally selfadjoint then $\mathcal{L}$ is symmetric on $C_0^\infty(R^r)$, that is $(\mathcal{L}\phi, \psi) = (\phi, \mathcal{L}\psi)$ $\forall \phi, \psi \in C_0^\infty(R^r)$, where (ϕ, ψ) is the inner product in $H = L^2(R^r)$. Sufficient conditions on $a_j(x)$ can be given for a formally selfadjoint differential expression to define a selfadjoint operator in H in the following way. Define a symmetric operator $\mathcal{L}_0$ with the domain $C_0^\infty(R^r)$ by the formula $\mathcal{L}_0 u = \mathcal{L}u$ for $u \in C_0^\infty(R^r)$. Under suitable conditions on $a_j(x)$ one can prove that $\mathcal{L}_0$ is essentially selfadjoint, that is, its closure is selfadjoint (see Chapter 8). In particular this is the case if $a_j = a_j^* = \text{const}$.

In what follows we assume that $R(x,y) \in \mathcal{R}$. Some generalizations will be considered later. The kernel $R(x,y)$ is the covariance function (1.3) of the random field $\mathcal{U}(x) = s(x) + n(x)$ observed in a bounded domain $D \subset R^r$.

A linear estimation problem can be formulated as follows: find a linear estimate

$$\hat{\mathcal{U}} := L\mathcal{U} := \int_D h(x,y)\mathcal{U}(y)dy \tag{2.7}$$

such that

$$\epsilon := \overline{|\hat{\mathcal{U}} - As|^2} = \min. \tag{2.8}$$

The kernel $h(x,y)$ in (2.7) is a distribution, so that, by L. Schwartz's theorem about kernels, estimate (2.7) is the most general linear estimate. The operator A in (2.8) is assumed to be known. It is an arbitrary operator, not necessarily a linear one. In the case when $A\mathcal{U} = \mathcal{U}$, that is $A = I$, where I is the identity operator, the estimation problem (2.8) is called the filtering problem. From (2.8) and (2.7) one obtains

$$\begin{aligned}
\epsilon &= \overline{\int_D h(x,y)\mathcal{U}(y)dy \int_D h^*(x,z)\mathcal{U}^*(z)dz} \\
&\quad - 2Re\overline{\int_D h(x,z)\mathcal{U}(z)dz(As)^*(x)} \quad + \quad \overline{|As(x)|^2} \\
&= \int_D \int_D h(x,y)h^*(x,z)R(z,y)dzdy \\
&\quad - 2Re \int_D h^*(x,z)f(z,x)dz \quad + \quad \overline{|As(x)|^2} \\
&= \min.
\end{aligned} \tag{2.9}$$

Here

$$f(y,x) := \overline{\mathcal{U}^*(y)(As(x))} = f^*(x,y), \tag{2.10}$$

the bar stands for the mean value and the star stands for complex conjugate. By the standard procedure one finds that a necessary condition for the minimum in (2.9) is:

$$\int_D R(z,y)h(x,y)dy = f(z,x), \quad x,z \in \overline{D} := D \cup \Gamma. \tag{2.11}$$

In order to derive (2.11) from (2.9) one takes $h + \alpha\eta$ in place of h in (2.9). Here α is a small number and $\eta \in C_0^\infty(D)$. The condition $\epsilon(h) \le \epsilon(h + \alpha\eta)$ implies $\frac{\partial\epsilon}{\partial\alpha}\big|_{\alpha=0} = 0$. This implies (2.11). Since h is a distribution, the left-hand side of (2.11) makes sense only if the kernel $R(z,y)$ belongs to the space of test functions on which the distribution h is defined. We will discuss this point later in detail. In (2.11) the variable x enters as a parameter. Therefore, the equation

$$Rh := \int_D R(x,y)h(y)dy = f(x), \quad x \in \overline{D}, \tag{2.12}$$

is basic for estimation theory. We have supressed the dependence on x in (2.11) and have written x in place of z in (2.12).

From this derivation it is clear that the operator A does not influence the theory in an essential way: if one changes A then f is changed but the kernel of the basic equation (2.12) remains the same. If $Au = \frac{\partial u}{\partial x_j}$, one has the problem of estimating the derivative of u. If $Au = u(x + x_0)$, where x_0 is a given point such that $x + x_0 \notin D$, then one has the extrapolation problem. Analytically these problems reduce to solving equation (2.12). If no assumptions are made about $R(x,y)$ except that $R(x,y)$ is a covariance function, then one cannot develop an analytical theory for equation (2.12). Such a theory will be developed below under the basic assumption $R(x,y) \in \mathcal{R}$.

Let us show that the class $\mathcal{R}$ of kernels, that is the class of random fields that we have introduced, is a natural one. To see this, recall that in the one-dimensional case, studied analytically in the literature, the covariance functions are of the form

$$R(x,y) = R(x-y), x, y \in R^1,$$

$$\tilde{R}(\lambda) := \int_{\infty}^{\infty} R(x)\exp(-i\lambda x)dx = P(\lambda)Q^{-1}(\lambda),$$

where $P(\lambda)$ and $Q(\lambda)$ are positive polynomials [Kai]. This case is a very particular case of the kernels in the class $\mathcal{R}$. Indeed, take

$$r = 1, \quad \mathcal{L} = -i\frac{d}{dx}, \quad \Lambda = (-\infty, \infty), \quad d\rho(\lambda) = d\lambda,$$

$$\Phi(x,y,\lambda) = (2\pi)^{-1}\exp\{i\lambda(x-y)\}.$$

Then formula (2.3) gives the above class of convolution covariance functions with rational Fourier transforms. If $p = q$, where p and q are defined in (2.4), then the basic equation (2.12) can be written as

$$Rh := \sigma^2 h(x) + \int_D R_1(x,y)h(y)dy = f(x), \quad x \in \overline{D}, \quad \sigma^2 > 0, \tag{2.13}$$

where

$$P(\lambda)Q^{-1}(\lambda) = \sigma^2 + P_1(\lambda)Q^{-1}(\lambda), \quad p_1 := \deg P_1 < q, \tag{2.14}$$

and $\sigma^2 > 0$ is interpreted as the variance of the white noise component of the observed signal $\mathcal{U}(x)$. If $p < q$, then the noise in $\mathcal{U}(x)$ is colored, it does not contain a white noise component.

Mathematically equation (2.13) is very simple. The operator R in (2.13) is of Fredholm type, selfadjoint and positive definite in H, $R \geq \sigma^2 I$, where I is the identity operator, and $A \geq B$ means $(Au, u) \geq (Bu, u)$ $\forall u \in H$. Therefore, if $p = q$ then equation (2.12) reduces to (2.13) and has a unique solution in H. This solution can be computed numerically without difficulties. There are many numerical methods which are applicable to equation (2.13). In particular (see Section 2.3.2), an iterative process can be constructed for solving (2.13) which converges as a geometrical series; a projection method can be constructed for solving (2.13) which converges and is stable computationally; one can solve (2.13) by collocation methods. However the important and practically interesting question is the following one: what happens with the solution h_σ to (2.13) as $\sigma \to 0$? This is a singular perturbation theory question: for $\sigma > 0$ the unique solution to equation (2.13) belongs to $L^2(D)$, while for $\sigma = 0$ the unique solution to (2.13) of minimal order of singularity is a distribution. What is the asymptotics of h_σ as $\sigma \to 0$? As we will show, the answer to this question is based on analytical results concerning the solution to (2.13).

The basic questions we would like to answer are:

1) **In what space of functions or distributions should one look for the solution to (2.12)?**
2) **When does a solution to (2.12) solve the estimation problem (2.8)?**

Note that (2.12) is only a necessary condition for $h(x, y)$ to solve (2.8). We will show that there is a solution to (2.12) which solves (2.8) and this solution to (2.12) is unique. The fact that estimation problem (2.8) has a unique solution follows from a Hilbert space interpretation of problem (2.8) as the problem of finding the distance from the element $(As)(x)$ to the subspace spanned by the values of the random field $u(y)$, $y \in D$. Since there exists and is unique an element in a subspace at which the distance is attained, problem (2.8) has a solution and the solution is unique. It was mentioned in the Introduction (see (1.9)) that equation (2.12) may have no solutions in $L^1(D)$ but rather its solution is a distribution. There can be several solutions to (2.12) in spaces of distributions, but only one of them solves the estimation problem (2.8). *This solution is characterized as the solution to (2.12) of minimal order of singularity.*

3) **What is the order of singularity and singular support of the solution to (2.12) which solves (2.8)?**
4) **Is this solution stable under small perturbations of the data, that is under small perturbations of $f(x)$ and $R(x,y)$? What is the appropriate notion of smallness in this case? What are the stability estimates for h?**
5) **How does one compute the solution analytically?**
6) **How does one compute the solution numerically?**
7) **What are the properties of the operator $R : L^2(D) \to L^2(D)$ in (2.12)? In particular, what is the asymptotics of its eigenvalues $\lambda_j(D)$ as $j \to +\infty$? What is the asymptotics of $\lambda_1(D)$ as $D \to R^r$, that is, as D expands uniformly in directions?**

These questions are of interest in applications. Note that if D is finite then the operator R is selfadjoint positive compact operator in $L^2(D)$, its spectrum is discrete, $\lambda_1 > \lambda_2 \geq \cdots > 0$, the first eigenvalue is nondegenerate by Krein-Rutman's theorem. However, if $D = R^r$ then the spectrum of R may be continuous, e.g., this is the case when $R(x,y) = R(x-y)$, $\tilde{R}(\lambda) = P(\lambda)Q^{-1}(\lambda)$. Therefore it is of interest to find $\lambda_{1\infty} := \lim \lambda_1(D)$ as $D \to R^r$. The quantity $\lambda_{1\infty}$ is used in some statistical problems.

8) **What is the asymptotics of the solution to (2.13) as $\sigma \to 0$?**

2.2 Formulation of the results (multidimensional case)

2.2.1 *Basic results*

We assume throughout that $R(x,y) \in \mathcal{R}$, $f(x)$ is smooth, more precisely, that $f \in H^\alpha$, $H^\alpha = H^\alpha(D)$is the Sobolev space, $\alpha := (q-p)s/2$, where s, q, p are the same as in (2.4), and the coefficients $a_j(x)$ of $\mathcal{L}$ (see (2.5)) are sufficiently smooth, say $a_s(x) \in C(R^r)$, $a_j(x) \in L^{r/(s-|\alpha|)}_{\text{loc}}$ if $s - |j| < r/2$, $a_j \in L^2_{\text{loc}}$ if $s - |j| > r/2$, $a_j \in L^{2+\epsilon}_{\text{loc}}$, $\epsilon > 0$, if $s - |j| = r/2$ (see [Hörmander (1983-85), Ch 17]).

If $q \leq p$, then the problem of finding the is simple: such a solution does exist, is unique, and belongs to $H^{m+2|\alpha|}$ if $f \in H^m$. This follows from the usual theory of elliptic boundary value problems [Berezanskij (1968)], since the operator $P(\mathcal{L})Q^{-1}(\mathcal{L})$ is an elliptic integral-differential operator of order $2|\alpha|$ if $q \leq p$. The solution satisfies the elliptic estimate:

$$\| h \|_{H^{m+2|\alpha|}} \leq c \| f \|_{H^m}, \quad q \leq p,$$

where c depends on $\mathcal{L}$ but does not depend on f.

If $q > p$, then the problem of finding the mos solution of (2.12) is more interesting and difficult because the order of singularity of h, is in general, positive, ord$h = \alpha$. The basic result we obtain is:

the mapping $R : \dot{H}^{-\alpha} \to H^{\alpha}$ is a linear isomorphism between the spaces $\dot{H}^{-\alpha}$ and H^{α}. The of h is $\partial D = \Gamma$, provided that f is smooth.

If h_1 is a solution to equation (2.12) and ord$h_1 > \alpha$ then $\epsilon = \infty$, where ϵ is defined in (2.8). Therefore if h_1 solves (2.12) and ord$h_1 > \alpha$ then h_1 does not solve the estimation problem (2.8).

The unique solution to (2.8) is the unique mos solution to (2.12). We give analytical formulas for the mos solution to (2.12). This solution is stable towards small perturbations of the data. We also give a stable numerical procedure for computing this solution.

In this section we formulate the basic results.

Theorem 2.1 *If $R(x,y) \in \mathcal{R}$, then the operator R in (2.12) is an isomorphism between the spaces $\dot{H}^{-\alpha}$ and H^{α}. The solution to (2.12) of minimal order of singularity, ord$h \leq \alpha$, can be calculated by the formula:*

$$h(x) = Q(\mathcal{L})G, \tag{2.15}$$

where

$$G(x) = \begin{cases} g(x) + v(x) & \text{in } D \\ u(x) & \text{in } \Omega := R^r \setminus D, \end{cases} \tag{2.16}$$

$g(x) \in H^{s(p+q)/2}$ is an arbitrary fixed solution to the equation

$$P(\mathcal{L})g = f \quad \text{in} \quad D, \tag{2.17}$$

and the functions $u(x)$ and $v(x)$ are the unique solution to the following (2.18)-(2.20):

$$Q(\mathcal{L})u = 0 \quad \text{in} \quad \Omega, \quad u(\infty) = 0, \tag{2.18}$$

$$P(\mathcal{L})v = 0 \quad \text{in} \quad D, \tag{2.19}$$

$$\partial_N^j u = \partial_N^j (v+g) \quad on \quad \Gamma, \quad 0 \le j \le \frac{s(p+q)}{2} - 1. \tag{2.20}$$

By $u(\infty) = 0$ we mean $\lim u(x) = 0$ as $|x| \to \infty$.

Corollary 2.1 *If $f \in H^{2\beta}$, $\beta \ge \alpha$, then*

$$sing\ supph = \Gamma. \tag{2.21}$$

Corollary 2.2 *If $P(\lambda) = 1$, then the transmission problem (2.18)-(2.20) reduces to the Dirichlet problem in Ω:*

$$Q(\mathcal{L})u = 0 \quad in \quad \Omega, \quad u(\infty) = 0, \tag{2.22}$$

$$\partial_N^j u = \partial_N^j f \quad on \quad \Gamma, \quad 0 \le j \le \frac{sq}{2} - 1, \tag{2.23}$$

and (2.15) takes the form

$$h = Q(\mathcal{L})F, \quad F = \begin{cases} f & in\ D \\ u & in\ \Omega. \end{cases} \tag{2.24}$$

Corollary 2.1 follows immediately from formulas (2.15) and (2.16) since $g(x)+v(x)$ and $u(x)$ are smooth inside D and Ω respectively. Corollary 2.2 follows immediately from Theorem 2.1: if $P(\lambda) = 1$ then $g = f$, $v = 0$, and $p = 0$.

Let $\omega(\lambda) \ge 0$, $\omega(\lambda) \in C(R^1)$, $\omega(\infty) = 0$,

$$\omega := \max_{\lambda \in \Lambda} \omega(\lambda), \tag{2.25}$$

$$R(x,y) = \int_\Lambda \omega(\lambda)\Phi(x,y,\lambda)d\rho(\lambda), \tag{2.26}$$

$\lambda_j = \lambda_j(D)$ be the eigenvalues of the operator $R : L^2(D) \to L^2(D)$ be the eigenvalues of the operator $R : L^2(D) \to L^2(D)$ with kernel (2.26), arranged so that

$$\lambda_1 \ge \lambda_2 \ge \lambda_3 \ge \cdots > 0. \tag{2.27}$$

Theorem 2.2 *If $D \subset D'$ then $\lambda_j \le \lambda'_j$, where $\lambda'_j = \lambda_j(D')$. If*

$$\sup_{x \in R^r} \int |R(x,y)|dy := A < \infty, \tag{2.28}$$

then

$$\lambda_{1\infty} = \omega, \tag{2.29}$$

where

$$\lim_{D \to R^r} \lambda_1(D) := \lambda_{1\infty}, \tag{2.30}$$

and ω is defined in (2.25).

Theorem 2.3 *If $\omega(\lambda) = |\lambda|^{-a}(1+o(1))$ as $|\lambda| \to \infty$, and $a > 0$, then the asymptotics of the eigenvalues of the operator R with kernel (2.26) is given by the formula:*

$$\lambda_j \sim cj^{-as/r} \quad as \quad j \to \infty, \quad c = const > 0, \tag{2.31}$$

where $c = \gamma^{as/r}$ and

$$\gamma := (2\pi)^{-r} \int_D \eta(x)dx, \tag{2.32}$$

with

$$\eta(x) := meas\{t : t \in R^r, \sum_{|\alpha|=|\beta|=s/2} a_{\alpha\beta}(x)t^{\alpha+\beta} \leq 1\}. \tag{2.33}$$

Here the form $a_{\alpha\beta}(x)$ generates the principal part of the selfadjoint elliptic operator $\mathcal{L}$:

$$\mathcal{L}u = \sum_{|\alpha|=|\beta|=s/2} \partial^\alpha(a_{\alpha\beta}(x))\partial^\beta u + \mathcal{L}_1, \quad ord\mathcal{L}_1 < s.$$

Corollary 2.3 *If $\omega(\lambda) = P(\lambda)Q^{-1}(\lambda)$ then $a = q - p$, where $q = \deg Q$, $p = \deg P$, and $\lambda_n \sim cn^{-(q-p)s/r}$, where λ_n are the eigenvalues of the operator in equation (2.12) .*

This Corollary follows immediately from Theorem 2.3. Theorems 2.1-2.3 answer questions 1)-5) and 7) in section 2.1. Answers to questions 6) and 8) will be given in Chapter 3. Proof of Theorem 2.3 is given in Section 8.3.2.10.

2.2.2 *Generalizations*

First, let us consider a generalization of the class $\mathcal{R}$ of kernels for the case when there are several commuting differential operators. Let $\mathcal{L}_1, \dots \mathcal{L}_m$ be a system of commuting selfadjoint differential operators in $L^2(R^r)$. There exists a $d\mu(\xi)$ and a spectral kernel $\Phi(x,y,\xi)$, $\xi = (\xi_1,\dots,\xi_m)$ such that a function $F(\mathcal{L}_1,\dots\mathcal{L}_m)$ is given by the formula

$$F(\mathcal{L}_1,\dots\mathcal{L}_m) = \int_M F(\xi)\phi(\xi)d\mu(\xi), \tag{2.34}$$

where $\Phi(\xi)$ is the operator with kernel $\Phi(x,y,\xi)$. The domain of definition of the operator $F(L_1,\dots L_m)$ is the set of all functions $u \in L^2(R^r)$ for which $\int_M |F(\xi)|^2(\Phi(\xi)u,u)d\mu < \infty$, M is the support of the spectral measure $d\mu$, and the parentheses denote the inner product in $L^2(R^r)$.

For example, let $m = r$, $\mathcal{L}_j = -i\frac{\partial}{\partial x_j}$. Then $\xi = (\xi_1,\dots\xi_r)$, $d\mu = d\xi_1\dots d\xi_r$, $\phi(x,y,\xi) = (2\pi)^{-r}\exp\{i\xi\cdot(x-y)\}$, where dot denotes the inner product in R^r.

If $F(\xi) = P(\xi)Q^{-1}(\xi)$, where $P(\xi)$ and $Q(\xi)$ are positive polynomials and the operators $P(\mathcal{L}) := P(\mathcal{L}_1,\dots\mathcal{L}_r)$ and $Q(\mathcal{L}) := Q(\mathcal{L}_1,\dots\mathcal{L}_r)$ are elliptic of orders m and n respectively, $m < n$, then the theorems, analogous to Theorems 2.1-2.2, hold with $sp = m$ and $sq = n$. Theorem 2.3 has also an analogue in which $as = n - m$ in formula (2.31).

Another generalization of the class $\mathcal{R}$ of kernels is the following one. Let $Q(x,\partial)$ and $P(x,\partial)$ be elliptic differential operators and

$$QR = P\delta(x-y) \quad \text{in} \quad R^r. \tag{2.35}$$

Note that the kernels $R \in \mathcal{R}$ satisfy equation (2.35) with $Q = Q(\mathcal{L})$, $P = P(\mathcal{L})$. Let ord$Q = n$, ord$P = m$, $n > m$. Assume that the transmission problem (2.18)-(2.20), with $Q(x,\partial)$ and $P(x,\partial)$ in place of $Q(\mathcal{L})$ and $P(\mathcal{L})$ respectively, and $ps = m$, $qs = n$, has a unique solution in $H^{(n+m)/2}$. Then Theorem 2.1 holds with $\alpha = (n-m)/2$.

The transmission problem (2.18)-(2.20) with $Q(x,\partial)$ and $P(x,\partial)$ in place of $Q(\mathcal{L})$ and $P(\mathcal{L})$ is uniquely solvable provided that, for example, $Q(x,\partial)$ and $P(x,\partial)$ are elliptic positive definite operators. For more details see Chapter 4 and Appendices A and B.

2.3 Formulation of the results (one-dimensional case)

In this section we formulate the results in the one-dimensional case, i.e., $r = 1$. Although the corresponding estimation problem is the problem for random processes (and not random fields) but since the method and the results are the same as in the multidimensional case, and because of the interest of the results in applications, we formulate the results in one-dimensional case separately.

2.3.1 Basic results for the scalar equation

Let $r = 1$, $D = (t - T, t)$, $R(x, y) \in \mathcal{R}$. The basic equation (2.12) takes the form

$$\int_{t-T}^{t} R(x,y)h(y)dy = f(x), \quad t - T \leq x \leq t. \tag{2.36}$$

Assume that $f \in H^\alpha$, $\alpha = s(q-p)/2$.

Theorem 2.4 *The solution to equation (2.36) in $\dot{H}^{-\alpha}$ exists, is unique, and can be found by the formula*

$$h = Q(\mathcal{L})G, \tag{2.37}$$

where

$$G(x) = \begin{cases} \sum_{j=1}^{sq/2} b_j^- \psi_j^-(x), & x \leq t - T \\ g(x), & t - T \leq t \leq t \\ \sum_{j=1}^{sq/2} b_j^+ \psi_j^+(x), & x \geq t. \end{cases} \tag{2.38}$$

Here $b_j^\pm$ are constants, the functions $\psi_j^\pm(x)$, $1 \leq j \leq sq/2$, form a fundamental system of solutions to the equation

$$Q(\mathcal{L})\psi = 0, \quad \psi_j^-(-\infty) = 0, \quad \psi_j^+(+\infty) = 0. \tag{2.39}$$

The function $g(x)$ is defined by the formula

$$g(x) = g_0(x) + \sum_{j=1}^{sp} c_j \phi_j(x), \tag{2.40}$$

where $g_0(x)$ is an arbitrary fixed solution to the equation

$$P(\mathcal{L})g = f, \quad t - T \leq x \leq t, \tag{2.41}$$

the functions ϕ_j, $1 \leq j \leq sp$, form a fundamental system of solutions to the equation

$$P(\mathcal{L})\phi = 0, \tag{2.42}$$

and c_j, $1 \le j \le sp$, are constants. The constants $b_j^{\pm}$, $1 \le j \le sq/2$, and c_j, $1 \le j \le sp$, are uniquely determined from the linear system:

$$D^k \left\{ \sum_{j=1}^{sq/2} b_j^- \psi_j^- \right\} \Bigg|_{x=t-T} = D^k \left\{ g_0 + \sum_{j=1}^{sp} c_j \phi_j \right\} \Bigg|_{x=t-T}, \tag{2.43}$$

$$D^k \left\{ \sum_{j=1}^{sq/2} b_j^+ \psi_j^+ \right\} \Bigg|_{x=t} = D^k \left\{ g_0 + \sum_{j=1}^{sp} c_j \phi_j \right\} \Bigg|_{x=t}, \tag{2.44}$$

where $D = d/dx$, $0 \le k \le \frac{1}{2}s(p+q) - 1$. The map $R^{-1} : f \to h$, where h is given by formula (2.37), is an isomorphism of the space H^α onto the space $\dot{H}^{-\alpha}$.

Remark 2.1 *This theorem is a complete analogue of Theorem 2.1. The role of $\mathcal{L}$ is played now by an ordinary differential selfadjoint operator $\mathcal{L}$ in $L^2(R^1)$. An ordinary differential operator is elliptic if and only if the coefficient in front of its senior (that is, the highest order) derivative does not vanish:*

$$\mathcal{L}u = \sum_{j=0}^{s} a_j(x) D^j u, \quad a_s(x) \ne 0, \tag{2.45}$$

One can assume that $a_s(x) > 0$, $x \in R^1$, and the condition of uniform ellipticity is assumed, that is

$$0 < c_1 \le a_s(x) \le c_2, \tag{2.46}$$

where c_1 and c_2 are positive constants which do not depend on x.

Corollary 2.4 *If $f \in H^\alpha$, then sing supp$h = \partial D$, where ∂D consists of two points t and $t - T$.*

Corollary 2.4 is a complete analogue of Corollary 2.1.

Corollary 2.5 *Let $Q(\lambda) = a_+(\lambda)a_-(\lambda)$, where $a_\pm(\lambda)$ are polynomials of degree $q/2$, the zeros of the polynomial $a_+(\lambda)$ lie in the upper half plane $Im\lambda > 0$, while the zeros of $a_-(\lambda)$ lie in the lower half-plane $Im\lambda < 0$. Since $Q(\lambda) > 0$ for $-\infty < \lambda < \infty$, the zeros of $a_+(\lambda)$ are complex conjugate of the corresponding zeros of $a_-(\lambda)$. Assume that $P(\lambda) = 1$. Then formula (2.37) can be written as*

$$h(x) = a_+(\mathcal{L})[\theta(x - t + T)a_-(\mathcal{L})f(x)] - a_-(\mathcal{L})[\theta(x - t)a_+(\mathcal{L})f(x)], \tag{2.47}$$

where $\theta(x) = \begin{cases} 1 & \text{if } x \geq 0 \\ 0 & \text{if } x < 0 \end{cases}$, *and the differentiation in (2.47) is understood in the sense of distributions.*

This Corollary is an analogue of Corollary 2.2.

Remark 2.2 *Formula (2.47) is convenient for practical calculations. Let us give a simple example of its application. Let* $\mathcal{L} = -i\partial$, $r = 1$, $P(\lambda) = 1$, $Q(\lambda) = (\lambda^2+1)/2$, $R(x,y) = \exp(-|x-y|)$, $\Phi(x,y,\lambda) = (2\pi)^{-1}\exp\{i\lambda(x-y)\}$, $d\rho(\lambda) = d\lambda$, $t = 1$, $t - T = -1$. *Equation (2.36) becomes*

$$\int_{-1}^{1} \exp(-|x-y|)h(y)dy = f(x), \quad -1 \leq x \leq 1, \tag{2.48}$$

$a_+(\lambda) = \frac{\lambda - i}{\sqrt{2}}$, $a_-(\lambda) = \frac{\lambda+i}{\sqrt{2}}$. *Formula (2.47) yields:*

$$\begin{aligned} h(x) &= \frac{1}{2}(-i\partial - i)[\theta(x+1)(-i\partial + i)f(x)] \\ &\quad -\frac{1}{2}(-i\partial + i)[\theta(x-1)(-i\partial - i)f(x)] \\ &= -\frac{1}{2}(\partial+1)[\theta(x+1)(\partial-1)f] + \frac{1}{2}(\partial-1)[\theta(x-1)(\partial+1)f] \\ &= -\frac{1}{2}\theta(x+1)(\partial^2-1)f - \frac{1}{2}\delta(x+1)(\partial-1)f \\ &\quad +\frac{1}{2}\theta(x-1)(\partial^2-1)f + \frac{1}{2}\delta(x-1)(\partial+1)f \\ &= \frac{-f''+f}{2} + \frac{f'(1)+f(1)}{2}\delta(x-1) \\ &\quad + \frac{-f'(-1)+f(-1)}{2}\delta(x+1). \end{aligned} \tag{2.49}$$

Here we have used the well known formula $\theta'(x-a) = \delta(x-a)$ *where* $\delta(x)$ *is the delta-function. Formula (2.49) is the same as formula (1.9). The term* $(-f''+f)/2$ *in (2.49) vanishes outside the interval* $[-1,1]$ *by definition.*

Remark 2.3 *If* $t = +\infty$ *and* $t - T = 0$, *so that equation (2.36) takes the form of Wiener-Hopf equation of the first kind*

$$\int_{0}^{\infty} R(x,y)h(y)dy = f(x), \quad x \geq 0, \tag{2.50}$$

then formula (2.47) reduces to

$$h(x) = a_+(\mathcal{L})[\theta(x)a_-(\mathcal{L})f(x)]. \tag{2.51}$$

If $\mathcal{L} = -i\partial$ formula (2.51) can be obtained by the well-known factorization method.

Example 2.1 Consider the equation

$$\int_0^\infty \exp(-|x-y|)h(y)dy = f(x), \quad x \geq 0. \tag{2.52}$$

By formula (2.51) one obtains

$$h(x) = \frac{-f'' + f}{2} + \frac{-f'(0) + f(0)}{2}\delta(x), \tag{2.53}$$

if one uses calculations similar to the given in formula (2.49).

2.3.2 *Vector equations*

In both cases $r = 1$ and $r > 1$ it is of interest to consider estimation problems for vector random processes and vector random fields. For vector random processes the basic equation is (2.36) with the positive kernel $\mathbf{R}(x, y)$ in the sense that $(\mathbf{Rh}, \mathbf{h}) > 0$ for $\mathbf{h} \not\equiv 0$, given by formula (2.3) in which $\tilde{\mathbf{R}}(\lambda) = P(\lambda)Q^{-1}(\lambda)$ is a matrix:

$$\tilde{\mathbf{R}}(\lambda) = (\tilde{R}_{ij}(\lambda)), \quad \tilde{R}_{ij}(\lambda) := P_{ij}(\lambda)Q_{ij}^{-1}(\lambda), \quad 1 \leq i, j \leq d, \tag{2.54}$$

where $P_{ij}(\lambda)$ and $Q_{ij}(\lambda)$ are relatively prime positive polynomials for each fixed pair of indices (ij), $1 \leq i,\ j \leq d$, d is the number of components of the random processes $\mathcal{U}$, s and n. Let $Q(\lambda)$ be the polynomial of minimal degree, $\deg Q(\lambda) = q$, for which any $Q_{ij}(\lambda)$, $1 \leq i,\ j \leq d$, is a divisor, $A_{ij}(\lambda) := R_{ij}(\lambda)Q(\lambda)$. Denote by $\mathbf{E}$ the unit $d \times d$ matrix and by $\mathbf{A}(\mathcal{L})$ the matrix differential operator with entries $A_{ij}(\mathcal{L})$. Assume that

$$\det |A_{ij}(\lambda)| > 0, \quad \forall \lambda \in R^1, \tag{2.55}$$

$$\det \mathbf{B}_m(x) \neq 0, \forall x \in R^1, \tag{2.56}$$

where $m := s \max_{1 \leq i,j \leq d} \deg A_{ij}(\lambda)$, $s = \text{ord}\mathcal{L}$,

$$\mathbf{A}(\mathcal{L}) := \sum_{j=0}^{m} \mathbf{B}_j(x)\partial^j, \quad \partial = \frac{d}{dx}. \tag{2.57}$$

Let $\mathbf{S}(x,y)$ denote the matrix kernel

$$\mathbf{S}(x,y) := \delta_{ij} \int_{\Lambda} Q^{-1}(\lambda)\Phi(x,y,\lambda)d\rho(\lambda), \quad \delta_{ij} = \begin{cases} 1 & i=j \\ 0 & i \neq j \end{cases} \tag{2.58}$$

of the diagonal operator $Q^{-1}(\mathcal{L})\mathbf{E}$. The operator $Q(\mathcal{L})\mathbf{E}$ is a diagonal matrix differential operator of order $n = sq$.

Let us write the basic equation

$$\int_{t-T}^{t} \mathbf{R}(x,y)\mathbf{h}(y)dy = \mathbf{f}(x), \quad t-T \leq x \leq t, \tag{2.59}$$

where $\mathbf{R}(x,y)$ is the $d \times d$ matrix with the spectral density (2.54), $\mathbf{h}$ and $\mathbf{f}$ are vector functions with d components, $\mathbf{f} \in \mathcal{H}^{\alpha}$, $\alpha = (n-m)/2$, $\mathcal{H}^{\alpha}$ denotes the space of vector functions $(f_1, \dots f_d)$ such that $\| \mathbf{f} \|_{\mathcal{H}^{\alpha}} := \left(\sum_{j=1}^{d} \| f_j \|_{H^{\alpha}}^2 \right)^{1/2}$.

Remark 2.4 *In the vector estimation problem* $\mathbf{h}$ *and* $\mathbf{f}$ *are* $d \times d$ *matrices, but for simplicity and without loss of generality we discuss the case when* $\mathbf{h}$ *is a vector. Matrix equation (2.59) is equivalent to* d *vector equations.*

Equation (2.59) one can write as

$$\mathbf{A}(\mathcal{L})\mathbf{v} = \mathbf{f}, \tag{2.60}$$

$$\mathbf{v} := Q^{-1}(\mathcal{L})\mathbf{E}h = \int_{D} \mathbf{S}(x,y)\mathbf{h}(y)dy. \tag{2.61}$$

Let $\mathbf{\Phi}_j$, $1 \leq j \leq m$, be a fundamental system of matrix solutions to the equation

$$\mathbf{A}(\mathcal{L})\phi = 0, \tag{2.62}$$

and $\mathbf{\Psi}_j^{\pm}$, $1 \leq j \leq n/2$ be the fundamental system of matrix solutions to the equation

$$Q(\mathcal{L})\mathbf{E}\mathbf{\Psi} = 0, \tag{2.63}$$

such that

$$\mathbf{\Psi}_j^{+}(+\infty) = 0, \quad \mathbf{\Psi}_j^{-}(-\infty) = 0. \tag{2.64}$$

The choice of the fundamental system of matrix solutions to (2.63) with properties (2.64) is possible if $\mathcal{L}$ is an elliptic ordinary differential operator, that is $a_s(x) \neq 0$, $x \in R^1$ (see Remark 2.1 and [Naimark (1969), p. 118]. Let us write equations (2.60) and (2.61) as

$$\int_D \mathbf{S}(x,y)\mathbf{h}(y)dy = \mathbf{g}_0(x) + \sum_{j=1}^{m} \mathbf{\Phi}_j(x)\mathbf{c}_j, \tag{2.65}$$

where $\mathbf{g}_0(x)$ is an arbitrary fixed solution to the equation (2.60), and $\mathbf{c}_j$, $1 \leq j \leq m$, are arbitrary linearly independant constant vectors.

Theorem 2.5 *If $\mathbf{R} \in \mathcal{R}$ with $\tilde{\mathbf{R}}$ given by (2.54), the assumptions (2.46), (2.55), (2.56) hold, and $\mathbf{f} \in \mathcal{H}^\alpha$, $\alpha = (n-m)/2$, then the matrix equation (2.59) has a solution in $\dot{\mathcal{H}}^{-\alpha}$, this solution is unique and can be found by the formula*

$$\mathbf{h} = Q(\mathcal{L})\mathbf{E}\mathbf{G}, \tag{2.66}$$

where the vector function $\mathbf{G}$ is given by

$$\mathbf{G}(x) = \begin{cases} \sum_{j=1}^{n/2} \mathbf{\Psi}_j^- \mathbf{b}_j^-, & x \leq t-T \\ \mathbf{g}_0(x) + \sum_{j=1}^{m} \mathbf{\Phi}_j \mathbf{c}_j, & t-T \leq x \leq t, \\ \sum_{j=1}^{n/2} \mathbf{\Psi}_j^+ \mathbf{b}_j^+, & x \geq t. \end{cases} \tag{2.67}$$

Here the functions $\mathbf{\Psi}_j^\pm$, $\mathbf{\Phi}_j$ and $\mathbf{g}_0(x)$ were defined above, and the constant vectors $\mathbf{b}_j^\pm$, $1 \leq j \leq n/2$ and $\mathbf{c}_j$, $1 \leq j \leq m$, can be uniquely determined from the linear system

$$D^k \left\{ \sum_{j=1}^{n/2} \mathbf{\Psi}_j^-(x)\mathbf{b}_j^- \right\}\Bigg|_{x=t-T} = D^k \left\{ \mathbf{g}_0(x) + \sum_{j=1}^{m} \mathbf{\Phi}_j(x)\mathbf{c}_j \right\}\Bigg|_{x=t-T} \tag{2.68}$$

$$D^k \left\{ \sum_{j=1}^{n/2} \mathbf{\Psi}_j^+(x)\mathbf{b}_j^+ \right\}\Bigg|_{x=t} = D^k \left\{ \mathbf{g}_0(x) + \sum_{j=1}^{m} \mathbf{\Phi}_j(x)\mathbf{c}_j \right\}\Bigg|_{x=t}, \tag{2.69}$$

where $0 \leq k \leq (n+m)/2 - 1$.

The map $\mathbf{R}^{-1} : \mathbf{f} \to \mathbf{h}$, given by formulas (2.66)-(2.69) is an isomorphism between the spaces $\mathcal{H}^\alpha$ and $\dot{\mathcal{H}}^{-\alpha}$, $\alpha = (n-m)/2$.

Remark 2.5 *The conditions (2.68), (2.69) guarantee that the function $\mathbf{G}(x)$, defined by formula (2.67), is maximally smooth so that the order of singularity of $\mathbf{G}$ and, therefore, of $\mathbf{h}$ (see formula (2.66)) is minimal.*

2.4 Examples of kernels of class $\mathcal{R}$ and solutions to the basic equation

1. If $r = 1$, $\mathcal{L} = -i\partial$, $\partial = d/dx$, $\Phi(x, y, \lambda) = (2\pi)^{-1}\exp\{i\lambda(x-y)\}$, $d\rho = d\lambda$, then $R(x, y) \in \mathcal{R}$ if

$$R(x,y) = (2\pi)^{-1}\int_{-\infty}^{\infty} \tilde{R}(\lambda)\exp\{i\lambda(x-y)\}d\lambda, \tag{2.70}$$

where

$$\tilde{R}(\lambda) = P(\lambda)Q^{-1}(\lambda) \tag{2.71}$$

and $P(\lambda)$, $Q(\lambda)$ are positive polynomials.

2. If $r > 1$, $\mathcal{L} = (\mathcal{L}_1, \dots \mathcal{L}_r)$, $\mathcal{L}_r = -i\partial_r$, $\partial_r = \partial/\partial x_r$, $\Phi(x, y, \lambda) = (2\pi)^{-r}\exp\{i\lambda \cdot (x-y)\}$, $\lambda = (\lambda_1, \dots \lambda_r)$, $d\rho(\lambda) = d\lambda = d\lambda_1 \dots d\lambda_r$, then

$$R(x,y) = (2\pi)^{-r}\int_{R^r} \tilde{R}(\lambda)\exp\{i\lambda \cdot (x-y)\}d\lambda, \tag{2.72}$$

where $\tilde{R}(\lambda)$ is given by (2.71) and

$$P(\lambda) = P(\lambda_1, \dots \lambda_r) > 0, \quad Q(\lambda) = Q(\lambda_1 \dots \lambda_r) > 0 \tag{2.73}$$

are polynomials. For the operators $P(\mathcal{L})$ and $Q(\mathcal{L})$ to be elliptic of orders p and q respectively, one has to assume that

$$0 < c_1 \le P(\lambda)|\lambda|^{-p} \le c_2, \quad 0 < c_3 \le Q(\lambda)|\lambda|^{-q} \le c_4, \quad \forall \lambda \in R^r \tag{2.74}$$

where $|\lambda| = (\lambda_1^2 + \dots + \lambda_r^2)^{1/2}$ and c_j, $1 \le j \le 4$, are positive constants.

3. If $r = 1$, $\mathcal{L} = -\frac{d^2}{dx^2}$, $D(\mathcal{L}) = \{u : u \in H^2(0, \infty),\ u'(0) = 0\}$, $D(\mathcal{L}) =$ domain of $\mathcal{L}$, then

$$R(x,y) = \frac{1}{2}[A(|x+y|) + A(|x-y|)], \quad x, y \ge 0 \tag{2.75}$$

where

$$A(x) = \pi^{-1}\int_0^{\infty} P(\lambda)Q^{-1}(\lambda)\cos(\sqrt{\lambda}x)\lambda^{-1/2}d\lambda \tag{2.76}$$

and $P(\lambda) > 0$, $Q(\lambda) > 0$ are polynomials.

Indeed, one has for $\mathcal{L}$

$$\Phi(x,y,\lambda)d\rho(\lambda) = \begin{cases} \pi^{-1}\cos(\sqrt{\lambda}x)\cos(\sqrt{\lambda}y)\lambda^{-1/2}d\lambda, & \lambda \ge 0 \\ 0, & \lambda < 0, \end{cases}$$

$0 \le x,\, y < \infty$. Since

$$\cos(kx)\cos(ky) = \frac{1}{2}[\cos(kx - ky) + \cos(kx + ky)], \quad k = \sqrt{\lambda}$$

one obtains (2.75) and (2.76).

If one put $\sqrt{\lambda} = k$ in (2.76) one gets

$$A(x) = \frac{2}{\pi}\int_0^\infty P(k^2)Q^{-1}(k^2)\cos(kx)dk, \tag{2.77}$$

which is a cosine transform of a positive rational function of k. The eigenfunctions of $\mathcal{L}$, normalized in $L^2(0,\infty)$, are $\left(\frac{2}{\pi}\right)^{1/2}\cos(kx)$ and $d\rho = dk$ in the variable k. If $\mathcal{L} = -\frac{d^2}{dx^2}$ is determined in $L^2(0,\infty)$ by the boundary condition $u(0) = 0$, then

$$R(x,y) = \frac{1}{2}[A(|x-y|) - A(x+y)], \quad x, y \ge 0, \tag{2.78}$$

where $A(x)$ is given by (2.77), the eigenfunctions of $\mathcal{L}$ with the Dirichlet boundary condition $u(0) = 0$ are $\sqrt{\frac{2}{\pi}}\sin(kx)$, $d\rho = dk$ in the variable k, and $\Phi(x,y,k)d\rho(k) = \frac{2}{\pi}\sin(kx)\sin(ky)dk$ one can compare this with the formula $\Phi(x,y,k)d\rho(k) = \frac{2}{\pi}\cos(kx)\cos(ky)dk$, which holds for $\mathcal{L}$ determined by the Neumann boundary condition $u'(0) = 0$.

4. If $\mathcal{L} = -\frac{d^2}{dx^2} + (\nu^2 - \frac{1}{4})x^{-2}$, $\nu \ge 0$, $x \ge 0$, then

$$\Phi(x,y,\lambda)d\rho(\lambda) = \begin{cases} \sqrt{x\lambda}J_\nu(x\lambda)\sqrt{y\lambda}J_\nu(y\lambda)d\lambda, & \text{if } \lambda \ge 0 \\ 0, & \text{if } \lambda < 0, \end{cases} \tag{2.79}$$

so that

$$R(x,y) = \sqrt{xy}\int_0^\infty P(\lambda)Q^{-1}(\lambda)J_\nu(\lambda x)J_\nu(\lambda y)\lambda d\lambda, \tag{2.80}$$

where $P(\lambda)$ and $Q(\lambda)$ are positive polynomials on the semiaxis $\lambda \ge 0$.

5. Let $R(x,y) = \exp(-a|x-y|)(4\pi|x-y|)^{-1}$, $x, y \in R^3$, $a = \text{const} > 0$. Note that $(-\Delta + a^2)R = \delta(x-y)$ in R^3. The kernel $R(x,y) \in \mathcal{R}$. One has $\mathcal{L} = (\mathcal{L}_1, \mathcal{L}_2, \mathcal{L}_3)$, $\mathcal{L}_j = -i\partial_j$, $P(\lambda) = 1$, $Q(\lambda) = \lambda^2 + a^2$, $\lambda^2 = \lambda_1^2 + \lambda_2^2 + \lambda_3^2$, $\Phi d\rho = (2\pi)^{-3}\exp\{i\lambda\cdot(x-y)\}d\lambda$,

$$R(x,y) = (2\pi)^{-3}\int_{R^3}\frac{\exp\{i\lambda\cdot(x-y)\}}{\lambda^2 + a^2}d\lambda. \tag{2.81}$$

6. Let $R(x,y) = R(xy)$. Put $x = \exp(\xi)$, $y = \exp(-\eta)$. Then $R(xy) = R(\exp(\xi - y)) := R_1(\xi - y)$. If $R_1 \in \mathcal{R}$ with $\mathcal{L} = -i\partial$, then one can solve

the equation

$$\int_a^b R(xy)h(y)dy = f(x), \quad a \le x \le b \tag{2.82}$$

analytically.

7. Let $K_0(a|x|)$ be the modified Bessel function which can be defined by the formula

$$K_0(a|x|) = (2\pi)^{-1} \int_{R^2} \frac{\exp(i\lambda \cdot x)}{\lambda^2 + a^2} d\lambda, \quad a > 0, \tag{2.83}$$

where $\lambda \cdot x = \lambda_1 x_1 + \lambda_2 x_2$. Then the kernel $R(x,y) := K_0(a|x-y|) \in \mathcal{R}$, $\mathcal{L} = (-i\partial_1, -i\partial_2)$, $r = 2$, $P(\lambda) = 1$, $Q(\lambda) = \lambda^2 + a^2$, $\Phi(x,y,\lambda)d\rho(\lambda) = (2\pi)^{-1} \exp\{i\lambda \cdot (x-y)\}d\lambda$.

8. Consider the equation

$$\int_D \frac{\exp(-a|x-y|)}{4\pi|x-y|} h(y)dy = f(x), \quad x \in \overline{D} \subset R^3, \quad a > 0, \tag{2.84}$$

with kernel (2.81). By formula (2.15), Theorem 2.1, one obtains the unique solution to equation (2.84) in $\dot{H}^{-1}(D)$:

$$h(x) = (-\Delta + a^2)f + \left(\frac{\partial f}{\partial N} - \frac{\partial u}{\partial N}\right)\delta_\Gamma, \tag{2.85}$$

where u is the unique solution to the Dirichlet problem in the exterior domain $\Omega := R^3 \setminus D$:

$$(-\Delta + a^2)u = 0 \quad \text{in} \quad \Omega, \quad u|_\Gamma = f|_\Gamma, \tag{2.86}$$

$\Gamma = \partial D = \partial\Omega$ is the boundary of D, and δ_Γ is the delta function with support Γ.

Let us derive formula (2.85). For the kernel (2.81) one has $r = 3$, $p = 0$, $P(\lambda) = 1$, $Q(\lambda) = \lambda^2 + a^2$, $s = 1$, $q = 2$, $\alpha = \frac{sq}{2} = 1$. Formula (2.15) reduces to

$$h(x) = (-\Delta + a^2)G, \tag{2.87}$$

with

$$G = \begin{cases} f & \text{in } D \\ u & \text{in } \Omega, \end{cases} \tag{2.88}$$

and u is the solution to (2.86). Indeed, since $P(\lambda) = 1$, one has $v = 0$ and $g = f$. In order to compute h by formula (2.87) one uses the definition of the derivative in the sense of distributions. For any $\phi \in C_0^\infty(R^r)$ one has:

$$\begin{aligned}
((-\Delta + a^2)G, \phi) &= (G, (-\Delta + a^2)\phi) \\
&= \int_D f(-\Delta + a^2)\phi dx + \int_\Omega u(-\Delta + a^2)\phi dx \\
&= \int_D (-\Delta + a^2) f\phi dx + \int_\Omega (-\Delta + a^2) u\phi dx \\
&\quad - \int_\Gamma \left(f\frac{\partial \phi}{\partial N} - \frac{\partial f}{\partial N}\phi \right) ds + \int_\Gamma \left(u\frac{\partial \phi}{\partial N} - \phi\frac{\partial u}{\partial N} \right) ds \\
&= \int_D (-\Delta + a^2) f\phi dx + \int_\Gamma \left(\frac{\partial f}{\partial N} - \frac{\partial u}{\partial N} \right) \phi ds, \qquad (2.89)
\end{aligned}$$

where the condition $u = f$ on Γ was used. Formula (2.89) is equivalent to (2.85).

9. Consider the equation

$$2\pi \int_D K_0(a|x - y|)h(y)dy = f(x), \quad x \in \overline{D} \subset R^2, \quad a > 0, \qquad (2.90)$$

where $D = \{x : x \in R^2, |x| \leq b\}$, and $K_0(x)$ is given by formula (2.83). The solution to (2.90) in $\dot{H}^{-1}(D)$ can be calculated by formula (2.85) in which $u(x)$ can be calculated explicitly

$$u(x) = \sum_{n=-\infty}^{\infty} f_n \frac{\exp(in\phi)K_n(ar)}{K_n(ab)}, \qquad (2.91)$$

where $x = (r, \phi)$, (r, ϕ) are polar coordinates in R^2,

$$f_n := (2\pi)^{-1} \int_0^{2\pi} f(b, \phi) \exp(-in\phi) d\phi, \qquad (2.92)$$

$K_n(r)$ is the modified Bessel function of order n which decays as $r \to +\infty$. One can easily calculate $\frac{\partial u}{\partial N} \big|_\Gamma$ in formula (2.85):

$$\frac{\partial u}{\partial N} \Big|_\Gamma = \frac{\partial u}{\partial r} \Big|_{r=b} = \sum_{n=-\infty}^{\infty} a f_n \frac{\exp(in\phi)K_n'(ab)}{K_n(ab)}. \qquad (2.93)$$

Formulas (2.85) and (2.93) give an explicit analytical formula for the solution to equation (2.90) in $\dot{H}^{-1}(D)$.

2.5 Formula for the error of the optimal estimate

In this section we give an explicit formula for the error of the optimal estimator. This error is given by formula (1.5). We assume for simplicity that $A = I$ in what follows. This means that we are discussing the filtering problem. In the same way the general estimation problem can be treated.

1. The error of the estimate can be computed by formula (2.9) with $A = I$. This yields

$$\epsilon = (Rh, h) - 2Re(h, f) + \epsilon_0(x), \quad \epsilon_0(x) := \overline{|s(x)|^2}, \tag{2.94}$$

where $(u, v) := \int_D u\bar{v}dx$, $(Rh, h) = \int_D \int_D R(z, y)h(y)h^*(z)dydz$, and $h(y) := h(x, y)$. The optimal estimate

$$\hat{\mathcal{U}}(x) = \int_D h(x, y)\mathcal{U}(y)dy \tag{2.95}$$

is given by the solution to the equation (2.11):

$$Rh = f, \tag{2.96}$$

and we assume that $R \in \mathcal{R}$. Since $(Rh, h) > 0$, it follows from (2.94) and (2.96) that

$$\epsilon(x) = \epsilon_0(x) - (Rh, h). \tag{2.97}$$

It is clear that the right side of (2.97) is finite if and only if the quadratic form (Rh, h) is finite. Our goal is to show that if and only if one takes the solution to (2.96) of minimal order of singularity, the mos solution to (2.96), that is the solution $h \in \dot{H}^{-\alpha}$, one obtains a finite value of (Rh, h). Therefore only the mos solution to (2.96) solves the estimation problem, and the error of the optimal estimate is given by formula (2.97) in which $h \in \dot{H}^{-\alpha}$ is the unique solution to (2.96) of minimal order of singularity.

2. In order to achieve our goal, let us write the form (Rh, h) using the Parseval equality and the basic assumption $R \in \mathcal{R}$:

$$(Rh, h) = \int_\Lambda P(\lambda)Q^{-1}(\lambda)|\tilde{h}(\lambda)|^2 d\rho(\lambda). \tag{2.98}$$

Here

$$\left|\tilde{h}(\lambda)\right|^2 = \sum_{j=1}^{N(\lambda)} \left|\int_D h(x)\phi_j^*(x, \lambda)dx\right|^2, \tag{2.99}$$

where $\phi_j(x, \lambda)$ are the eigenfunctions of $\mathcal{L}$ which are used in the expansion of the spectral kernel:

$$\Phi(x, y, \lambda) = \sum_{j=1}^{N_\lambda} \phi_j(x, \lambda)\phi_j^*(y, \lambda), \tag{2.100}$$

and $N_\lambda \le \infty$ (see Section 8.2). One has

$$h \in \dot{H}^{-bs} \Leftrightarrow \int_\Lambda \left|\tilde{h}(\lambda)\right|^2 (1 + \lambda^2)^{-b} d\rho(\lambda) < \infty, \tag{2.101}$$

where $b > 0$ is an arbitrary number. By the assumption (see 2.74)

$$0 < c_1 \le P(\lambda)(1 + \lambda^2)^{-p/2} \le c_2, \tag{2.102}$$

$$0 < c_3 \le Q(\lambda)(1 + \lambda^2)^{-q/2} \le c_4, \tag{2.103}$$

where c_j, $1 \le j \le 4$, are positive constants. Thus

$$0 < c_5 \le PQ^{-1}(1 + \lambda^2)^{(q-p)/2} \le c_6. \tag{2.104}$$

From (2.98),(2.101) and (2.104) it follows that

$$(Rh, h) < \infty \Leftrightarrow h \in \dot{H}^{-(q-p)s/2} = \dot{H}^{-\alpha}. \tag{2.105}$$

In particular, if $m(h) := \text{ord} h > \alpha$, then $(Rh, h) = \infty$.

Let $\mathcal{L}$ be an operator with constant coefficients in $L^2(R^r)$ so that $R(x, y) = R(x - y)$, $\mathcal{L} = (\mathcal{L}_1, \dots \mathcal{L}_r)$, $\mathcal{L}_r := -i\partial/\partial x_r$, and formula (2.98) takes the form

$$(Rh, h) = \int_{R^r} P(\lambda)Q^{-1}(\lambda) \left|\tilde{h}(\lambda)\right|^2 d\lambda, \tag{2.106}$$

with $\lambda = (\lambda_1, \dots \lambda_r)$ and

$$\tilde{h}(\lambda) := (2\pi)^{-r/2} \int_{R^r} h(x) \exp(i\lambda \cdot x) dx. \tag{2.107}$$

Then $P(\mathcal{L})$ is elliptic of order p if and only if $P(\lambda)$ satisfies (2.102) and $Q(\mathcal{L})$ is elliptic of order q if and only if $Q(\lambda)$ satisfies (2.103). The integral (2.106) is finite if and only if $\tilde{h}(\lambda)(1+\lambda^2)^{(p-q)/4} \in L^2(R^r)$, where $\tilde{h}(\lambda)$ is the usual Fourier transform. This is equivalent to $h \in H^{-\alpha}(R^r)$, $\alpha = (q-p)/2$. Since we assume that $\text{supp} h \subset D$, we conclude in the case described that $h \in \dot{H}^{-\alpha}(D)$.

3. Formula (2.97) can be written as

$$\epsilon(x) = \epsilon_0(x) - (f, h) = \epsilon_0(x) - (f, R^{-1}f). \tag{2.108}$$

If $\mathcal{U}$ is a vector random field then $\mathbf{h}(x, y)$ is a $d \times d$ matrix and formulas (2.97) and (2.108) take the form

$$\epsilon(x) = \epsilon_0(x) - tr(\mathbf{f}, \mathbf{h}) = \epsilon_0(x) - \int_D tr\{\mathbf{f}(x, y)\mathbf{h}^*(y, x)\}dy, \tag{2.109}$$

where $tr\mathbf{A}$ is the trace of the matrix $\mathbf{A}$.

Chapter 3

Numerical Solution of the Basic Integral Equation in Distributions

3.1 Basic ideas

It is convenient to explain the basic ideas using the equation

$$Rh = \int_{-1}^{1} \exp(-|x-y|)h(y)dy = f(x), \quad -1 \le x \le 1 \tag{3.1}$$

as an example, which contains all of the essential features of the general equation

$$Rh = \int_D R(x,y)h(y)dy = f(x), \quad x \in \overline{D} \subset R^r \tag{3.2}$$

with the kernel $R \in \mathcal{R}$.

The first idea that might come to mind is that equation (3.1) is Fredholm's equation of the first kind, so that the regularization method can yield a numerical solution to (3.1). On second thought one realizes that, according to Theorem 3.1, the solution to (3.1) does not belong to $L^2(-1,1)$ in general, and that the mapping $R^{-1} : H^\alpha \to \dot{H}^{-\alpha}$, $\alpha = 1$ for equation (3.1), is an isomorphism. Therefore the problem of numerical solution of equation (3.1) is not an ill-posed but a well-posed problem. The solution to (3.1) is a distribution in general, which is clear from formula (2.49). For the solution of (3.1) to be an integrable function it is necessary and sufficient that the following boundary conditions hold

$$f'(1) + f(1) = 0, \quad f'(-1) = f(-1). \tag{3.3}$$

This follows immediately from the formula (2.49). The problem is to develop a numerical method for solving equations (3.1) and (3.2) in the space of distributions $\dot{H}^{-\alpha}$.

Much work has been done on the effective numerical inversion of the Toeplitz matrices which one obtains by discretizing the integral equation

$$\epsilon h_\epsilon + Rh_\epsilon = f, \quad \epsilon > 0 \tag{3.4}$$

with R, for example, given by (3.1). If the nodes of discretization are equidistant, equation (3.4), after discretizing, reduces to a linear algebraic system with Toeplitz matrix $t_{ij} = t_{i-j}$. Discussion of this, however, is not in the scope of our work. The question of principal interest is the question about asymptotic behavior of the solution to (3.4) as $\epsilon \to +0$. Note that, for any $\epsilon > 0$, equation (3.4) is an equation with a selfadjoint positive definite operator $\epsilon I + R$ in $L^2(-1,1)$. Therefore, for any $\epsilon > 0$, equation (3.4) has a solution in $L^2(-1,1)$ for any $f \epsilon L^2(-1,1)$, and this solution is unique. Numerical solution of equation (3.4) by the above mentioned discretization (or collocation) method becomes impossible as $\epsilon \to +0$ because the condition number of the matrix of the discretized problem grows quickly as $\epsilon \to +0$. The nature of singularity of the solution to the limiting equation ($\epsilon = 0$) (3.1) is not clear from the discretization method described above.

Numerical solution of equation (3.1) requires therefore a new approach which we wish to describe.

The basic idea is to take into account theoretical results obtained in Theorems 2.1 and 2.4 . According to the Theorem 2.4, the solution to equation (3.1) with a smooth right hand side $f(x)$ has the following structure:

$$h = A\delta(x-1) + B\delta(x+1) + h_{sm}, \tag{3.5}$$

where A and B are constants and h_{sm} is a smooth function. The order of singularity of the solution to equation (3.1) is 1 since $\alpha = 1$ for this equation.

Let us assume that f is smooth, $f \in H^2$, for example, so that

$$\frac{-f'' + f}{2} \in H^0 = L^2(D).$$

Let us look for an approximate solution to equation (3.1) of the form

$$h_n(x) = \sum_{j=1}^{n} c_j \phi_j(x) + c_0 \delta(x-1) + c_{-1}\delta(x+1), \tag{3.6}$$

where c_j, $j = -1, 0, 1, \ldots$, are constants, $\{\phi_j\}$, $1 \le j < \infty$, is a basis in $H^\circ = L^2(-1,1)$. The constants can be found, for example, from the least squares method:

$$\| Rh_n - f \|_1 = \min, \tag{3.7}$$

where $\| f \|_\alpha$ is the norm of f in the space H^α.

The variational problem (3.7) can be written as

$$\begin{aligned} \epsilon := & \int_{-1}^{1} \{ |c_0' \exp(x) + c_{-1}' \exp(-x) + \textstyle\sum_{j=1}^{n} c_j \psi_j(x) - f(x)|^2 \\ & + |c_0' \exp(x) - c_{-1}' \exp(-x) + \textstyle\sum_{j=1}^{n} c_j \psi_j'(x) - f'(x)|^2 \} dx \\ = & \min, \end{aligned} \tag{3.8}$$

where

$$c_0' := c_0 e^{-1}, \quad c_{-1}' = c_{-1} e^{-1}, \quad \psi_j(x) = R\phi_j, \quad 1 \le j \le n. \tag{3.9}$$

The linear system for finding the c_j, $1 \le j \le n$ and c_0', c_{-1}' is

$$\frac{\partial \epsilon}{\partial c_j} = 0, \quad 1 \le j \le n, \quad \frac{\partial \epsilon}{\partial c_0'} = 0, \quad \frac{\partial \epsilon}{\partial c_{-1}'} = 0. \tag{3.10}$$

The matrix of this system is

$$a_{ij} := (\psi_j, \psi_i)_1, \quad -1 \le i, \quad j \le n, \tag{3.11}$$

where $\psi_0 = \exp(x)$, $\psi_{-1} = \exp(-x)$, ψ_j for $1 \le j < \infty$ is defined in (3.9), the system $\{\psi_j\}$, $-1 \le j \le n$ is assumed to be linearly independent for any n, and the inner product is taken in the space H^1: $(u,v)_1 := \int_{-1}^{1} (u\bar{v} + u'\bar{v}')dx$. Matrix a_{ij} is positive definite for any n so that the system

$$\sum_{j=-1}^{n} a_{ij} c_j = b_i, \quad -1 \le i \le n, \quad b_i := (f, \psi_i)_1 \tag{3.12}$$

is uniquely solvable for any n.

Convergence of the suggested numerical method is easy to prove. One wishes to prove that

$$\| h_n - h \|_{-1} \to 0 \quad \text{as} \quad n \to \infty. \tag{3.13}$$

Since $R : \dot{H}^{-1} \to H^1$ is an isomorphism, it is sufficient to prove that

$$\| Rh_n - Rh \|_1 \to 0 \quad \text{as} \quad n \to \infty. \tag{3.14}$$

Since $Rh = f$, equation (3.14) reduces to

$$\| Rh_n - f \|_1 \to 0 \quad \text{as} \quad n \to \infty. \tag{3.15}$$

Equation (3.15) holds if the set of functions $\{\psi_j\}$, $-1 \le j < \infty$ is complete in H^1. Here

$$\psi_j = R\phi_j, \quad 1 \le j \le \infty, \quad \psi_{-1} = \exp(-x), \psi_0 = \exp(x). \tag{3.16}$$

Therefore, if one chooses a system $\psi_j \in H^1$, $1 \le j < \infty$, such that the system $\{\psi_j\}$, $-1 \le j < \infty$, forms a basis of H^1 (or just a complete system in H^1) then (3.15) holds. Since for practical calculations one need only know the matrix a_{ij} and the vector b_i (see (3.12)), and both these quantities can be computed if the system $\{\psi_j\}$, $-1 \le j < \infty$, and f are known, it is not necessary to deal with the system $\{\phi_j\}$, $1 \le j \le \infty$.

We have proved the following

Proposition 3.1 *If $\{\psi_j\}$, $-1 \le j < \infty$, $\psi_0 = \exp(x)$, $\psi_{-1} = \exp(-x)$, is a complete system in H^1 then, for any n, the system (3.12) is uniquely solvable. Let $c_j^{(n)}$ be its solution. Then the function $h_n = \sum_{j=-1}^{n} c_j^{(n)} \phi_j(x)$, where $\phi_0 = \delta(x-1)$, $\phi_{-1} = \delta(x+1)$, $\phi_j = R^{-1}\psi_j$, $1 \le j \le n$ converges in H^{-1} to the solution h of equation (3.1):*

$$\| h - h_n \|_{-1} \to 0, \quad n \to \infty.$$

There are some questions of a practical nature:

1) how does one choose the system ψ_j, $1 \le j < \infty$, so that the matrix a_{ij} in equation (3.12) is easily invertible?
2) how does one choose ψ_j, $1 \le j < \infty$, so that the functions ϕ_j are easily computable?

The first question is easy to answer: it is sufficient that the condition number of the matrix a_{ij}, $-1 \le j \le n$ for any n is bounded. This will be the case if the system $\{\psi_j\}$, $-1 \le j < \infty$, forms a Riesz basis. Let us recall that the system $\{\psi_j\}$ is called a Riesz basis of a Hilbert space if and only if there exists an orthonormal basis $\{f_j\}$ of H and a linear isomorphism B of H onto H such that $Bf_j = \psi_j$, $\forall j$. The system $\{\psi_j\}$ forms a Riesz basis of the Hilbert space H if and only if the Gram matrix $(\psi_i, \psi_j) := \Gamma_{ij}$ defines a linear isomorphism of ℓ^2 onto itself.

If $\{\psi_j\}$, $1 \le j < \infty$, in (3.6) is a basis of H° then $\{\psi_j\}$, $1 \le j < \infty$, $\psi_j = R\phi_j$, is a complete set in H^1. Indeed, suppose that $f \in H^1$ and $(f, R\phi_j)_1 = 0\ \forall j$. Then $0 = (f, R\phi_j)_+ = (I^{-1}f, R\phi_j)_0 = (RI^{-1}f, \phi_j)_0\ \forall j$, where the operator I has been introduced in Section 8.1.2, and $H_+ = H^1$.

Since the system $\{\phi_j\}$ is complete in H_0 ($H_0 = H^\circ$ in our case) by the assumption, one concludes that $RI^{-1}f = 0$. Since $I^{-1}f \in H_-$ and $R(x,y)$ is positive so that $(Rg, g)_0 > 0$ for $g \neq 0$, $g \in H_-$, one concludes that $I^{-1}f = 0$. Since I^{-1} is an isometry between H_+ and H_-, one concludes that $f = 0$. Therefore, by Proposition 3.1, if the system $\{\phi_j\}$ forms a basis of H° then $\| h_n - h \|_{-1} \to 0$ as $n \to \infty$, where h_n is the solution to (3.7) of the form (3.6).

If $\{\phi_j\}$ is a basis of H° then the system (3.12) is uniquely solvable for all n if and only if the system $\{\psi_j\}$, $-1 \leq j \leq n$ is linearly independent in H^1 for all n. Here the system ψ_j is defined by formula (3.16).

3.2 Theoretical approaches

1. Let us consider equation (3.2) as an equation with the operator $R : H_- \to H_+$ which is a linear isomorphism between the spaces H_- and H_+. The general theory of the triples of spaces $H_+ \subset H_0 \subset H_-$ is given in Section 8.1, and we will use the results proved in Section 8.1.

In our case $H_+ = H^\alpha$, $H_0 = H^0$, $H_- = \dot{H}^{-\alpha}$. In general, $H_+ \subset H_0$ and H_0 are Hilbert spaces, H_+ is dense in H_0, $\| u \|_0 \leq \| u \|_+$, and H_- is the dual space to H_+ with respect to H_0. It is proved in Section 8.1 that there exist linear isometries $p_+ : H_0 \to H_+$ and $p_- : H_- \to H_0$, and $(u,v)_+ = (qu, qv)_0$, where $q = p_+^{-1}$. The operator q^*, the adjoint of q in H_0, is an isometry of H_0 onto H_-.

Let us rewrite equation (3.2) in the equivalent form

$$Ah_0 := qRq^*h_0 = f_0, \tag{3.17}$$

where

$$f_0 := qf, \quad h_0 := (q^*)^{-1}h, \quad f_0 \in H_0, \quad h_0 \in H_0. \tag{3.18}$$

The linear operator qRq^* is bounded, selfadjoint, and positive definite:

$$(qRq^*\phi, \phi)_0 = (Rq^*\phi, q^*\phi)_0 \geq c_1 \| q^*\phi \|_-^2 = c_1 \| \phi \|_0^2, \quad c_1 > 0. \tag{3.19}$$

Moreover

$$(R_1^*\phi, q^*\phi)_0 \leq c_2 \| q^*\phi \|_-^2 = c_2 \| \phi \|_0^2, \quad c_2 > 0. \tag{3.20}$$

Here we used the isometry of q^*: $\| q^*\phi \|_- = \| \phi \|_0$, and the inequality

$$c_1 \| h \|_-^2 \leq (Rh, h)_0 \leq c_2 \| h \|_-^2, \quad c_2 \geq c_1 > 0. \tag{3.21}$$

This inequality is proved in Section 3.4, Lemma 3.5, below.

Equation (3.17) with a linear positive definite operator A on a Hilbert space H_0 is uniquely solvable in H_0 and its solution can be obtained by iterative or projection methods. If the solution h_0 to equation (3.17) is found then the desired function $h = q^* h_0$. Let us describe these methods.

Let us start with an iterative method. Assume that A is a bounded positive definite operator on a Hilbert space:

$$0 < m \leq A \leq M. \tag{3.22}$$

This means that $m \parallel \phi \parallel^2 \leq (A\phi, \phi) \leq M \parallel \phi \parallel^2$, $\forall \phi \in H$. Let

$$Au = f. \tag{3.23}$$

Consider the iterative process

$$u_{n+1} = (I - aA)u_n + af, \quad a := \frac{2}{M+m}, \tag{3.24}$$

where $u_0 \in H$ is arbitrary.

Lemma 3.1 *There exists* $\lim_{n\to\infty} u_n = u$. *This limit solves equation (3.23). One has*

$$\parallel u_n - u \parallel \leq cq^n, \quad q := \frac{M-m}{M+m}, \quad 0 < q < 1, \quad c = const > 0. \tag{3.25}$$

This is a well known result (see e.g. [Kantorovich and Akilov (1980)]). We give a proof for convenience of the reader.

Proof. If $u_n \to u$ in H then, passing to the limit in (3.24), one concludes that the limit u solves equation (3.23). In order to prove convergence and the estimate (3.25), it is sufficient to check that

$$\parallel I - aA \parallel \leq q, \tag{3.26}$$

where q is defined in (3.25). This follows from the spectral representation:

$$\parallel I - aA \parallel = \sup_{m \leq \lambda \leq M} \left| 1 - \frac{2\lambda}{m+M} \right| = \frac{M-m}{M+m} = q. \tag{3.27}$$

Lemma 3.1 is proved. □

If A is not positive definite but only nonnegative, and $f \in R(A)$, where $R(A)$ is the range of A, then consider the following iterative process

$$u_{n+1} + Au_{n+1} = u_n + f \tag{3.28}$$

$u_0 \in H$ is arbitrary.

Lemma 3.2 *If $A \geq 0$ and $f \in R(A)$ then there exists*

$$\lim_{n\to\infty} u_n = u, \tag{3.29}$$

where u_n is defined by (3.28) and u solves equation (3.23).

Proof. If $u_n \to u$ in H then, passing to the limit in (3.28) yields equation (3.23) for u. In order to prove that $u_n \to u$ one writes equation (3.28) as

$$u_{n+1} = Bu_n + h, \tag{3.30}$$

where

$$B := (I + A)^{-1}, \quad h := Bf. \tag{3.31}$$

Since $A \geq 0$ one has $0 \leq B \leq I$, where I is the identity operator in H. Under this condition ($0 \leq B \leq I$) one can prove [Krasnoselskii *et al.* (1972), p. 71] that $u_n \to u$. Lemma 3.2 is proved. □

Remark 3.1 *If A satisfies assumptions (3.22), then $0 < (M+1)^{-1} \leq B \leq (m+1)^{-1}$, and the iterative process (3.30) converges as a geometrical series with $q = (m+1)^{-1}$.*

2. Let us consider the projection methods for solving equation (3.23) under the assumption (3.22).

First, consider the least squares method which is a variant of the projection method. The least squares method can be described as follows. Take a complete linearly independent system $\{\phi_j\}$ in H. Look for a solution

$$u_n = \sum_{j=1}^{n} c_j \phi_j. \tag{3.32}$$

Find the constants c_j from the condition

$$\| Au_n - f \| = \min. \tag{3.33}$$

This leads to the linear system for c_j:

$$\sum_{j=1}^{n} a_{ij} c_j = f_i, \quad 1 \leq i \leq n \tag{3.34}$$

where

$$a_{ij} := (A\phi_j, A\phi_i), \quad f_i = (f, A\phi_i). \tag{3.35}$$

Since the system ϕ_j, $1 \le j \le n$, is linearly independent for any n, and A is an isomorphism of H onto H, the system $\{A\phi_j\}$, $1 \le j \le n$, is linearly independent for any n. Therefore $\det a_{ij} \ne 0$, $1 \le i, j \le n$, and the system (3.34) is uniquely solvable for any right hand sides and any n. Let $c_j^{(n)}$, $1 \le j \le n$, be the unique solution to system (3.34) and

$$u_n := \sum_{j=1}^{n} c_j^{(n)} \phi_j. \tag{3.36}$$

Let us prove that $u_n \to u$ as $n \to \infty$. It is sufficient to prove that the system $\{A\phi_j\}$, $1 \le j < \infty$, is complete in H. Indeed, if this is so, then $\| Au_n - f \| \to 0$ as $n \to \infty$, where u_n is given by (3.36). Therefore

$$\| u_n - u \| = \| A^{-1}(Au_n - f) \| \le m^{-1} \| Au_n - f \| \to 0. \tag{3.37}$$

Here we used the estimate $\| A^{-1} \| \le m^{-1}$. It is easy to check that the system $\{A\phi_j\}$, $1 \le j < \infty$, is complete in H. Indeed, suppose $(h, A\phi_j) = 0$, $1 \le j < \infty$, for some $h \in H$. Then $(Ah, \phi_j) = 0$, $1 \le j < \infty$. Thus $Ah = 0$ since by the assumption the system $\{\phi_j\}$, $1 \le j < \infty$, is complete in H. Since A^{-1} exists, equation $Ah = 0$ implies $h = 0$. We have proved the following lemma.

Lemma 3.3 *If A satisfies condition (3.22) and $\{\phi_j\}$, $1 \le j < \infty$, is a complete linearly independent system in H, then the least squares method of solving equation (3.23) converges. Namely: a) for any n the system (3.34) is uniquely solvable and the aproximate solution u_n is uniquely determined by formula (3.36), and b) $\| u_n - u \| \to 0$ as $n \to \infty$, where u is the unique solution to equation (3.23).*

The general projection method can be described as follows. Pick two complete linearly independent systems in H $\{\phi_j\}$ and $\{\psi_j\}$, $1 \le j < \infty$. Look for an approximate solution to equation (3.23) of the form (3.32). Find the coefficients c_j, $1 \le j \le n$, from the condition

$$(Au_n - f, \psi_i) = 0, \quad 1 \le i \le n. \tag{3.38}$$

Geometrically this means that the vector $Au_n - f$ is orthogonal to the linear span of the vectors ψ_i, $1 \le i \le n$. Equations (3.38) can be written as

$$\sum_{j=1}^{n} b_{ij} c_j = f_i, \quad 1 \le i \le n \tag{3.39}$$

where

$$b_{ij} = (A\phi_j, \psi_i), \quad f_i = (f, \psi_i). \tag{3.40}$$

The least squares method is the projection method with $\psi_i = A\phi_i$. In [Krasnoselskii, M. (1972)] one can find a detailed study of the general projection method.

Let us give a brief argument which demonstrates convergence of the projection method. Let $\{\phi_j\}$ be a complete linearly independent system in H, $L_n := \text{span}\{\phi_1, \dots \phi_n\}$, P_n is the orthogonal projection on L_n in H. An infinite system $\{\phi_j\}$ is called linearly independent in H, if, for any n, the system $\{\phi_j\}$, $1 \leq j \leq n$, is linearly independent in H. Take $\psi_j = \phi_j$ and write equation (3.38) as

$$P_n A u_n = P_n f, \quad u_n \in L_n. \tag{3.41}$$

Since $u_n = P_n u_n$ and the operator $P_n A P_n$ is selfadjoint positive definite on the subspace $L_n \subset H$, equation (3.41) is uniquely solvable for any $f \in H$ and any n. Note that $P_n A u_n = P_n A P_n u_n$ and A satisfies assumes (3.22). To prove that $\| u_n - u \| \to 0$ as $n \to \infty$, let us subtract from (3.41) the equation

$$P_n A h = P_n f. \tag{3.42}$$

The result is

$$P_n A P_n (u_n - u) = P_n A (u - P_n u). \tag{3.43}$$

Since $\{\phi_j\}$ is complete, $P_n \to I$, as $n \to \infty$, strongly, where I is the identity operator. This and the boundedness of A imply

$$\| P_n A (u - P_n u) \| \leq c \| u - P_n u \| \to 0, \quad n \to \infty. \tag{3.44}$$

Multiply (3.43) by $P_n(u - u_n)$ and use the positive definiteness of A to obtain

$$c \| P_n (u - u_n) \|^2 \leq \| u - P_n u \| \| P_n (u - u_n) \|,$$

or

$$c \| P_n (u - u_n) \| \leq \| u - P_n u \| \to 0, \quad n \to \infty. \tag{3.45}$$

Since $P_n u_n = u_n$, equations (3.44) and (3.45) imply

$$\| u - u_n \| \leq \| u - P_n u \| + \| P_n u - u_n \| \to 0, \quad n \to \infty. \tag{3.46}$$

We have proved

Lemma 3.4 *If (3.22) holds and $\{\phi_j\}$ is a complete linearly independent system in H, then the projection method (3.41) for solving equation (3.23) converges, and equation (3.41) is uniquely solvable for any n.*

3. If one applies the projection method with $\psi_j = \phi_j$ to equation (3.17) and chooses a Schauder basis $\{\phi_j\}$ of H, then the matrix

$$a_{ij} := (qRq^*\phi_j, \phi_i) = (Rq^*\phi_j, q^*\phi_i) \tag{3.47}$$

and the numbers

$$f_{0i} = (f_0, \phi_i) = (f, q^*\phi_i) \tag{3.48}$$

can be computed. The parentheses here denote the inner product in $H^0 = H_0$, $(u, v) = (u, v)_0$. As $q^*\phi_i := w_i$ one can take a basis of H_-: since q^* is an isometry between H_0 and H_- it sends a basis $\{\phi_i\}$ of H_o onto a basis $\{w_i\}$ of H_-. Let us suggest a system $\{w_i\}$ for computational purposes. Let B be a ball which contains the domain D, $\{v_j\}$ be the orthonormal in $L^2(B)$ system of eigenfunctions of the Dirichlet Laplacian in B:

$$-\Delta v_j = \lambda_j v_j, \quad v_j = 0 \quad \text{on} \quad \partial B. \tag{3.49}$$

For any $-\infty < \beta < \infty$, the system $\{v_j\}$, $1 \leq j < \infty$, forms a basis of $H^\beta(B)$. Indeed the norm in $H^\beta(B)$ is equivalent to the norm

$$\| (-\Delta)^{\beta/2} u \|_{L^2(B)} := \| u \|_s$$

and

$$\| u \|_\beta = \left(\sum_{j=1}^{\infty} |u_j|^2 \lambda_j^\beta \right)^{1/2}, \quad u_j := (u, v_j). \tag{3.50}$$

Therefore, $u \in H^\beta(B)$ is a necessary and sufficient condition for the Fourier series

$$u = \sum_{j=1}^{\infty} u_j v_j \tag{3.51}$$

to converge in $H^\beta(B)$, so that the system $\{v_j\}$ is a basis of $H^\beta(B)$ for any β, $-\infty < \beta < \infty$. Moreover, this basis is orthogonal in $H^\beta(B)$ for any β, although it is not normalized in $H^\beta(B)$ for $\beta \neq 0$. In order to check the orthogonality property, note that

$(v_j, v_i)_\beta = \left((-\Delta)^\beta v_j, v_i\right)_0 = \lambda_j^\beta (v_j, v_i)_0 = \lambda_j^\beta \delta_{ji}$,
where δ_{ji} is the Kronecker delta.

The basis $\{v_j\}$ can be used therefore as a basis of $H_- = \dot{H}^{-\beta}$ for any $\beta \geq 0$. In the case of equation (3.17) with $R \in \mathcal{R}$, one has $H_- = \dot{H}^{-\alpha}$, $\alpha = s(q-p)/2$. Although the system $\{v_j\}$, $1 \leq j < \infty$, is a basis of $\dot{H}^{-\alpha}$, it is not very convenient for representation of singular functions, such as δ_Γ, for example. The situation is similar to the one arising when $\delta(x-y)$ is represented as

$$\delta(x-y) = \sum_{j=1}^{\infty} \phi_j(x)\phi_j^*(y), \tag{3.52}$$

where $\{\phi_j\}$ is an orthonormal basis in $L^2(D)$. Formula (3.52) is valid in the sense that for any $f \in L^2(D)$ one has

$$f(x) = \sum_{j=1}^{\infty} (f, \phi_j)\phi_j. \tag{3.53}$$

The sequence $\delta_n(x-y) := \sum_{j=1}^{n} \phi_j(x)\phi_j^*(y)$ is a delta-sequence, that is

$$\left\| \int_D f\delta_n(x-y)dy - f(x) \right\|_{L^2(D)} \to 0 \quad \text{as} \quad n \to \infty \quad \forall f \in L^2(D). \tag{3.54}$$

The series (3.52) does not converge in the classical sense.

3.3 Multidimensional equation

In this section we describe the application of the basic idea presented in Section 3.1 to the multidimensional equation of random fields estimation theory

$$\int_D R(x,y)h(y)dy = f(x), \quad x \in \overline{D} \subset R^r. \tag{3.55}$$

We assume that $R \in \mathcal{R}$ and Γ is smooth. Let $j = (j_1, \ldots j_r)$ be a multiindex, $|j| = j_1 + j_2 + \cdots + j_r$. By $\{b(s)\delta_\Gamma\}^{(j)}$ we mean the distribution with support on $\Gamma = \partial D$ which acts on a test function $\phi \in C_0^\infty(R^r)$ by the formula

$$\left(\{b(x)\delta_\Gamma\}^{(j)}, \phi\right) = (-1)^{|j|} \int_\Gamma b(s)\phi^{(j)}(s)ds. \tag{3.56}$$

Here $b(s)$ is a smooth function on Γ. Let us look for an approximate solution to equation (3.55) in $\dot{H}^{-\alpha}$ of the form

$$h_n = \sum_{j=1}^{n} c_j\phi_j(x) + \sum_{i=0}^{\alpha-1}\sum_{m=1}^{n}(-1)^{|i|}a_{mi}\{b_{mi}(s)\delta_\Gamma\}^{(i)} := h_{on} + h_{sn} \quad (3.57)$$

where a_{mi} and c_j are constants, the system $\{\phi_j\}$, $1 \le j < \infty$, forms a basis of $H_0 = L^2(D)$, and the systems $\{b_{mi}(s)\}$, $1 \le m < \infty$, form a basis of $L^2(\Gamma)$, $0 \le |i| \le \alpha - 1$. Here h_{on} stands for an approximation of h_0, the ordinary part of the solution $h = h_o + h_s$ to equation (3.55), and h_{sn} stands for an approximation of the singular part, h_s, of this solution. If $f \in H^{2qs-ps}$, then $G(x)$, defined by formula (2.16), belongs to H^{2qs} and h, the solution to (3.55) given by formula (2.15), belongs to $H^0 = L^2(D)$ in the interior of D, $h_o = Q(\mathcal{L})G|_D$, where the symbol $h|_D$ denotes the restriction of the distribution h to the interior of D. For example, if $D = (t - T, t)$, h is given by formula (2.49), then $h|_D = \frac{-f''+f}{2}$. The term

$$h_{on} := \sum_{j=1}^{n} c_j\phi_j(x) \quad (3.58)$$

in (3.57) can approximate h_o with an arbitrary accuracy in H^0, if n is large enough, because the system $\{\phi_j\}$, $1 \le j < \infty$, forms a basis of H^0. It would be sufficient to assume that $\{\phi_j\}$ is a complete in H^0 linearly independent system. The term

$$h_{sn} := \sum_{i=0}^{\alpha-1}\sum_{m=1}^{n}(-1)^{i}a_{mi}\{b_{mi}(s)\delta_\Gamma\}^{(i)} \quad (3.59)$$

can approximate h_s in $\dot{H}^{-\alpha}$ with an arbitrary accuracy, if n is large enough, because the systems $\{b_{mi}(s)\}$, $1 \le m < \infty$, are complete in $L^2(\Gamma)$ and h_s is of the form (see formula (2.16))

$$h_s = \sum_{i=0}^{\alpha-1}(-1)^{|i|}\{b_i(s)\delta\}^{(i)}, \quad (3.60)$$

where the coefficients $b_i(s)$ are the traces on Γ of certain derivatives of $f(x)$ (see formula (2.85), for example). If $f(x)$ is sufficiently smooth then the functions $b_i(s)$ are in $L^2(\Gamma)$ and can be approximated in $L^2(\Gamma)$ with an arbitrary accuracy, if m is large enough, by linear combinations of functions $b_{mi}(s)$ because the systems $\{b_{mi}(s)\}$, $1 \le m < \infty$, are assumed to be complete in $L^2(\Gamma)$ for any $0 \le |i| \le \alpha - 1$.

Choose coefficients c_j and b_{mi} in (3.57) so that

$$\| Rh_n - f \|_1 = \min. \tag{3.61}$$

The variational problem (3.61) leads to the linear algebraic system for the coefficients c_j and b_{mi}. The arguments given in Section 3.1 below formula (3.7) remain valid without essential changes. Rather than formulate some general statements, consider an example. Let

$$\int_D \frac{\exp(-|x-y|)}{4\pi|x-y|} h dy = f(x), \quad D \subset R^3. \tag{3.62}$$

This is equation (2.84) with $a = 1$. Look for its approximate solution in $\dot{H}^{-1}$ of the form

$$h_n = \sum_{j=1}^{n} c_j \phi_j + \sum_{m=1}^{n} a_m b_m(s) \delta_\Gamma. \tag{3.63}$$

Here $\alpha = 1$, so that the double sum in (3.59) reduces to the second sum in (3.63), c_j and a_m are the coefficients to be determined by the least squares method (3.61). One has

$$Rh_n = \sum_{j=1}^{n} c_j \eta_j(x) + \sum_{m=1}^{n} a_m w_m(x) \tag{3.64}$$

where

$$g(x,s) = \frac{\exp(-|x-s|)}{4\pi|x-s|}, \quad \eta_j := R\phi_j(x) \tag{3.65}$$

$$w_m(x) := \int_\Gamma b_m(s) g(x,s) ds. \tag{3.66}$$

Therefore (3.61) yields:

$$\| \sum_{j=1}^{n} c_j \eta_j(x) + \sum_{m=1}^{n} a_m w_m(x) - f \|_1 = \min. \tag{3.67}$$

This leads to the linear system for the $2n$ coefficients c_j and a_m:

$$\sum_{j=1}^{2n} a_{ij} \beta_j = \gamma_i, \quad 1 \le i \le 2n \tag{3.68}$$

where

$$\begin{aligned}
\beta_j &= c_j, 1 \le j \le n, \quad \beta_j = a_m, \quad j = n+m, \quad 1 \le m \le n, && (3.69)\\
\gamma_i &= (f, \eta_i)_1, 1 \le i \le n, \gamma_i = (f, w_i)_1, \quad i = n+m, 1 \le m \le n, && (3.70)\\
a_{ij} &:= (v_j, v_i)_1, && (3.71)\\
v_i &= \eta_i, \quad 1 \le i \le n, \quad v_i = w_i, \quad i = n+m, j \quad 1 \le m \le n. && (3.72)
\end{aligned}$$

Exercise. Under what assumptions can one prove that
$\| Rh_n - f \|_1 \to 0$ as $n \to \infty$ implies $\| h_{on} - h_o \|_o \to 0$ and
$\| h_{sn} - h_s \|_{-1} \to 0$ as $n \to \infty$?

3.4 Numerical solution based on the approximation of the kernel

Consider the basic equation

$$Rh := \int_D R(x,y)h(y)dy = f(x), \quad x \in \overline{D} \subset R^r \tag{3.73}$$

with kernel

$$R(x,y) = \int_\Lambda \tilde{R}(\lambda)\Phi(x,y,\lambda)d\rho(\lambda), \quad \tilde{R}(\lambda) > 0, \tag{3.74}$$

where $\tilde{R}(\lambda)$ is a positive continuous function vanishing at infinity. Let us call this function the spectral density corresponding to the kernel $R(x,y)$. Assume that, for any $\epsilon > 0$, one can find polynomials $P_\epsilon(\lambda) > 0$ and $Q_\epsilon(\lambda) > 0$, such that the kernel $\tilde{R}_\epsilon(\lambda) := P_\epsilon(\lambda)Q_\epsilon^{-1}(\lambda)$ approximates $\tilde{R}(\lambda)$ in the following sense:

$$\sup_{\lambda\in\Lambda}\{|\tilde{R}_\epsilon - \tilde{R}|(1+\lambda^2)^\beta\} :=\| \tilde{R}_\epsilon - \tilde{R} \|_\beta < \epsilon. \tag{3.75}$$

We assume that for all sufficiently small ϵ, $0 < \epsilon < \epsilon_o$, ϵ_o is a small number, one has

$$\deg Q_\epsilon(\lambda) - \deg P_\epsilon(\lambda) = 2\beta > 0, \tag{3.76}$$

where β does not depend on ϵ. For example, if
$\tilde{R}(\lambda) = \left(\frac{\lambda^2+3}{\lambda^2+2}\right)^{1/2} \frac{1}{\lambda^2+1}$, $\Lambda = (-\infty, \infty)$,

then $\beta = 1$. We also assume that

$$\inf_{0<\epsilon<\epsilon_o} \left\{ \inf_{\lambda\in\Lambda} |\tilde{R}_\epsilon(\lambda)|(1+\lambda^2)^\beta \right\} := \gamma_o > 0, \tag{3.77}$$

and

$$\inf_{\lambda\in\Lambda} \{\tilde{R}(\lambda)(1+\lambda^2)^\beta\} := \gamma_1 > 0, \quad \sup_{\lambda\in\Lambda}\{\tilde{R}(\lambda)(1+\lambda^2)^\beta\} := \gamma_2 < \infty. \tag{3.78}$$

The basic idea of this section is this: if the operator $R_\epsilon : \dot{H}^{-\beta s} \to H^{\beta s}$, with the spectral density $\tilde{R}_\epsilon(\lambda)$ is for all $\epsilon \in (0, \epsilon_o)$ an isomorphism and the asumptions (3.75)-(3.78) hold with constants β, c which do not depend on ϵ, then R, the operator with the kernel $R(x, y)$, is also an isomorphism between $\dot{H}^{-\beta s}$ and $H^{\beta s}$. Therefore the properties of the operator R will be expressed in terms of the properties of the rational approximants of its spectral density.

We will need some preparations. First, let us prove a general lemma.

Lemma 3.5 *Let $H_+ \subset H_0 \subset H_-$ be a rigged triple of Hilbert spaces, where H_- is the dual space to H_+ with respect to H_0. Assume that $R : H_- \to H_+$ is a linear map such that*

$$c_1 \parallel h \parallel_-^2 \leq (Rh, h) \leq c_2 \parallel h \parallel_-^2, \quad \forall h \in H_-, \tag{3.79}$$

where $0 < c_1 < c_2$ are constants, and (f, h) is the value of the functional $h \in H_-$ on the element $f \in H_+$. Then

$$\parallel R \parallel \leq c_2, \quad \parallel R^{-1} \parallel \leq c_1^{-1}, \tag{3.80}$$

so that R is an isomorphism of H_- onto H_+. Here $\parallel R \parallel$ is the norm of the mapping

$$R : H_- \to H_+.$$

Proof. One has

$$c_1 \parallel h \parallel_-^2 \leq (Rh, h) \leq \parallel Rh \parallel_+ \parallel h \parallel_-,$$

so that

$$\parallel Rh \parallel_+ \geq c_1 \parallel h \parallel_- \quad \forall h \in H_-. \tag{3.81}$$

Therefore R^{-1} is defined on the range of R and

$$\parallel R^{-1} \parallel \leq c_1^{-1}. \tag{3.82}$$

Let us prove that the map R is surjective, that is, the range of R is all of H_+. If it is not, then there exists a $\phi \in H_-$, $\phi \neq 0$, such that

$$(Rh, \phi) = 0 \quad \forall h \in H_-. \tag{3.83}$$

It follows from (3.83) that

$$(h, R\phi) = 0 \quad \forall h \in H_-. \tag{3.84}$$

Therefore $R\phi = 0$ and

$$0 = (R\phi, \phi) \geq c_1 \parallel \phi \parallel_-^2 . \tag{3.85}$$

Thus $\phi = 0$ contrary to the assumption. Therefore the map R is surjective.

Let us now prove the first inequality (3.80). One has

$$\begin{aligned} \parallel R \parallel &= \sup |(Rg, h)| = \sup |Re(Rg, h)| \\ &= \sup \frac{|(R(h+g), h+g) - (R(h-g), h-g)|}{4} \\ &\leq \frac{c_2}{4} \sup \left\{ \parallel h+g \parallel_-^2 + \parallel h-g \parallel_-^2 \right\} \leq c_2, \end{aligned} \tag{3.86}$$

where the supremum is taken over all $h, g \in H_-$ such that

$$\parallel h \parallel_- \leq 1, \quad \parallel g \parallel_- \leq 1. \tag{3.87}$$

□

Remark 3.2 *The surjectivity of the map $R : H_- \to H_+$ follows also from the fact that R is a coercive, monotone and continuous mapping (see, e.g., [Deimling (1985), p. 100]).*

Let us now prove

Lemma 3.6 *Let $R_\epsilon : H_- \to H_+$ be an isomorphism for all $\epsilon \in (0, \epsilon_o)$, where $\epsilon_o > 0$ is a small fixed number. Let $R : H_- \to H_+$ be a linear map defined on all of H_-. Assume that*

$$\parallel R_\epsilon^{-1} \parallel \leq M \tag{3.88}$$

and

$$\parallel R - R_\epsilon \parallel < \epsilon, \tag{3.89}$$

where $M = const > 0$ does not depend on $\epsilon \in (0, \epsilon_o)$. Then $R : H_- \to H_+$ is an isomorphism and

$$\| R^{-1} \| \le M(1 - \epsilon M)^{-1} \quad \textit{for} \quad \epsilon M < 1. \tag{3.90}$$

Proof. One has

$$R = R_\epsilon + R - R_\epsilon = R_\epsilon[I + R_\epsilon^{-1}(R - R_\epsilon)], \tag{3.91}$$

where I is the identity operator on H_-. The operator $R_\epsilon^{-1}(R - R_\epsilon)$ is an operator from H_- into H_- and

$$\| R_\epsilon^{-1}(R - R_\epsilon) \| \le \epsilon M \tag{3.92}$$

because of (3.88) and (3.89). If $\epsilon M < 1$, then the operator $I + R_\epsilon^{-1}(R - R_\epsilon)$ is an isomorphism of H_- onto H_-, and

$$\| [I + R_\epsilon^{-1}(R - R_\epsilon)]^{-1} \| \le (1 - \epsilon M)^{-1}. \tag{3.93}$$

Therefore, the operator R is an isomorphism of H_- onto H_+,

$$R^{-1} = [I + R_\epsilon^{-1}(R - R_\epsilon)]^{-1} R_\epsilon^{-1}, \tag{3.94}$$

and

$$\| R^{-1} \| \le M(1 - \epsilon M)^{-1}, \epsilon M < 1. \tag{3.95}$$

Lemma 3.6 is proved. □

Let us choose $H_- = \dot{H}^{-\beta s}$, $H_0 = H^0 = L^2(D)$, and $H_+ = H^{\beta s}$, where $s = \text{ord}\mathcal{L}$, and $\mathcal{L}$ is the elliptic operator which defines the kernel $R \in \mathcal{R}$ of equation (3.73). Note that, by Parseval's equality, one has:

$$(Rh, h) = \int_\Lambda \tilde{R}(\lambda)|\tilde{h}(\lambda)|^2 d\rho(\lambda) \le \gamma_2 \int_\Lambda (1+\lambda^2)^{-\beta}|\tilde{h}(\lambda)|^2 d\rho(\lambda) = \gamma_2 \| h \|_{-\beta s}^2 \tag{3.96}$$

and, similarly,

$$\gamma_1 \| h \|_{-\beta s}^2 \le (Rh, h), \tag{3.97}$$

where γ_1 and γ_2 are constants from condition (3.78). From (3.96), (3.97) and Lemma 3.5, one obtains

Lemma 3.7 *If the spectral density $\tilde{R}(\lambda)$ of the kernel (3.74) satisfies conditions (3.78) with some $\beta > 0$, then the operator R with kernel $R(x, y)$,* *defined by formula (3.74), is an isomorphism between the spaces* $\dot{H}^{-\beta s}(D)$ *and* $H^{\beta s}(D)$ *and*

$$\| R \| \leq \gamma_2, \quad \| R^{-1} \| \leq \gamma_1^{-1}, \tag{3.98}$$

where γ_1 *and* γ_2 *are constants from condition (3.78).*

Let us discuss briefly the approximation problem. Let $\tilde{R}(\lambda)$ be a continuous positive function such that conditions (3.78) hold and

$$\lim_{\lambda \to \pm\infty} \tilde{R}(\lambda)\lambda^{2\beta} = \gamma_3, \tag{3.99}$$

where β is a positive integer, and let $\lambda = tg(\phi/2)$. Then λ runs through the real axis, $-\infty < \lambda < \infty$, if ϕ runs through the unit circle, $-\pi \leq \phi \leq \pi$. Because of the assumption (3.99), one can identify $+\infty$ and $-\infty$, and consider the function $\tilde{R}(\lambda)$ as a function

$$R(\phi) := \tilde{R}\left(tg\frac{\phi}{2} \right), \quad -\pi \leq \phi \leq \pi, \tag{3.100}$$

which is 2π periodic function defined on a unit circle. Since

$$\sin\phi = \frac{1-\lambda^2}{1+\lambda^2}, \quad \cos\phi = \frac{2\lambda}{1+\lambda^2} \tag{3.101}$$

and $\cos(m\phi)$ is a polynomial of degree m of $\cos\phi$, while $\frac{\sin(m+1)\phi}{\sin\phi}$ is also a polynomial of degree m of $\cos\phi$, a trigonometric polynomial

$$S_n(\phi) := a_o + \sum_{j=1}^{n} a_j \cos(j\phi) + b_j \sin(j\phi), \tag{3.102}$$

where a_j are b_j are constants, can be written as

$$S_n(\phi) := \tilde{R}_n(\lambda) = \frac{\sum_{m=0}^{2n} c_m \lambda^m}{(1+\lambda^2)^n}, \tag{3.103}$$

where c_m are some constants. Therefore, if one wishes to approximate a function $\tilde{R}(\lambda)$ on the whole axis $(-\infty, \infty)$ by a rational function, one can approximate the function $R(\phi)$, defined by formula (3.100) on the unit circle, by a trigonometric polynomial $S_n(\phi)$ and then write this polynomial as a rational function of λ as in formula (3.103). The function $\tilde{R}_n(\lambda)$, defined by formula (3.103), satisfies the condition

$$\tilde{R}_n(\lambda) \sim \gamma_3 \lambda^{-2\beta} \quad \text{as} \quad \lambda \to \pm\infty \tag{3.104}$$

if and only if β, $0 < \beta < n$, is an integer and

$$0 = c_{2n} = c_{2n-1} = \ldots = c_{2n-2\beta+1} = 0. \tag{3.105}$$

If (3.105) holds then the constant γ_3 in formula (3.104) equals $c_{2n-2\beta}$. The theory of approximation by trigonometric polynomials is well developed (see [Akhieser (1965)]). In particular, if condition (3.99) holds, then approximation in the norm

$$\sup_{\lambda \in R^1} (1+\lambda^2)^\beta |\tilde{R}(\lambda) - \tilde{R}_\epsilon(\lambda)| :=\| \tilde{R} - \tilde{R}_\epsilon \|_\beta \tag{3.106}$$

is possible. The norm (3.106) is the norm (3.75) with $\Lambda = R^1$. We keep the same notation for these two norms since there will be no chance to make an error by confusing these two norms: in all our arguments they are interchangeable.

Lemma 3.8 *If $\tilde{R}(\lambda)$ is a continuous function defined on $R^1 = (-\infty, \infty)$ which satisfies condition (3.99), then, for any $\epsilon > 0$, there exists a rational function $\tilde{R}_\epsilon = P_\epsilon(\lambda) Q_\epsilon^{-1}(\lambda)$ such that*

$$\| \tilde{R}(\lambda) - \tilde{R}_\epsilon(\lambda) \|_\beta < \epsilon \tag{3.107}$$

and condition (3.76) holds. If $\tilde{R}(\lambda) > 0$ then the polynomials $P_\epsilon(\lambda)$ and $Q_\epsilon(\lambda)$ can be chosen positive.

Proof. The function $\psi(\lambda) := (1+\lambda^2)^\beta \tilde{R}(\lambda)$ is continuous on R^1 and

$$\lim_{\lambda \to \pm\infty} (1+\lambda^2)^\beta \tilde{R}(\lambda) = \gamma_3 \tag{3.108}$$

because of (3.99). The function $\psi\left(tg\frac{\phi}{2}\right)$, $tg\frac{\phi}{2} = \lambda$, is a continuous function of ϕ on the interval $[-\pi, \pi]$. Therefore, there exists, for any given $\epsilon > 0$, a trigonometric polynomial such that

$$\max_{-\pi \le \phi \le \pi} \left| \psi\left(tg\frac{\phi}{2}\right) - S_n(\phi) \right| < \epsilon, \quad \lambda = tg\frac{\phi}{2}, \tag{3.109}$$

where $n = n(\epsilon)$ depends on ψ. From (3.109) and (3.103) it follows that

$$\max_{-\infty < \lambda < \infty} \left| (1+\lambda^2)^\beta \tilde{R}(\lambda) - \sum_{m=0}^{2n} c_m \lambda^m (1+\lambda^2)^{-n} \right| < \epsilon. \tag{3.110}$$

Therefore (3.107) holds with

$$P_\epsilon(\lambda) := \sum_{m=0}^{2n} c_m \lambda^m, \quad Q_\epsilon(\lambda) := (1+\lambda^2)^{-n-\beta}. \tag{3.111}$$

In order to prove the last statement of Lemma 3.8, one approximates first the continuous function $(1+\lambda^2)^{\beta/2}\tilde{R}^{1/2}(\lambda)$ by a rational function $T(\lambda)$ in the uniform norm on R^1 with accuracy $\leq \epsilon$, where $\epsilon > 0$ is a given number. Then the square of this rational function approximates the function $(1+\lambda^2)^{\beta}\tilde{R}(\lambda)$ with accuracy const $\cdot\, \epsilon$, where the constant does not depend on ϵ. Indeed, if $|f - T| < \epsilon$, then

$$|f^2 - T^2| \leq |f - T|(\max|f| + \max|T|) \leq \epsilon \cdot \text{const}. \tag{3.112}$$

If T is a rational function then $T^2(\lambda) = P(\lambda)Q^{-1}(\lambda)$ where $P(\lambda)$ and $Q(\lambda)$ are positive. Lemma 3.8 is proved. □

Let us summarize the results in the following theorem.

Theorem 3.1 *Let $\tilde{R}(\lambda)$ be a continuous positive function on R^1 and suppose condition (3.99) holds with a positive integer β. Then:*

a) for any $\epsilon > 0$, there exists a positive rational function $\tilde{R}_\epsilon(\lambda)$ such that conditions (3.75) and (3.76) hold;
b) for all sufficiently small ϵ, $0 \leq \epsilon \leq \epsilon_0$, the operator R_ϵ, with the kernel defined by the spectral density $\tilde{R}_\epsilon(\lambda)$, is an isomorphism of the space $\dot{H}^{-\beta s}(D) := \dot{H}^{-\beta s}$ onto $H^{\beta s}(D) := H^{\beta s}$;
c) the operator $R : \dot{H}^{-\beta s} \to H^{\beta s}$ is an isomorphism;
d) there exist positive constants γ_0, γ_1, and γ_2, such that conditions (3.77) and (3.78) hold;
e) the following estimates hold:

$$\| R \| \leq \gamma_2, \quad \| R^{-1} \| \leq \gamma_1^{-1}, \tag{3.113}$$

where γ_1 and γ_2 are the constants in formula (3.78);
f) if $\gamma_1^{-1}\epsilon < 1$, then

$$\| R_\epsilon^{-1} \| \leq \gamma_1^{-1}(1 - \gamma_1^{-1}\epsilon)^{-1} \tag{3.114}$$

and

$$\| R^{-1} - R_\epsilon^{-1} \| \leq \epsilon\gamma_1^{-2}(1 - \gamma_1^{-1}\epsilon)^{-1}. \tag{3.115}$$

Proof. The statements (a) to (e) of Theorem 3.1 follow from Lemmas 3.5-3.8. The statement (3.114) is analogous to (3.95) and can be proved similarly. The last statement (3.115) follows immediately from the identity:

$$R^{-1} - R_\epsilon^{-1} = R_\epsilon^{-1}(R_\epsilon - R)R^{-1} \tag{3.116}$$

and estimate (3.114), second estimate (3.113), and the estimate

$$\| R_\epsilon - R \| \leq \epsilon, \tag{3.117}$$

which is a consequence of (3.75). Let us explain why (3.75) implies (3.117). One has

$$\begin{aligned}
\| R \| &= \sup_{\|h\|_- \leq 1} |(Rh, h)| = \sup_{\|h\|_- \leq 1} \int \tilde{R}(\lambda) \left|\tilde{h}(\lambda)\right|^2 d\rho(\lambda) \\
&\leq \sup_{\lambda \in R^1} \left\{(1+\lambda^2)^\beta \left|\tilde{R}(\lambda)\right|\right\} \sup_{\|h\|_- \leq 1} \int (1+\lambda^2)^{-\beta} |\tilde{h}|^2 d\rho(\lambda) \\
&= \sup_{\lambda \in R^1} \left\{(1+\lambda^2)^\beta \left|\tilde{R}(\lambda)\right|\right\} \sup_{\|h\|_- \leq 1} \| h \|_-^2 \\
&= \sup_{\lambda \in R^1} \left\{(1+\lambda^2)^\beta \left|\tilde{R}(\lambda)\right|\right\}.
\end{aligned} \tag{3.118}$$

Here $H_- = \dot{H}^{-\beta s}$. Thus

$$\| R \| \leq \sup_{\lambda \in R^1} \left\{(1+\lambda^2)^\beta \left|\tilde{R}(\lambda)\right|\right\}. \tag{3.119}$$

From (3.75) and (3.119) one obtains (3.117). Theorem 3.1 is proved. □

It is now easy to study the stability of the numerical solution of equation (3.73) based on the approximation of $R(x, y)$.

Consider the equation

$$Rh_\delta = f_\delta, \quad f_\delta \in H_+, \tag{3.120}$$

where f_δ is the noisy data:

$$\| f_\delta - f \|_+ \leq \delta. \tag{3.121}$$

This means that, in place of the exact data $f \in H_+$, an approximate data f_δ is given, where $\delta > 0$ is the accuracy with which the given data approximates in H_+ the exact data. Suppose that $\tilde{R}(\lambda)$, the spectral density of the given kernel, satisfies condition (3.99) with $\beta > 0$ an integer. Take a kernel $R_\epsilon \in \mathcal{R}$, such that the estimate (3.117) holds with $\epsilon > 0$ sufficiently small. This is possible by Theorem 3.1. Then estimate (3.115) holds. Consider the equation

$$R_\epsilon h_{\epsilon\delta} = f_\delta. \tag{3.122}$$

By Theorem 3.1 the operator $R_\epsilon : H_- \to H_+$ is an isomorphism if $\epsilon > 0$ is sufficiently small, $0 < \epsilon < \epsilon_0$. Therefore, for such ϵ equation (3.122) is uniquely solvable in H_-. We wish to estimate the error of the approximate solution:

$$\begin{aligned} \| h - h_{\epsilon\delta} \|_- &= \| R^{-1} f - R_\epsilon^{-1} f_\delta \|_- \\ &\le \| R^{-1}(f - f_\delta) \|_- + \| (R^{-1} - R_\epsilon^{-1}) f_\delta \|_- \\ &\le \| R^{-1} \| \| f - f_\delta \|_+ + \| R^{-1} - R_\epsilon^{-1} \| \| f_\delta \|_+ \\ &\le \gamma_1^{-1}\delta + \epsilon\gamma_1^{-2}(1 - \epsilon\gamma_1^{-1})^{-1} \| f_\delta \|_+ . \end{aligned} \tag{3.123}$$

In our case $H_- = \dot{H}^{-\beta s}$, $H_+ = H^{\beta s}$. Estimate (3.123) proves that the error of the approximate solution goes to zero as the accuracy of the data increases, that is $\delta \to 0$. Indeed, one can choose $\epsilon > 0$ sufficiently small, so that the second term on the right side of the inequality (3.123) will be arbitrarily small, say less than δ. Then the right side of (3.123) is not more than $(\gamma_1^{-1} + 1)\delta$. We have proved

Lemma 3.9 *The error estimate of the approximate solution $h_{\epsilon\delta}$ is given by the inequality:*

$$\| h - h_{\epsilon\delta} \|_- \le \gamma_1^{-1}\delta + \epsilon\gamma_1^{-1}(1 - \epsilon\gamma_1^{-1})^{-1} \| f_\delta \|_+ \tag{3.124}$$

where γ_1 is the constant in condition (3.78).

3.5 Asymptotic behavior of the optimal filter as the white noise component goes to zero

Consider the equation

$$\epsilon h_\epsilon + R h_\epsilon = f, \quad \epsilon > 0, \tag{3.125}$$

where $R \in \mathcal{R}$, or, more generally, R is an isomorphism between H_- and H_+, where $H_+ \subset H_0 \subset H_-$ is a rigged triple of Hilbert spaces. We wish to study the behavior as $\epsilon \to 0$ of h_ϵ, the optimal filter. This question is of theoretical and practical interest as was explained in the Introduction. It will be discussed in depth in Chapter 5. We assume that the estimate

$$c_1 \parallel h \parallel_-^2 \leq (Rh, h) \leq c_2 \parallel h \parallel_-^2 \quad \forall h \in H_- \tag{3.126}$$

holds, where $c_1 > 0$ and $c_2 > 0$ are constants and the parentheses denote the pairing between H_- and H_+.

From (3.125) it follows that

$$\epsilon \parallel h_\epsilon \parallel_0^2 + (Rh_\epsilon, h_\epsilon) = (f, h_\epsilon), \tag{3.127}$$

where the parentheses denote the inner product in H_0, which is the pairing between H_- and H_+ (see Section 8.1). It follows from (3.126) and (3.127) that

$$c_1 \parallel h_\epsilon \parallel_-^2 \leq (Rh_\epsilon, h_\epsilon) \leq (f, h_\epsilon) \leq \parallel f \parallel_+ \parallel h_\epsilon \parallel_- . \tag{3.128}$$

Thus

$$\parallel h_\epsilon \parallel_- \leq c \parallel f \parallel_+, \quad c = c_1^{-1}, \tag{3.129}$$

where the constant $c > 0$ does not depend on ϵ. Since H_- is a Hilbert space, and bounded sets are weakly compact in Hilbert spaces, inequality (3.129) implies that there is a weakly convergent subsequence of h_ϵ, which we denote again h_ϵ, so that

$$h_\epsilon \rightharpoonup h \quad \text{in} \quad H_- \quad \text{as} \quad \epsilon \to 0. \tag{3.130}$$

Here $\rightharpoonup$ denotes weak convergence in H_- which means that for any $f \in H_+$ one has

$$(f, h_\epsilon) \to (f, h) \quad \text{as} \quad \epsilon \to 0, \quad \forall f \in H_+. \tag{3.131}$$

Let $\phi \in H_+$ be arbitrary. It follows from (3.125) that

$$\epsilon(h_\epsilon, \phi) + (Rh_\epsilon, \phi) = (f, \phi),$$

or, since $(Rh, \phi) = (h, R\phi)$, $R\phi \in H_+$,

$$\epsilon(h_\epsilon, \phi) + (h_\epsilon, R\phi) = (f, \phi), \quad \forall \phi \in H_+. \tag{3.132}$$

One has

$$|\epsilon(h_\epsilon, \phi)| \leq \epsilon \parallel h_\epsilon \parallel_- \parallel \phi \parallel_+ \to 0 \quad \text{as} \quad \epsilon \to 0, \tag{3.133}$$

where we used estimate (3.129). Therefore one can pass to the limit $\epsilon \to 0$ in equation (3.132) and obtain

$$(h, R\phi) = (f, \phi) \quad \forall \phi \in H_+, \tag{3.134}$$

or

$$(Rh - f, \phi) = 0 \quad \forall \phi \in H_+, \tag{3.135}$$

where $h \in H_-$ is the weak limit (3.130). Since $H_+ \subset H_-$ is dense in H_- in the norm of H_-, one concludes from (11) that

$$Rh = f. \tag{3.136}$$

We have proved the following theorem.

Theorem 3.2 *Let $H_+ \subset H_0 \subset H_-$ be a triple of rigged Hilbert spaces. If $R : H_- \to H_+$ is an isomorphism, and (3.126) holds, then the unique solution to equation (3.125) converges weakly in H_- to the unique solution to the limit equation (3.136).*

Remark 3.3 *The weak convergence in H_- is exactly what is natural in the estimation theory. Indeed, the estimate*

$$\hat{u} = \int_D h(x,y)f(y)dy = (h_x, f) \tag{3.137}$$

is the value of the functional (3.137) at the element f, $h_x = h(x,y) \in H_-$, $f \in H_+$, $x \in R^r$ being a parameter. The error of the optimal estimate (see e.g. formulas (2.96) and (2.108) are also expressed as the values of a functional of the form (3.137)).

One can prove that actually h_ϵ converges strongly to H_-. Indeed, equation (3.125) implies $(h_\epsilon, h_\epsilon)_- = (Rh_\epsilon, h_\epsilon) \le (Rh, h_\epsilon) = (h, h_\epsilon)_-$. Thus $||h_\epsilon||_- \le ||h||_-$. Choose a weakly convergent in H_- sequence $h_n := h_{\epsilon_n}$, $\lim_{n\to\infty} \epsilon_n = 0$. Then $h_n \rightharpoonup h$ in H_-, $||h||_- \le \underline{\lim}_{n\to\infty}||h_n||_-$ and $\overline{\lim}_{n\to\infty}||h_n|| \le ||h||_-$. Consequently, $\lim_{n\to\infty} ||h_n||_- = ||h||_-$. This and the weak convergence $h_n \rightharpoonup h$ in H_- imply strong convergence in H_-, so $\lim_{n\to\infty} ||h - h_n||_- = 0$.

3.6 A general approach

In this section we outline an approach to solving the equation

$$Rh = f \tag{3.138}$$

which is based on the theory developed in Section 8.1.

Assume that $R : H \to H$ is a compact positive operator in the Hilbert space H, that is

$$(Rh, h) > 0 \quad \forall h \in H, \quad h \neq 0. \tag{3.139}$$

The parentheses denote the inner product in H. The inner product

$$(h, g)_- := (Rh, g) \tag{3.140}$$

induces on H a new norm

$$\| h \|_- = (Rh, h)^{1/2} = \| R^{1/2} h \| . \tag{3.141}$$

Let H_- be the Hilbert space with the inner product (3.140) which is the completion of H in the norm (3.141). By H_+ we denote the dual to H_- space with respect to $H = H_0$ (see Section 8.1). One has

$$H_+ \subset H \subset H_- \tag{3.142}$$

where H_+ is dense in H and H is dense in H_-. The inner product in H_+ is

$$(u, v)_+ = (R^{-1}u, v) = (R^{-1/2}u, R^{-1/2}v), \quad u, v \in \text{Dom}(R^{-1/2}). \tag{3.143}$$

Therefore

$$\| u \|_+ = \| R^{-1/2} u \| . \tag{3.144}$$

One can see that H_+ is the range of $R^{1/2}$.

Indeed, $\text{Ran}(R^{1/2}) \subset H_+$ by definition and is closed in H_+ norm. Indeed, let $f_n = R^{1/2}u_n$ and assume that $\| f_n - f_m \|_+ \to 0$ as $n, m \to \infty$. Then, by (3.144), $\| u_n - u_m \| \to 0$ as $n, m \to \infty$. Therefore there exists a $u \in H$ such that $\| u_n - u \| \to 0$ as $n \to \infty$. Let $f := R^{1/2}u$. Then $f \in H_+$ and $\| f - f_n \|_+ \to 0$ as $n \to \infty$. Thus $\text{Ran}(R^{1/2})$ is closed in H_+, where $\text{Ran}(A)$ is the range of an operator A. Since H_+ is the completion of $\mathrm{Ran}(R^{1/2})$ in H_+ norm, it follows that $H_+ = \mathrm{Ran}(R^{1/2})$. One can also define the norm in H_+ as

$$\| u \|_+ = \sup_{\substack{h \in H \\ h \neq 0}} \frac{|(u, R^{-1/2}h)|}{\| h \|}. \tag{3.145}$$

If and only if the right side of (3.145) is finite one concludes that $u \in \mathrm{Dom}(R^{-1/2})$ and obtains from (3.145) equation (3.144). The triple (3.142) is a triple of rigged Hilbert spaces (see Section 8.1), and $R : H_- \to H_+$ is an isomorphism. Therefore, a general approach to stable solving equation (3.138) can be described as follows. Suppose an operator $A : H \to H$ is found such that $A > 0$ and the norm $(Au, u)^{1/2}$ is equivalent to the norm (3.141). In this case the spaces H_+ and H_- constructed with the help of A will consist of the same elements as the spaces H_+ and H_- constructed above, and the norms of these spaces are equivalent, so that one can identify these spaces. Suppose that one can construct the mapping $A^{-1} : H_+ \to H_-$. This mapping is an isomorphism between H_+ and H_-. Then equation (3.138) can be written as

$$Bh := A^{-1}Rh = A^{-1}f := g, \tag{3.146}$$

where $B : H_- \to H_-$ is an isomorphism. Therefore equation (3.138) is reduced to the equation

$$Bh = g \tag{3.147}$$

which is an equation in the Hilbert space H_- with a linear bounded operator B which is an isomorphism of H_- onto H_-. The operator B in equation (3.148) is selfadjoint and positive in H_-. Indeed

$$(Bh, v)_- = (RBh, v) = (RA^{-1}Rh, v) = (h, RA^{-1}Rv) = (h, Bv)_-. \tag{3.148}$$

Moreover

$$(Bh, h)_- = (A^{-1}Rh, Rh) > 0 \quad \text{for} \quad h \neq 0 \tag{3.149}$$

since A and R are positive.

Equation (3.148) with positive, in the sense (3.149), isomorphism B from H_- onto H_- can be easily solved numerically by iterative or projection methods described in Section 3.2.

Let us now describe some connections between the concepts of this section and the well known concept of a reproducing kernel.

Definition 3.1 *A kernel $K(x,y)$, $x,y \in D \subset R^r$, is called a reproducing kernel for a Hilbert space H_+ of functions defined on D, where D is a (not necessarily bounded) domain in R^r, if for any $u \in H_+$ one has*

$$(K(x,y), u(y))_+ = u(x). \tag{3.150}$$

It is assumed that, for every $x \in D$, $K(x,y) \in H_+$, and H_+ consists of functions for which their values at a point are well defined.

From (3.150) it follows that $(K(x,y)u, u)_+ \geq 0$, where $K(x,y)u := (K(x,y), u)_+$, and

$$K(x,x) \geq 0, \quad K(x,y) = K^*(y,x) \quad |K(x,y)|^2 \leq K(x,x)K(y,y) \tag{3.151}$$

as we will prove shortly. The reproducing kernel, if it exists for a Hilbert space H_+, is unique. Indeed, if K_1 is another reproducing kernel then

$$(K(x,y) - K_1(x,y), u(y))_+ = 0 \quad \forall u \in H_+. \tag{3.152}$$

Therefore $K(x,y) = K_1(x,y)$.

The reproducing kernel exists if and only if the estimate

$$|u(x)| \leq c \parallel u \parallel_+ \quad \forall u \in H_+ \tag{3.153}$$

holds with a positive constant c which does not depend on u. Indeed

$$|u(x)| \leq \left|(K(x,y), \quad u(y))_+\right| \leq \parallel K(x,y) \parallel_+ \parallel u \parallel_+ . \tag{3.154}$$

Thus (3.153) holds with $c = \parallel K(x,y) \parallel_+$. Note that

$$\parallel K(x,y) \parallel_+^2 = (K(x,y), K(x,y))_+ = K(x,x) \tag{3.155}$$

because of (3.150). Conversely, if (3.153) holds then, by Riesz's theorem about linear functionals on a Hilbert space, there exists a $K(x,y)$ such that (3.150) holds.

Since (3.150) implies that, for any numbers t_j, $1 \leq j \leq n$, one has

$$\sum_{i,j=1}^{n} K(x_i, x_j) t_i t_j^* = \left(\sum_{i=1}^{n} K(x_i, y) t_j, \quad \sum_{i=1}^{n} K(x,y) t_j \right) \geq 0, \tag{3.156}$$

one sees that the matrix $K(x_i, x_j)$ is nonnegative definite and therefore (3.151) holds.

Lemma 3.10 *Assume that $D \subset R^r$ is a bounded domain and the kernel $R(x,y)$ of the operator $R : H \to H$, $H = L^2(D)$ is nonnegative definite and* *continuous in $x, y \in \overline{D}$. Then the Hilbert space H_+ generated by R (see formula (3.143)) is a Hilbert space with reproducing kernel $R(x, y)$.*

Proof. If $u \in H_+$ then $u \in \mathrm{Ran}(R^{1/2})$ so that there is a v such that

$$u = R^{1/2}v, \quad v \in H. \tag{3.157}$$

If we prove that the operator $R^{1/2}$ is an integral operator:

$$u(x) = R^{1/2}v = \int_D T(x, y)v(y)dy \tag{3.158}$$

such that the function

$$t(x) := \left(\int_D |T(x, y)|^2 \, dy\right)^2 \tag{3.159}$$

is continuous in $\overline{D}$, then (3.158) and (3.144) imply

$$|u(x)| \le t(x) \parallel v \parallel = t(x) \parallel R^{-1/2}u \parallel = t(x) \parallel u \parallel_+ . \tag{3.160}$$

This is an estimate identical with (3.153), and we have proved that this estimate implies that H_+ has the reproducing kernel $K(x, y)$.

To finish the proof one has to prove (3.158) and (3.159). Since D is bounded, the operator $R : H \to H$ with continuous kernel is in the trace class. This means that

$$R(x, y) = \sum_{j=1}^{\infty} \lambda_j \phi_j(x) \phi_j^*(y) \tag{3.161}$$

where $\lambda_1 \ge \lambda_2 \ge \cdots > 0$ are the eigenvalues of R counted according to their multiplicities, ϕ_j are the normalized eigenfunctions

$$\int_D R(x, y)\phi_j(y)dy = \lambda_j \phi_j(x), \quad (\phi_j, \phi_i) = \delta_{ij}, \tag{3.162}$$

and

$$TrR = \sum_{j=1}^{\infty} \lambda_j = \int_D R(x, x)dx < \infty. \tag{3.163}$$

We will explain the second equality (3.163) later. The operator $R^{1/2}$ has the kernel

$$T(x, y) = \sum \lambda_j^{1/2} \phi_j(x) \phi_j^*(y), \tag{3.164}$$

which can be easily checked: the kernel of R is the composition:

$$T \circ T := \int_D T(x,z)T(z,y)dz = R(x,y). \tag{3.165}$$

Therefore

$$\int_D |T(x,y)|^2 dy = \sum_{j=1}^{\infty} \lambda_j |\phi_j(x)|^2 = R(x,x). \tag{3.166}$$

Therefore (3.158) and (3.159) are proved, and

$$t(x,x) = [R(x,x)]^{1/2}. \tag{3.167}$$

Let us finally sketch the proof of the second equality (3.163). This equality is well known and the proof is sketched for convenience of the reader.

It is sufficient to use

Mercer's theorem : If $R(x,y)$ is continuous and nonnegative definite, that is

$$\int_D \int_D R(x,y)h(y)h^*(x)dydx \geq 0 \quad \forall h \in L^2(D), \tag{3.168}$$

then the series (3.161) converges absolutely and uniformly in $\overline{D} \times \overline{D}$.

If Mercer's theorem is applied to the series (3.161) with $x = y$, then

$$\int_D R(x,x)dx = \sum_{j=1}^{\infty} \lambda_j \int_D |\phi_j|^2 dx = \sum_{j=1}^{\infty} \lambda_j. \tag{3.169}$$

Thus, (3.163) holds.

To prove Mercer's theorem, note that the kernel $R_n(x,y) := R(x,y) - \sum_{j=1}^{n} \lambda_j \phi_j(x)\phi_j^*(y)$ is nonnegative definite for every n. Therefore $R_n(x,x) \geq 0$ so that

$$\sum_{j=1}^{n} \lambda_j |\phi_j(x)|^2 \leq R(x,x) \quad \forall n. \tag{3.170}$$

Therefore the series $\sum_{j=1}^{\infty} \lambda_j |\phi_j(x)|^2 \leq R(x,x) \leq c$ converges and c does not depend on x because $R(x,x)$ is a continuous function in $\overline{D}$. Thus, the series

$$R(x,y) = \sum_{j=1}^{\infty} \lambda_j \phi_j(x)\phi_j^*(y) \tag{3.171}$$

converges uniformly in x for each $y \in \overline{D}$. Indeed:

$$\left|\sum_m^n \lambda_j \phi_j(x)\phi_j^*(y)\right|^2 \leq \sum_m^n \lambda_j |\phi_j(x)|^2 \cdot \sum_m^n \lambda_j |\phi_j(y)|^2$$
$$\leq c\sum_m^n \lambda_j |\phi_j(y)|^2 \to 0 \quad \text{as} \quad m,n \to \infty. \tag{3.172}$$

Take $y = x$ in (3.171) and get

$$R(x,x) = \sum_{j=1}^{\infty} \lambda_j |\phi_j(x)|^2. \tag{3.173}$$

Since $R(x,y)$ is continuous in $\overline{D} \times \overline{D}$, the functions $\phi_j(x)$ are continuous in $\overline{D}$. By Dini's lemma the series (3.173) converges uniformly in $x \in \overline{D}$. Therefore the series (3.173) can be termwise integrated which gives (3.169). Lemma 3.10 is proved. □

Exercise. Prove Dini's lemma: *if a monotone sequence of continuous functions on a compactum $\overline{D} \subset R^r$ converges to a continuous function, then it converges uniformly.*

In Definition 3.1 we assume that the space H_+ with reproducing kernel consists of functions $u(x)$ whose values at a point are well defined. This excludes spaces $L^2(D)$, for example. If the definition (3.150) is understood in the sense that both sides of (3.150) are equal as elements of H_+ (and not pointwise) then the spaces of the type L^2 can be included. However, in general, for such kind of spaces the reproducing kernel is not necessarily an element of the space. For example, if $H_+ = L^2(D)$ then (3.150) implies that $K(x,y)$ is the kernel of the identity operator in $L^2(D)$. But the identity operator in $L^2(D)$ does not have a kernel in the set of locally integrable functions. In the set of distributions, however, it has kernel $\delta(x-y)$, where $\delta(x)$ is the delta-function. As the kernel of the identity operator in $L^2(D)$ the delta-function is understood in weak sense:

$$\int_D \int_D \delta(x-y) f\overline{g}\,dxdy = \int_D f(x)\overline{g}(x)dx, \quad \forall f,g \in L^2(D). \tag{3.174}$$

Remark 3.4 *We have seen that the operator R in $L^2(D)$, $D \subset R^r$ is bounded, with continuous nonnegative definite kernel $R(x,y)$, belongs to the trace class. Therefore $R^{1/2}$ is a Hilbert-Schmidt operator. One can prove that such an operator is an integral operator without assuming that $R^{1/2} \geq 0$. A linear operator $A : H \to H$ on a Hilbert space is called a*

Hilbert-Schmidt operator if $\sum_{j=1}^{\infty} \| A\phi_j \|^2 < \infty$ *where* $\{\phi_j\}$, $1 \leq j < \infty$, *is an orthonormal basis of* H. *Pick an arbitrary* $f \in H$, $f = \sum_{j=1}^{\infty}(f, \phi_j)\phi_j$. *Consider*

$$Af = \sum_{i=1}^{\infty}(Af, \phi_i)\phi_i = \sum_{i,j=1}^{\infty}(f, \phi_j)(\phi_j, A^*\phi_i)\phi_i. \tag{3.175}$$

Let $H = L^2(D)$. *Then (3.175) can be written as*

$$Af = \int_D A(x,y)f(y)dy \tag{3.176}$$

with

$$A(x,y) := \sum_{i,j=1}^{\infty} a_{ji}\phi_j^*(y)\phi_i(x), \quad a_{ji} := (\phi_j, A^*\phi_i). \tag{3.177}$$

One can check that the series (3.177) converges in $L^2(D) \times L^2(D)$ *and*

$$\int_D \int_D |A(x,y)|^2 dx\, dy = \sum_{i,j=1}^{\infty} |a_{ji}|^2 < \infty. \tag{3.178}$$

Indeed, by Parseval's equality one has

$$\sum_{i=1}^{\infty}\sum_{j=1}^{\infty} |a_{ji}|^2 = \sum_{j=1}^{\infty} \| A\phi_j \|^2 < \infty. \tag{3.179}$$

One can prove that the sum (3.179) does not depend on the choice of the orthonormal basis of H.

Chapter 4

Proofs

In this chapter we prove all of the theorems formulated in Chapter 2 except Theorem 2.3, the proof of which is given in Section 8.3.2.10 as a consequence of an abstract theory we develop.

4.1 Proof of Theorem 2.1

In order to make it easier for the reader to understand the basic ideas, we give first a proof of Corollary 2.2, which is a particular case of Theorem 2.1. This case corresponds to the assumption $P(\lambda) = 1$, and in this case the transmission problem (2.18 - 2.20) reduces to the exterior Dirichlet boundary problem (2.22 - 2.23).

Proof of Corollary 2.2 Equation (2.12)

$$\int_D R(x,y)h(y)dy = f(x), \quad x \in \overline{D} \tag{4.1}$$

holds if and only if

$$(Rh, \phi) = (f, \phi) \quad \forall \phi \in \dot{H}^{-\alpha}. \tag{4.2}$$

Since smooth functions with compact support in D are dense in $\dot{H}^{-\alpha}$ equation (4.2) holds if and only if

$$(Rh, \phi) = (f, \phi) \quad \forall \phi \in H_0^m(D), \quad m \geq -\alpha, \tag{4.3}$$

where $H_0^m(D)$ is the Sobolev space of functions defined in the domain D with compact support in D. Let us take $\phi = Q(\mathcal{L})\psi$, $\psi \in C_0^\infty(D)$. The function $\phi \in H_0^m(D)$ if the coefficients of the operator $\mathcal{L}$ belong to $H^m(D)$.

We will assume that these coefficients are sufficiently smooth. Sharp conditions of smoothness on the coefficients of the operator $\mathcal{L}$ are formulated in the beginning of Section 2.2.

Let us write (4.3) as

$$(Q(\mathcal{L})Rh, \psi) = (Q(\mathcal{L})f, \psi) \quad \forall \psi \in C_0^\infty(D). \tag{4.4}$$

By the assumption

$$Q(\mathcal{L})R = \delta(x-y), \tag{4.5}$$

equation (4.4) reduces to

$$(h, \psi) = (Q(\mathcal{L})f, \psi) \quad \forall \psi \in C_0^\infty(D). \tag{4.6}$$

This means that the distribution h equals to $Q(\mathcal{L})f$ in the domain D (which is an open set). If f is smooth enough, say $f \in H^{qs}$, then the obtained result says that

$$\text{sing supp } h = \partial D = \Gamma, \tag{4.7}$$

since in D the distribution h is equal to a regular function $Q(\mathcal{L})f$ in D. In order to find h in $\overline{D}$, we extend f from D to R^r so that the extension F has two properties:

$$F \quad \text{is maximally smooth} \tag{4.8}$$

and

$$Q(\mathcal{L})F = 0 \quad \text{in} \quad \Omega. \tag{4.9}$$

Requirement (4.9) is necessary because the function $h = Q(\mathcal{L})F$ has to have support in $\overline{D}$. Requirement (4.8) is natural from two points of view. The first, purely mathematical, is: the requirement (4.8) selects the unique solution to equation (4.1) of minimal order of singularity, the mos solution to (4.1). The second point of view is of statistical nature: only the mos solution to equation (4.1) gives the (unique) solution to the estimation problem we are interested in (see formula (2.105)).

Let $F = u$ in Ω. Then (4.9) says that

$$Q(\mathcal{L})u = 0 \quad \text{in} \quad \Omega. \tag{4.10}$$

Since $F = f$ in D, condition (4.8) requires that

$$\partial_N^j u = \partial_N^j f \quad \text{on} \quad \Gamma = \partial D, \quad 0 \le j \le \frac{qs}{2} - 1, \tag{4.11}$$

where N is the outer normal to Γ, and one cannot impose more than $\frac{qs}{2}$ boundary conditions on u since the Dirichlet problem in Ω allows one to impose not more than $\frac{qs}{2}$ conditions on Γ.

Finally one has to impose the condition

$$u(\infty) = 0. \tag{4.12}$$

Indeed, one can consider F as the left hand side of equation (4.1) with $h \in \dot{H}^{-\alpha}$. In this case it is clear that condition (4.12) holds since $R(x,y) \to 0$ as $|x| \to \infty$.

The Dirichlet problem (4.10)-(4.12) is uniquely solvable in $H^\infty(\overline{\Omega})$ if $f \in C^\infty(\overline{D})$, $\Gamma \in C^\infty$, and the coefficients of $\mathcal{L}$ are C^∞. If Γ and the coefficients $a_j(x)$ of $\mathcal{L}$ are C^∞, but $f \in H^m(\overline{D})$, $m \geq \alpha = \frac{qs}{2}$, then the solution u to the Dirichlet problem (4.10)-(4.12) belongs to $H^m(\overline{\Omega}) \cap H^\infty(\Omega)$. We assume that $a_j(x) \in C^\infty$ and $\Gamma \in C^\infty$. This is done for simplicity and in order to avoid lengthy explanations of the results on elliptic regularity of the solutions and the connection between smoothness of Γ and of the coefficients of $\mathcal{L}$ with the smoothness of the solution u to the problem (4.10)-(4.12).

The uniqueness of the solution to the Dirichlet problem (4.10)-(4.12) follows from the positivity of $Q(\lambda)$: the quadratic form $(Q(\mathcal{L})u,u)_{L^2(\Omega)} = 0$ if and only if $u = 0$ provided that u satisfies conditions (4.11) with $f = 0$. If u is the unique solution to the problem (4.10)-(4.12) then

$$F(x) = \begin{cases} f(x) & \text{in } D, \\ u(x) & \text{in } \Omega, \end{cases} \qquad F \in H^\alpha(R^r). \tag{4.13}$$

Indeed, $F \in H^\alpha_{\text{loc}}(R^r)$, and $F \in H^\alpha(R^r)$ since u decays at infinity. Since $F \in H^\alpha(R^r)$ and ord$Q(\mathcal{L}) = qs = 2\alpha$, one has

$$h(x) = Q(\mathcal{L})F \in \dot{H}^{-\alpha}(D). \tag{4.14}$$

Corollary 2.2 is proved.

□

Remark 4.1 *Consider the operator $\mathcal{L} = -\Delta + a^2$, $a > 0$, in $L^2(R^3)$ with the domain of definition $H^2(R^3)$. The operator $\mathcal{L}$ is elliptic selfadjoint and positive definite in $H_0 = L^2(R^3)$:*

$$(\mathcal{L}u,u) \geq a^2(u,u), \quad a > 0, \quad \forall u \in H_0. \tag{4.15}$$

The Green function of $\mathcal{L}$, which is the kernel of the operator $\mathcal{L}^{-1}$, is

$$G(x, y) = \frac{\exp(-a|x-y|)}{4\pi|x-y|}. \tag{4.16}$$

It decays exponentially as $|x| \to \infty$.

Exponential decay as $|x| \to \infty$ holds for the Green function of $\mathcal{L} = -\Delta + q$ in H_0 if $\mathcal{L}$ has property (4.15), in particular if $q(x) \geq q_0 > 0$.

Exercise. Is it true that if $\mathcal{L}$ is an elliptic selfadjoint operator in $H_0 = L^2(R^r)$ and $Q(\lambda) > 0$ for all $\lambda \in R^1$ is a polynomial, then the kernel of the operator $[Q(\mathcal{L})]^{-1}$ decays exponentially as $|x| \to \infty$?

Proof of Theorem 2.1 Let us start with rewriting equation (4.1) with $R \in \mathcal{R}$ in the form

$$P(\mathcal{L}) \int_D S(x, y)h(y)dy = f(x), \quad x \in \overline{D}, \tag{4.17}$$

where

$$S(x, y) := \int_\Lambda Q^{-1}(\lambda)\Phi(x, y, \lambda)d\rho(\lambda). \tag{4.18}$$

Equation (4.17) can be written as

$$\int_D S(x, y)h(y)dy = g + v(x), \tag{4.19}$$

where g is a fixed solution to the equation

$$P(\mathcal{L})g = f \quad \text{in} \quad D \tag{4.20}$$

and v is an arbitrary solution to the equation

$$P(\mathcal{L})v = 0 \quad \text{in} \quad D. \tag{4.21}$$

Equation (4.19) is of the form considered in the proof of Corollary 2.2 with $g+v$ in place of f. Applying the result proved in this corollary, one obtains the following formula:

$$h = Q(\mathcal{L})G, \tag{4.22}$$

where

$$G = \begin{cases} g + v & \text{in } D, \\ u & \text{in } \Omega, \end{cases} \tag{4.23}$$

and

$$Q(\mathcal{L})u = 0 \quad \text{in} \quad \Omega, \quad u(\infty) = 0. \tag{4.24}$$

Here g is a particular solution of (4.20) and v is an arbitrary solution to (4.21). Formula (4.22) gives the unique solution of minimal order of singularity, mos solution, to equation (4.1) if and only if G is maximally smooth. If f and Γ are sufficiently smooth, the maximal smoothness of G is guaranteed if and only if the following transmission boundary conditions hold on Γ:

$$\partial_N^j u = \partial_N^j (v+g) \quad \text{on} \quad \Gamma, \quad 0 \le j \le \frac{s(p+q)}{2} - 1. \tag{4.25}$$

Given the orders of the elliptic operators

$$\text{ord } P(\mathcal{L}) = ps, \quad \text{ord } Q(\mathcal{L}) = qs, \tag{4.26}$$

one cannot impose, in general, more than $\frac{ps+qs}{2}$ conditions of the form (4.25). We will prove that if one imposes $\frac{ps+qs}{2}$ conditions then the transmission problem (4.20), (4.21), (4.24), (4.25) is uniquely solvable and $G \in H^{s(p+q)/2}(R^r)$. Therefore the mos solution h to equation (4.1), given by formula (4.22) in which G has maximal smoothness: $G \in H^{s(p+q)/2}(R^r)$, has the minimal order of singularity:

$$h \in \dot{H}^{-\alpha}(D), \quad \alpha = -\left(\frac{s(p+q)}{2} - qs\right) = \frac{(q-p)s}{2}.$$

In order to complete the proof one has to prove that the transmission problem (4.20), (4.21), (4.24), (4.25) has a solution and its solution is unique. This problem can be written as

$$P(\mathcal{L})G = f \quad \text{in} \quad D \tag{4.27}$$

$$Q(\mathcal{L})G = 0 \quad \text{in} \quad \Omega \tag{4.28}$$

$$G(\infty) = 0 \tag{4.29}$$

$$\left(\partial_N^j G\right)_+ = (\partial^j G)_-, \quad 0 \le j \le \frac{s(p+q)}{2} - 1, \tag{4.30}$$

where $+$ and $-$ in (4.30) denote the limiting values on Γ from D and from Ω respectively.

First let us prove uniqueness of the solution to the problem (4.27)-(4.30). Suppose $f = 0$. The problem (4.27)-(4.30) is equivalent to finding the mos solution of equation (4.1).

Indeed, we have proved that if h is the mos solution to (4.1) then h is given by formula (4.22) where G solves (4.27)-(4.30). Conversely, if G solves (4.27)-(4.30)) then h given by formula (4.22) solves equation (4.1) and has minimal order of singularity. This is checked by a straightforward calculation: for any $\phi \in C_0^\infty(D)$ one has

$$\begin{aligned}(RQ(\mathcal{L})G, \phi) &= (G, Q(\mathcal{L})R\phi) = (G, P(\mathcal{L})\phi) = (P(\mathcal{L})G, \phi) \\ &= (f, \phi), \quad \forall \phi \in C_0^\infty(D),\end{aligned} \tag{4.31}$$

where we have used the formula

$$Q(\mathcal{L})R(x, y) = P(\mathcal{L})\delta(x - y) \tag{4.32}$$

and the selfadjointness of $P(\mathcal{L})$. Formula (4.31) implies that

$$Rh = RQ(\mathcal{L})G = f \quad \text{in} \quad D. \tag{4.33}$$

Equations (4.22)-(4.24) imply that supp$h \subset \overline{D}$. It follows from (4.22) and the inclusion $G \in H^{s(p+q)/2}(R^r)$ that $h \in \dot{H}^{-\alpha}(D)$, $\alpha = (q-p)s/2$. Thus, we have checked that h given by (4.22) with G given by (4.27)-(4.30) solves equation (4.1) and belongs to $\dot{H}^{-\alpha}(D)$, that is h is the mos solution to (4.1).

Exercise. Prove uniqueness of the solution to problem (4.27)-(4.30) in $H^{s(p+q)/2}(R^r)$ by establishing equivalence of this problem and the problem of solving equation (4.1) in $\dot{H}^{-\alpha}(D)$, and then proving that equation (4.1) has at most one solution in $\dot{H}^{-\alpha}(D)$.

Hint: If $h \in \dot{H}^{-\alpha}$ and $Rh = 0$ in D, then

$$0 = (Rh, h) \geq c_1 \parallel h \parallel_{-\alpha}^2, \quad c_1 > 0, \tag{4.34}$$

so that $h = 0$.

Let us prove the existence of the solution to the problem (4.27)-(4.30) in $H^{s(p+q)/2}(R^r)$. Consider the bilinear form

$$[\phi, \psi] := \int_\Lambda P(\lambda)Q(\lambda)\tilde{\phi}(\lambda)\tilde{\psi}^*(\lambda)d\rho(\lambda) \tag{4.35}$$

defined on the set $V := H^{s(p+q)/2}(R^r) \cap H^{sq}(\Omega)$ of functions which satisfy the equation:

$$Q(\mathcal{L})\phi = 0 \quad \text{in} \quad \Omega. \tag{4.36}$$

Since $P(\lambda)Q(\lambda) \geq c > 0 \ \forall\lambda \in R^1$, one has the norm

$$[\phi, \phi]^{1/2} = \left\{ \int_\Lambda P(\lambda)Q(\lambda)|\tilde{\phi}|^2 d\rho(\lambda) \right\}^{1/2}, \tag{4.37}$$

which is equivalent to the norm of $H^{s(p+q)/2}(R^r)$. Indeed

$$0 < d_1 \leq P(\lambda)Q(\lambda)(1+\lambda^2)^{-(p+q)/2} \leq d_2. \tag{4.38}$$

Therefore

$$\begin{aligned} d_1 \int_\Lambda (1+\lambda^2)^{(p+q)/2} |\tilde{\phi}|^2 d\rho &\leq \int_\Lambda P(\lambda)Q(\lambda)|\tilde{\phi}|^2 d\rho \\ &\leq d_2 \int_\Lambda (1+\lambda^2)^{(p+q)/2} |\tilde{\phi}|^2 d\rho(\lambda). \end{aligned} \tag{4.39}$$

On the other hand,

$$\int_\Lambda (1+\lambda^2)^\beta |\tilde{\phi}|^2 d\rho(\lambda) = \| \phi \|^2_{H^{\beta s}(R^r)}\,. \tag{4.40}$$

This proves that the norm (4.37) is equivalent to the norm of the space $H^{s(p+q)/2}(R^r)$. Consider the form

$$\begin{aligned} [G, \psi] &= \int_\Lambda P(\lambda)Q(\lambda)\tilde{G}(\lambda)\tilde{\psi}^*(\lambda)d\rho(\lambda) \\ &= \int_{R^r} P(\mathcal{L})G(y)Q(\mathcal{L})\psi^*(y)dy = \int_D P(\mathcal{L})G\{Q(\mathcal{L})\psi\}^* dy \\ &= \int_D f\{Q(\mathcal{L})\psi\}^* dy \quad \forall\psi \in v, \end{aligned} \tag{4.41}$$

where Parseval's equality was used. Let W be the Hilbert space which is the completion of V in the norm (4.37). For any $f \in H^\alpha(D)$, $\alpha = s(q-p)/2$, the right-hand side of (4.41) is a linear bounded functional on W. Indeed, extend f to all of R^r so that $f \in H^\alpha(R^r)$ and use Parseval's equality and the equation $Q(\mathcal{L})\psi = 0$ to obtain

$$\begin{aligned}
\left|\int_D f\{Q(\mathcal{L})\psi\}^* dy\right| &= \left|\int_{R^r} f\{Q(\mathcal{L})\psi\}^* dy\right| \\
&= \left|\int_\Lambda Q(\lambda)\tilde{f}(\lambda)\tilde{\psi}^*(\lambda)d\rho(\lambda)\right| \\
&\leq \left\{\int_\Lambda |\tilde{f}|^2 \frac{Q(\lambda)}{P(\lambda)} d\rho(\lambda)\right\}^{1/2} \left\{\int_\Lambda Q(\lambda)P(\lambda)|\tilde{\psi}|^2 d\rho(\lambda)\right\}^{1/2} \\
&\leq \| f \|_\alpha \| \psi \|_W \leq c \| \psi \|_W, \qquad (4.42)
\end{aligned}$$

where $c = \| f \|_\alpha$ is a positive constant which does not depend on $\psi \in W$. According to Riesz's theorem about linear functionals one concludes from (4.41) and (4.42) that

$$[G, \psi] = [Tf, \psi] \quad \forall \psi \in W, \qquad (4.43)$$

where $T : H^\alpha \to W$ is a bounded linear mapping. The function

$$G = Tf, \quad G \in W, \qquad (4.44)$$

is the solution to problem (4.27)-(4.30) in $H^{s(p+q)/2}(R^r)$. The last statement can be proved as follows. Suppose that $G \in W$ satisfies equation (4.41) for all $\psi \in V$. Then $G \in H^{s(p+q)/2}(R^r)$ so that equations (4.30) and (4.29) hold, and G solves equation (4.28). In order to check that equation (4.27) holds, let $\psi \in C_0^\infty(R^r)$ in (4.41). This is possible since $C_0^\infty(R^r) \subset V$. Then

$$\int_D \{P(\mathcal{L})G - f\}\, \eta^* dy = 0 \quad \forall \eta = Q(\mathcal{L})\psi, \quad \psi \in C_0^\infty(R^r). \qquad (4.45)$$

If the set of such η is complete in $L^2(D)$, one can conclude from equation (4.45) that equation (4.27) holds. To finish the proof of Theorem 2.1, let us prove that the set $\{Q(\mathcal{L})\psi\}$ $\forall \psi \in C_0^\infty(R^r)$ is complete in $L^2(D)$. But this is clear because even the smaller set of functions

$$\{Q(\mathcal{L})\psi\} \quad \forall \psi \in C^\infty(D), \quad \partial_N^j \psi = 0 \quad \text{on} \quad \Gamma, \quad 0 \leq j \leq \frac{qs}{2} - 1 \qquad (4.46)$$

is dense in $L^2(D)$. Indeed, the operator $Q(\mathcal{L})$ is essentially selfadjoint in $L^2(D)$ on the set

$$\left\{\psi : \psi \in C^\infty(D), \quad \partial_N^j \psi = 0 \quad \text{on} \quad \Gamma, \quad 0 \leq j \leq \frac{qs}{2} - 1\right\}. \qquad (4.47)$$

That is, the closure of $Q(\mathcal{L})$ with the domain (4.47) is selfadjoint in $L^2(D)$: it is the Dirichlet operator $Q(\mathcal{L})$ in $L^2(D)$. Since $Q(\mathcal{L})$ is positive definite on the set (4.47), its closure is also a positive definite selfadjoint operator in $L^2(D)$. Therefore, the range of the closure of $Q(\mathcal{L})$ is the whole space $L^2(D)$, and the range of the operator $Q(\mathcal{L})$ with the domain of definition (4.47) is dense in $L^2(D)$. This completes the proof of Theorem 2.1.

□

4.2 Proof of Theorem 2.2

Let us first prove a lemma of a general nature.

Lemma 4.1 *Let*

$$R\phi := \int_D R(x,y)\phi(y)dy, \quad R(x,y) = R^*(y,x) \tag{4.48}$$

and assume that the kernel $R(x,y)$ defines a compact selfadjoint operator in $H = L^2(D)$ for any bounded domain $D \subset R^r$. Let $\lambda_j(D)$ be the eigenvalues of $R : L^2(D) \to L^2(D)$,

$$R\phi_j = \lambda_j(D)\phi_j \tag{4.49}$$

and $\lambda_j^+(D)$ be the positive eigenvalues, ordered so that

$$\lambda_1^+(D) \geq \lambda_2^+ \geq \cdots \tag{4.50}$$

and counted according to their multiplicities. Then

$$\lambda_j^+(D_2) \geq \lambda_j^+(D_1) \quad \forall j \quad \text{provided that } D_2 \supset D_1. \tag{4.51}$$

Proof. By the well-known minimax principle one has

$$\lambda_j^+(D_2) = \min_{\psi_1,\dots,\psi_{j-1}} \max_{\substack{(\phi,\psi_i)_2=0, \quad 1\leq i\leq j-1 \\ (\phi,\phi)_2=1}} (R\phi,\phi)_2 := \min_{\psi} \mu_j(\psi). \tag{4.52}$$

Here $(u,v)_m := \int_{D_m} uv^* dx$, $m = 1,2$, and

$$\mu_j(\psi) := \max_{\substack{(\phi,\psi_i)_2=0, \quad 1\leq i\leq j-1 \\ (\phi,\phi)_2=1}} (R\phi,\phi)_2. \tag{4.53}$$

Let $D_3 := D_2 \setminus D_1$. If we assume that an additional restriction is imposed on ϕ in formula (4.52), namely

$$\phi = 0 \quad \text{in} \quad D_3 \tag{4.54}$$

then $\mu_j(\psi)$ cannot increase:

$$\nu_j(\psi) := \max_{\substack{(\phi,\psi_i)_2=0, \quad 1\le i\le j-1 \\ (\phi,\phi)_2=1, \phi=0 \quad \text{in} \quad D_3}} (R\phi,\phi)_2 \le \mu_j(\psi). \tag{4.55}$$

On the other hand,

$$\nu_j(\psi) = \max_{\substack{(\phi,\psi_i)_1=0, \quad 1\le i\le j-1 \\ (\phi,\phi)_1=1}} (R\phi,\phi)_1. \tag{4.56}$$

Since

$$\lambda_j^+(D_1) = \min_\psi \nu_j(\psi) \le \min_\psi \mu_j(\psi) = \lambda_j(D_2), \tag{4.57}$$

one obtains (4.51). Lemma 4.1 is proved. □

Assume now that D expands uniformly in directions, $D \to R^r$. We wish to find out if the limit

$$\lim_{D\to R^r} \lambda_1(D) := \lambda_{1\infty} \tag{4.58}$$

exists and is finite. We assume that

$$R(x,y) = \int_\Lambda \tilde{R}(\lambda)\Phi(x,y,\lambda)d\rho(\lambda), \tag{4.59}$$

where $\tilde{R}(\lambda) > 0$ is a continuous function which vanishes at infinity. This implies that the operator $R : L^2(D) \to L^2(D)$ with kernel (4.59) is compact, as follows from

Lemma 4.2 *If $\tilde{R}(\lambda)$ is a continuous function such that*

$$\lim_{\lambda\to\pm\infty} \tilde{R}(\lambda) = 0, \tag{4.60}$$

then the operator $R : L^2(D) \to L^2(D)$, where $D \subset R^r$ is a bounded domain, with the kernel (4.59) is compact in $H = L^2(D)$.

Proof. Given a number $\epsilon > 0$, find a continuous function $r_\epsilon(\lambda) = 0$ for $|\lambda| > N$ such that

$$\max_{|\lambda|\le N} |\tilde{R}(\lambda) - r_\epsilon(\lambda)| < \epsilon. \tag{4.61}$$

Here the number N is chosen so large that

$$|\tilde{R}(\lambda)| < \epsilon \quad \text{for} \quad |\lambda| > N. \tag{4.62}$$

Denote by R_ϵ the operator in $L^2(D)$ whose kernel has the spectral density $r_\epsilon(\lambda)$. Let $\mathcal{P}$ denote the orthoprojection in $L^2(R^r)$ onto $L^2(D)$. If R_∞ is the operator with kernel (4.59) (considered as an operator in $L^2(R^r)$), then

$$R = \mathcal{P} R_\infty \mathcal{P} \tag{4.63}$$

and

$$R_\epsilon = \mathcal{P} R_{\epsilon\infty} \mathcal{P} \tag{4.64}$$

with the same notation for $R_{\epsilon\infty}$. One has

$$\| R - R_\epsilon \| \leq \| R_\infty - R_{\epsilon\infty} \| \leq \epsilon. \tag{4.65}$$

Here one has used the fact that the norm of the operator $r : L^2(R^r) \to L^2(R^r)$ with the kernel

$$r(x, y) = \int_\Lambda \tilde{r}(\lambda) \Phi(x, y, \lambda) d\rho(\lambda) \tag{4.66}$$

is given by the formula

$$\| r \| = \max_{\lambda \in \Lambda} |\tilde{r}(\lambda)|. \tag{4.67}$$

Indeed

$$\begin{aligned} \| r \| &= \sup_{\|\phi\|_{L^2(R^r)}=1} |(R\phi, \phi)| = \sup_{\|\phi\|_{L^2(R^r)}=1} \left| \int_\Lambda \tilde{r}(\lambda) |\tilde{\phi}|^2 d\rho(\lambda) \right| \\ &\leq \max_{\lambda \in \Lambda} |\tilde{r}(\lambda)| \sup_{\|\phi\|_{L^2(R^r)}=1} \| \phi \|^2_{L^2(R^r)} = \max_{\lambda \in \Lambda} |\tilde{r}(\lambda)|. \end{aligned} \tag{4.68}$$

This proves the inequality

$$\| r \| \leq \max_{\lambda \in \Lambda} |\tilde{r}(\lambda). \tag{4.69}$$

In order to establish the equality (4.67), take the point λ_0 at which the function $|r(\lambda)|$ attains its maximum. Such a point does exist since the function $|r(\lambda)|$ is continuous and vanishes at infinity. Then find a $\phi(x)$, $\| \phi \|_{L^2(R^r)} = 1$, such that $\tilde{\phi} = 0$ for $|\lambda - \lambda_0| > \delta$, where $\delta > 0$ is an arbitrary small number. Then, using continuity of $\tilde{r}(\lambda)$, one obtains

$$\begin{aligned} \| r \| &= \sup_{\|\phi\|_{L^2(R^r)}=1} \left| \int_{|\lambda - \lambda_0| \leq \delta} \tilde{r}(\lambda) |\tilde{\phi}|^2 d\rho(\lambda) \right| \geq (|\tilde{r}(\lambda_0)| - \eta(\delta)) \\ &= \max_{\lambda \in \Lambda} |\tilde{r}(\lambda)| - \eta(\delta) \end{aligned} \tag{4.70}$$

where $\eta(\delta)$ is arbitrarily small if $\delta > 0$ is sufficiently small. From (4.69) and (4.70) formula (4.67) follows. From (4.67) and the obvious inequality $\| \mathcal{P} \| \leq 1$ one obtains (4.65). If one can prove that the operator R_ϵ is compact in $L^2(D)$ then Lemma 4.2 is proved, because R can be approximated in the norm by compact operators R_ϵ with arbitrary accuracy according to (4.65).

Let us prove that the operator R_ϵ is compact in $L^2(D)$. One has

$$w := R_\epsilon f = \int_D \left(\int_{-N}^{N} r_\epsilon(\lambda)\phi(x, y, \lambda)d\rho(\lambda) \right) f(y)dy. \tag{4.71}$$

Let $\| f \|_{L^2(R^r)} \leq 1$. Taking into account that $\mathcal{L}\Phi = \lambda\Phi$ and using Parseval's equality, one obtains

$$\begin{aligned} \| \mathcal{L}R_\epsilon f \|^2_{L^2(R^r)} &= \left\| \int_D \left(\int_{-N}^{N} \lambda r_\epsilon(\lambda)\Phi(x, y, \lambda)d\rho(\lambda) \right) f(y)dy \right\|_{L^2(R^r)} \\ &= \int_{-N}^{N} \lambda^2 |r_\epsilon(\lambda)|^2 |\tilde{f}|^2 d\rho(\lambda) \leq N^2 \max_{-N\leq\lambda\leq N} |r_\epsilon(\lambda)|^2 \| f \|^2_{L^2(R^r)} \\ &\leq c(N). \end{aligned} \tag{4.72}$$

Let us now recall the well-known elliptic estimate ([Hörmander (1983-85)]):

$$\| w \|_{H^s(D_1)} \leq c(D_1, D_2) \left(\| \mathcal{L}w \|_{L^2(D_2)} + \| w \|_{L^2(D_2)} \right), \tag{4.73}$$

which holds for the elliptic operator $\mathcal{L}$, ord$\mathcal{L} = s$, and for arbitrary bounded domains $D_1 \subset D_2 \subset R^r$, D_1 is a strictly inner subdomain of D_2. From (4.72) and (4.73) it follows that

$$\| w \|_{H^s(D)} \leq c \quad \forall f \in B_1 := \left\{ f : \| f \|_{L^2(R^r)} \leq 1 \right\}, \tag{4.74}$$

where $c > 0$ is a constant which does not depend on $f \in B_1$. Indeed, the estimate for $\| \mathcal{L}w \|_{L^2(R^r)}$ is given by formula (4.72), and the estimate for $\| w \|_{L^2(R^r)}$ is obtained in the same way. Inequality (4.74) and the imbedding theorem (see Theorem 8.1) imply that the set $\{R_\epsilon f\}$ is relatively compact in $L^2(D)$. Therefore the operator R_ϵ maps a unit ball in $L^2(D)$ into a relatively compact set in $L^2(D)$. This means that R_ϵ is compact in $L^2(D)$. Lemma 4.2 is proved. □

In order to proceed with the study of the behavior of $\lambda_1(D)$ as $D \to R^r$ let us assume that

$$\sup_{x\in R^r} \int_{R^r} |R(x, y)|dy := A < \infty. \tag{4.75}$$

Lemma 4.3 *If the kernel $R(x, y)$ defines, for any bounded domain $D \subset R^r$, a selfadjoint compact nonnegative operator in $L^2(D)$, and condition*

(4.75) holds, then the limit (4.58) exists and

$$\lambda_{1\infty} \le A. \tag{4.76}$$

Proof. From Lemma 4.1 one knows that $\lambda_1(D)$ grows monotonically as D increases in the sense (4.51). Therefore, existence of the limit (4.58) and the estimate (4.76) will be established if one proves that

$$\lambda_1(D) \le A \tag{4.77}$$

for all $D \subset R^r$. Let $R\phi_1 = \lambda_1(D)\phi_1$. One has

$$\begin{aligned}\lambda_1(D) \sup_{x\in D} |\phi_1(x)| &\le \sup_{x\in D} \int_D |R(x,y)|dy \sup_{y\in D} |\phi_1(y)| \\ &\le \sup_{x\in R^r} \int_{R^r} |R(x,y)|dy \sup_{y\in D} |\phi_1(y)|. \end{aligned} \tag{4.78}$$

Therefore inequality (4.77) is obtained. Lemma 4.3 is proved. □

Let us now prove Theorem 2.2.

Proof of Theorem 2.2 We need only prove formula (2.29). Take ϕ_1 in Lemma 4.3 with $\| \phi_1 \|_{L^2(D)} = 1$. Extend ϕ_1 to all of R^r by setting $\phi_1 = 0$ in Ω. Then, using Parseval's equality, one obtains:

$$\begin{aligned}\lambda_1(D) &= \int_D \left(\int_D R(x,y)\phi_1(y)dy \right) \phi_1^*(x)dx \\ &= \int_{R^r}\int_{R^r} R(x,y)\phi_1(y)\phi_1^*(x)dydx = \int_\Lambda \tilde{R}(\lambda) \left|\tilde{\phi}_1(\lambda)\right|^2 d\rho(\lambda) \\ &\le \max_{\lambda\in\Lambda} \tilde{R}(\lambda) \| \phi_1 \|_{L^2(R^r)} = \max_{\lambda\in\Lambda} \tilde{R}(\lambda). \end{aligned} \tag{4.79}$$

Choose $\tilde{\phi}_1$ with support in a small neighborhood of the point λ_0 at which $\tilde{R}(\lambda)$ attains its maximum. Then $\lambda(D) \ge \max_\lambda \tilde{R}(\lambda) - \epsilon$, where $\epsilon > 0$ is an arbitrary small number.

This proves formula (2.29) and Theorem 2.2 in which $\omega(\lambda)$ stands for $\tilde{R}(\lambda)$. □

We now discuss some properties of the eigenvalues $\lambda_j(D)$.

Suppose that $R(x,y) = R(x-y)$, $R(-x) = R(x)$, and the domain D is centrally symmetric with respect to the origin, that is, if $x \in D$ then $-x \in D$. Let us recall that an eigenvalue is called simple if the corresponding eigenspace is one dimensional.

Lemma 4.4 *If λ is a simple eigenvalue of the operator $R : L^2(D) \to L^2(D)$ with the kernel $R(x-y)$, $R(-x) = R(x)$, and D is centrally symmetric, then the corresponding eigenfunction ϕ*

$$R\phi = \lambda\phi \tag{4.80}$$

is either even or odd.

Proof. One has

$$\lambda\phi(-x) = \int_D R(-x-y)\phi(y)dy = \int_D R(x+y)\phi(y)dy$$
$$= \int_D R(x-z)\phi(-z)dz. \tag{4.81}$$

Here we set $y = -z$ in the second integral and used the assumption of central symmetry. Therefore $\phi(-x)$ is the eigenfunction corresponding to the same eigenvalue λ. Since this eigenvalue is simple, one has $\phi(-x) = c\phi(x)$, $c =$ const. This implies $\phi(x) = c\phi(-x)$, so that $c^2 = 1$. Thus $c = \pm 1$. If $c = 1$ then $\phi(x)$ is even. Otherwise it is odd. Lemma 4.4 is proved. □

Remark 4.2 *If λ is not simple, the corresponding eigenfunction may be neither even nor odd.*

For example the operator

$$R\phi := \int_{-\pi}^{\pi} \phi(y)dy \tag{4.82}$$

has an eigenvalue $\lambda = 0$. The corresponding eigenspace is infinite dimensional: it consists of all functions orthogonal to 1 in $L^2(-\pi, \pi)$. In particular, the function $\cos y + \sin y$ is an eigenfunction which is neither even nor odd and it corresponds to the eigenvalue $\lambda = 0$.

Suppose one has a family of domains D_t, $0 < t < \infty$, such that $D_1 = D$, and $D_t = \{x : x = t\xi, \xi \in D_1\}$. Then the eigenvalues $\lambda_j(D_t) := \lambda_j(t)$ depend on the parameter t and one can study this dependence. If one writes

$$\int_{D_t} R(x,y)\phi(y)dy = t^r \int_{D_1} R(t\xi, t\eta)\phi(t\eta)d\eta = \lambda_j(t)\phi(t\xi)$$

then one sees that $\lambda_j(t)$ are the eigenvalues of the operator $R(t)$ in $L^2(D)$ with the kernel $R(\xi, \eta, t) := t^r R(t\xi, t\eta)$, where $D = D_1$ does not depend on

t. This implies immediately that $\lambda_j(t)$ depend on t continuously provided that

$$\| R(t') - R(t) \| \to 0 \quad \text{as} \quad t \to t'. \tag{4.83}$$

Indeed,

$$\max_j |\lambda_j(t') - \lambda_j(t)| \leq \| R(t) - R(t') \| . \tag{4.84}$$

Estimate (4.84) follows from the minimax principle and is derived in Section 8.3. Condition (4.83) holds, for example, if

$$\int_D \int_D |R(x,y)|^2 dxdy \leq c(D) < \infty \tag{4.85}$$

for any bounded domain $D \subset R^r$.

One can also study differentiability of the eigenvalues $\lambda_j(t)$ with respect to parameter t using, for example, methods given in [K]. This and the study of the eigenfunctions as functions of the parameter t would lead us astray.

4.3 Proof of Theorems 2.4 and 2.5

Proof of Theorem 2.4

This proof can be given in complete analogy to the proof of Theorem 2.1. On the other hand, there is a special feature in the one-dimensional ($r = 1$) theory which is the subject of Theorem 2.4. Namely the spaces of all solutions to homogeneous equations

$$P(\mathcal{L})\phi = 0 \tag{4.86}$$

and

$$Q(\mathcal{L})\psi = 0 \tag{4.87}$$

are finite dimensional (in contrast to the case when $r > 1$). The system (2.43)-(2.44), for example, is a linear algebraic system. Therefore, existence of the solution to this system follows from the uniqueness of this solution by Fredholm's alternative.

□

Let us briefly describe the basic steps of the proof.

Step 1. Consider first the case when $P(\lambda) = 1$.

Lemma 4.5 *The set of all solutions of the equation*

$$Rh := \int_{t-T}^{t} R(x,y)h(y)dy = f(x), \quad t-T \le x \le t \tag{4.88}$$

with the kernel $R(x,y) \in \mathcal{R}$ and $P(\lambda) = 1$ in the space $\dot{H}^{-sq} := \dot{H}^{-sq}(D)$, $D = (t-T,t)$, is in one-to-one correspondence with the set of the solutions of the equation

$$\int_{-\infty}^{\infty} R(x,y)h(y)dy = f(x), \quad x \in R^1, \tag{4.89}$$

where

$$h = Q(\mathcal{L})F, \tag{4.90}$$

$h \in H^{-sq}(R^1)$ and supph $\subset \overline{D} = [t-T,t]$. Here

$$F(x) = \begin{cases} \sum_{j=1}^{qs/2} b_j^- \psi_j^-(x), & x \le t-T, \\ f(x), & t-T \le x \le t, \\ \sum_{j=1}^{qs/2} b_j^+ \psi_j^+(x), & t \le x, \end{cases} \tag{4.91}$$

$b_j^\pm$ are arbitrary constants, and the functions $\{\psi_j^\pm\}$, $1 \le j \le qs/2$, form a fundamental system of solutions to equation (4.87) such that

$$\psi_j^+(+\infty) = 0, \quad \psi_j^-(-\infty) = 0. \tag{4.92}$$

Proof. Let $h \in \dot{H}^{-qs}$ be a solution to (4.88). This means that for any $\phi \in C_0^\infty(R^1)$ one has

$$(Rh, \phi) = (F, \phi), \quad \forall \phi \in C_0^\infty(R^1), \tag{4.93}$$

where the parentheses in (4.93) denote the $L^2(R^1)$ inner product and F is given by (4.91). One can say that F is defined to be the integral Rh for $x > t$ and $x < t-T$ (see formula (4.88)). This integral for $h \in \dot{H}^{-qs}$ can be considered as the value of the functional h on the test function $R(x,y)$, since, for a fixed x, $R(x,y) \in H^{qs}_{\text{loc}}$ outside of an arbitrarily small neighborhood of the point x. Since $Q(\mathcal{L})R = \delta(x-y)$, the kernel R does not belong to H^{qs}, however, $\delta(x-y)$ can be interpreted as a kernel of the identity operator in L^2, so that $(\delta(x-y), \phi) = \phi$ in the sense that $(\delta(x-y)\phi, \psi) = (\phi, \psi)\ \forall \phi, \psi \in L^2$.

Since $h \in \dot{H}^{-qs}$ one can consider it as an element of $H^{-qs}(R^1)$. Equation (4.93) then is equivalent to (4.89). The correspondence mentioned in Lemma 4.5 can be described as follows. If $h \in H^{-qs}(R^1)$,

$\operatorname{supp} h \in [t-T, t]$, and h solves equation (4.89), then $h = h$ solves equation (4.88). Conversely, if $h \in \dot{H}^{-qs}$ solves equation (4.88), then $h = h$ is an element of $H^{-qs}(R^1)$ with $\operatorname{supp} h \in [t-T, t]$, and h solves equation (4.89). The constants $b_j^{\pm}$ in (4.91) can be uniquely determined by the given h from the formula

$$\int_{t-T}^{t} R(x,y) h dy = F(x).$$

Lemma 4.5 is proved. □

Example 4.1 Let

$$\int_{-1}^{1} e^{-|x-y|} h(y) dy = f(x), \quad -1 \le x \le 1. \tag{4.94}$$

Here $\mathcal{L} = -i\frac{d}{dx}$, $r = 1$, $P(\lambda) = 1$, $Q(\lambda) = \frac{\lambda^2+1}{2}$, $Q(\mathcal{L}) = (-\partial^2 + 1)/2$, $\partial = \frac{d}{dx}$. (See Section 2.4.) Equation (4.87) has the solutions

$$\psi_- = \exp(x), \quad \psi_+ = \exp(-x). \tag{4.95}$$

Choose F by formula (4.91) with some $b^{\pm}$, $qs = 1$, $t = 1$, $t - T = -1$. Then

$$\begin{aligned} h &= \frac{1}{2}(-\partial^2 + 1)F \\ &= \frac{-f'' + f}{2} + \delta'(x-1)\left[f(1) - b^+ e\right] + \delta(x-1)\left[f'(1) - b^+ e\right] \\ &+ \delta'(x+1)[b^- e - f(-1)] + \delta(x+1)[-f'(-1) - b^- e]. \end{aligned} \tag{4.96}$$

For convenience of the reader let us give all the details of the calculation. By definition of the distributional derivative one has

$$\begin{aligned} (Q(\mathcal{L})F, \phi) = (F, Q(\mathcal{L})\phi) = \frac{b^+}{2} \int_1^{\infty} \exp(x)(-\phi'' + \phi) dx \\ + \frac{1}{2}\int_{-1}^{1} f(x)(-\phi'' + \phi) dx + \frac{b^-}{2} \int_{-\infty}^{-1} \exp(-x)(-\phi'' + \phi) dx. \end{aligned}$$

After integrating by parts two times one obtains

$$\begin{aligned} (Q(\mathcal{L})F, \phi) &= \frac{1}{2}\int_{-1}^{1} (-f'' + f)\phi dx + b^+\left[\phi'(1)e - \phi(1)e\right] \\ &- \phi'(1)f(1) + \phi(1)f'(1) + \phi'(-1)f(-1) - \phi(-1)f'(-1) \\ &+ b^-\left[-\phi'(-1)e - \phi(-1)e\right]. \end{aligned} \tag{4.97}$$

Formula (4.97) is equivalent to (4.96). Corollary 2.5 follows from Lemma 4.5 immediately. Indeed, the solution of minimal order of singularity one obtains from formula (4.90) if and only if the constants $b_j^{\pm}$ are chosen so that F has minimal order of singularity, that is, if F is maximally smooth. This happens if and only if the following conditions hold:

$$\left[F^{(j)}(t)\right] = 0, \quad \left[F^{(j)}(t-T)\right] = 0, \quad 0 \leq j \leq \frac{qs}{2} - 1. \tag{4.98}$$

Here, for example,

$$\left[F^{(j)}(t)\right] := F^{(j)}(t+0) - F^{(j)}(t-0) \tag{4.99}$$

is the jump of $F^{(j)}(x)$ across point t. If conditions (4.98) hold, one can rewrite formula (4.90) as formula (2.47).

Exercise. Check the last statement.

Hint: The calculations are similar to the calculations given in Example 4.1.

Step 2. Assume now that $P(\lambda) \not\equiv 1$. Write equation (4.88) as

$$P(\mathcal{L}) \int_{t-T}^{t} S(x,y)h(y)dy = f(x), \quad t-T \leq x \leq t, \tag{4.100}$$

where

$$S(x,y) = \int_{\Lambda} Q^{-1}(\lambda)\Phi(x,y,\lambda)d\rho(\lambda). \tag{4.101}$$

Rewrite equation (4.100) as

$$\int_{t-T}^{t} S(x,y)h(y)dy = g_0(x) + \sum_{j=1}^{ps} c_j\phi_j. \tag{4.102}$$

Here $\{\phi_j\}$, $1 \leq j \leq ps$, is the fundamental system of solutions to equation (4.86), g_0 is a particular solution to the equation

$$P(\mathcal{L})g = f, \quad t-T \leq x \leq t, \tag{4.103}$$

and c_j, $1 \leq j \leq ps$, are arbitrary constants.

Apply to equation (4.102) the result of Lemma 4.5, in particular, use formula (4.90) to get the conclusion of Theorem 2.4. Equations (2.43)-(2.44) are necessary and sufficient for G, given by formula (2.38), to have minimal order of singularity and, therefore, for h, given by formula (2.37), to have minimal order of singularity. As we have already noted, the solvability of the system (2.43)-(2.44) for the coefficients c_j, $1 \le j \le ps$, and $b_j^{\pm}$, $0 \le j \le \frac{qs}{2} - 1$, follows from the fact that the homogeneous system (2.43)-(2.44) has only the trivial solution and from Fredholm's alternative.

The fact that for $f(x) = 0$ the system (2.43)-(2.44) has only the trivial solution can be established as in the proof of Theorem 2.1.

Corollary 2.1 follows from formulas (2.15)-(2.20) immediately.

Exercise. Give a detailed proof of the uniqueness of the solution of the homogeneous system (2.43)-(2.44).

Hint: A solution to this system generates a solution h to equation (4.88) with $f = 0$, $h \in \dot{H}^{-\alpha}$. Use Parseval's equality to derive from

$$Rh = 0 \quad t - T \le x \le t, \quad h \in \dot{H}^{-\alpha} \tag{4.104}$$

that $h = 0$. If $h = 0$ then $c_j = b_j^{\pm} = 0$ for all j.

Proof of Theorem 2.5 The proof is similar to the proof of Theorem 2.4. Equation (2.65)

$$\int_{t-T}^{t} S(x,y)\mathbf{h}(y)dy = g(x), \quad t - T \le x \le t, \tag{4.105}$$

where $g(x)$ is the right hand side of equation (2.65), can be written as

$$\int_{-\infty}^{\infty} S(x,y)\mathbf{h}(y)dy = \mathbf{G}(x), \quad -\infty < x < \infty, \tag{4.106}$$

where $\mathcal{G}(x)$ is given by (2.67) and

$$\text{supp}\mathbf{h} \subseteq [t - T, t]. \tag{4.107}$$

There is a one-to-one correspondence between solutions to equation (4.105) in $\dot{\mathcal{H}}^{-\alpha}$ and solutions to equation (4.106) in $\mathcal{H}^{-\alpha}(R^1)$ with property (4.107). Equation (4.106) can be solved by the formula

$$\mathbf{h} = Q(\mathcal{L})E\mathbf{G}, \tag{4.108}$$

and $\mathbf{h}$ given by formula (4.108) has property (4.107). This $\mathbf{h}$ has minimal order of singularity $\le \alpha = \frac{n-m}{2}$. The constant vectors $b_j^{\pm}$ and c_j solve linear algebraic system (2.68-2.69). That this system is solvable follows from Fredholm's alternative and the fact that this system has at most one solution. The fact that this system has at most one solution follows, as in the proof of Theorem 2.4, from Parseval's equality and the positivity of the kernel $R(x,y)$.

Theorem 2.5 is proved.

□

4.4 Another approach

The remark we wish to make in this section concerns the case of the positive definite kernel $R(x,y)$ which satisfies the equation

$$Q(x,\partial)R(x,y) = P(x,\partial)\delta(x-y), \tag{4.109}$$

where $Q(x,\partial)$ and $P(x,\partial)$ are elliptic positive definite in $L^2(R^r)$ operators, not necessarily commuting, $\text{ord}Q(x,\partial) = n$, $\text{ord}P(x,\partial) = m$, $m < n$, and $R(x,y) \to 0$ as $|x-y| \to \infty$. As in Section 4.1, we wish to study the equation

$$Rh = f, \qquad x \in \overline{D} \tag{4.110}$$

with $f \in H^\alpha(D)$, $\alpha := \frac{n-m}{2}$, to prove that the operator $R : \dot{H}^{-\alpha}(D) \to H^\alpha(D)$ is an isomorphism, and to give analytical formulas for the solution h of minimal order of singularity. Consider equation (4.110) as an equation in $L^2(R^r)$

$$Rh = F, \qquad x \in R^r, \tag{4.111}$$

with

$$F = \begin{cases} f & \text{in } D \\ u & \text{in } \Omega := R^r \setminus D, \end{cases} \tag{4.112}$$

where

$$Q(x,\partial)u = 0 \quad \text{in} \quad \Omega. \tag{4.113}$$

Equation (4.113) follows from equation (4.109) and the assumption that $\text{supp}h \subset \overline{D}$. The function F has minimal order of singularity if and only if u solves (4.113) and satisfies the boundary conditions

$$\partial_N^j u = \partial_N^j f \quad \text{on} \quad \Gamma, \quad 0 \le j \le \frac{n}{2} - 1, \quad u(\infty) = 0. \tag{4.114}$$

The problem (5)-(6) is the exterior Dirichlet problem which is uniquely solvable if Q is positive definite. If u solves problem (4.113)-(4.114) then $F \in H^{(n-m)/2}(R^r)$, $\text{supp}Q(x,\partial)F \subset \overline{D}$, $QF \in H^{-(n+m)/2}(R^r)$. Apply $Q(x,\partial)$ to both sides of equation (4.111) and use equation (4.109) to get

$$P(x,\partial)h = Q(x,\partial)F, \quad QF \in H^{-(n+m)/2}(R^r), \quad \text{supp}QF \subset \overline{D}. \quad (4.115)$$

The solution to (4.115) in the space $\dot{H}^{-(n-m)/2}(D)$ exists, is unique, and is the solution to equation (4.110) of the minimal order of singularity.

More details are given in Appendix B.

Chapter 5

Singular Perturbation Theory for a Class of Fredholm Integral Equations Arising in Random Fields Estimation Theory

A basic integral equation of random fields estimation theory by the criterion of minimum of variance of the estimation error is of the form $Rh = f$, where $Rh = \int_D R(x,y)h(y)\,dy$, and $R(x,y)$ is a covariance function. The singular perturbation problem we study consists of finding the asymptotic behavior of the solution to the equation $\varepsilon h(x,\varepsilon)+Rh(x,\varepsilon) = f(x)$, as $\varepsilon \to 0$, $\varepsilon > 0$. The domain D can be an interval or a domain in R^n, $n > 1$. The class of operators R is defined by the class of their kernels $R(x,y)$ which solve the equation $Q(x,D_x)R(x,y) = P(x,D_x)\delta(x-y)$, where $Q(x,D_x)$ and $P(x,D_x)$ are elliptic differential operators. The presentation in this chapter is based on [Ramm and Shifrin (2005)].

5.1 Introduction

Consider the equation

$$\varepsilon h(x,\varepsilon) + Rh(x,\varepsilon) = f(x)\,, \quad x \in D \subset R^n\,, \tag{5.1}$$

where D is a bounded domain with a sufficiently smooth boundary ∂D, and

$$Rg(x) := \int_D R(x,y)g(y)\,dy\,.$$

In this paper we study the class $\mathcal{R}$ of kernels $R(x,y)$ which satisfy the equation

$$Q(x,D_x)R(x,y) = P(x,D_x)\delta(x-y) \quad \text{in } R^n$$

and tend to zero as $|x-y| \to \infty$, where $Q(x, D_x)$ and $P(x, D_x)$ are elliptic differential operators with smooth coefficients, and $\delta(x-y)$ is the delta-function.

For technical reasons below we use the kernels $R(x, y)$ of the same class, but written in a slightly different form (see (5.5)).

Specifically, we write

$$R(x, y) = P(y, D_y)G(x, y) \tag{5.2}$$

where

$$P(y, D_y) = \sum_{|\alpha| \le p} a_\alpha(y) D_y^\alpha , \quad Q(x, D_x) = \sum_{|\beta| \le q} b_\beta(x) D_x^\beta , \quad p < q, \tag{5.3}$$

and

$$Q(x, D_x)G(x, y) = \delta(x-y). \tag{5.4}$$

Note that

$$Q(x, D_x)R(x, y) = P(y, D_y)\delta(x-y) . \tag{5.5}$$

In this paper all the functions are assumed to be real-valued. We assume that the coefficients $a_\alpha(x)$, $b_\beta(x)$ and $f(x)$ are sufficiently smooth functions in R^n, $\alpha = (\alpha_1, \cdots, \alpha_n)$ and $\beta = (\beta_1, \cdots, \beta_n)$ are multiindices, $|\alpha| = \sum_{i=1}^n \alpha_i$, $|\beta| = \sum_{j=1}^n \beta_j$, $D_y^\alpha = \frac{\partial^{|\alpha|}}{\partial y_1^{\alpha_1} \cdots \partial y_n^{\alpha_n}}$, $D_x^\beta = \frac{\partial^{|\beta|}}{\partial x_1^{\beta_1} \cdots \partial x_n^{\beta_n}}$. Sufficient smoothness of the coefficients means that the integrations by parts we use are justified.

The following assumptions hold throughout the paper:

$$\text{A1)} \quad (Q(x, D_x)\varphi, \varphi) \ge c_1(\varphi, \varphi) , \quad c_1 = \text{const} > 0 , \quad \forall \varphi(x) \in C_0^\infty(R^n) , \tag{5.6}$$

$$(P(x, D_x)\varphi, \varphi) \ge c_2(\varphi, \varphi) , \quad c_2 = \text{const} > 0 , \quad \forall \varphi(x) \in C_0^\infty(R^n) , \tag{5.7}$$

where $(\cdot, \cdot)$ is the $L^2(R^n)$ inner product, and L^2 is the real Hilbert space.

By $Q^*(x, D_x)$ and $P^*(x, D_x)$ the operators formally adjoint to $Q(x, D_x)$ and $P(x, D_x)$ are denoted.

If (5.6) holds, then $q > 0$ is an even integer, and (5.7) implies that p is an even integer, $0 \le p < q$. Define $\mathbf{a} := (q-p)/2$. Let $H^\lambda(D)$ be the usual Sobolev space and $\dot{H}^{-\lambda}(D)$ be its dual with respect to $L^2(D) = H^0(D)$. Denote $\|\varphi\|_\lambda = \|\varphi\|_{H^\lambda(D)}$ for $\lambda > 0$ and $\|\varphi\|_\lambda = \|\varphi\|_{\dot{H}^\lambda(D)}$ for $\lambda < 0$. Let us denote, for the special value $\lambda = \mathbf{a}$, $H^{\mathbf{a}}(D) = H_+$, $\dot{H}^{-\mathbf{a}}(D) = H_-$. Denote by $(h_1, h_2)_-$ and by $(\cdot, \cdot)$ the inner products in H_- and, respectively, in $L^2(D)$. As in Chapter 8, let us assume that

$$\text{A2)}\ c_3\|\varphi\|_-^2 \leq (R\varphi, \varphi) \leq c_4\|\varphi\|_-^2\,,\ c_3 = \text{const} > 0\,, \forall\varphi(x) \in C_0^\infty(R^n)\,. \tag{5.8}$$

This assumption holds, for example (see Chapter 8) if

$$c_5\|\varphi\|_{(p+q)/2} \leq \|Q^*\varphi\|_{-\mathbf{a}} \leq c_6\|\varphi\|_{(p+q)/2}\,,\ c_5 = \text{const} > 0\,, \quad \forall\varphi(x) \in C_0^\infty(R^n)\,, \tag{5.9}$$

and

$$c_7\|\varphi\|_{(p+q)/2} \leq (PQ^*\varphi, \varphi) \leq c_8\|\varphi\|_{(p+q)/2}\,,\ c_7 = \text{const} > 0\,, \quad \forall\varphi(x) \in C_0^\infty(R^n)\,. \tag{5.10}$$

The following result is proved in Chapter 8.

Theorem 5.1 *If (5.8) holds, then the operator $R : H_- \to H_+$ is an isomorphism. If $QR = P\delta(x - y)$ and (5.9) and (5.10) hold, then (5.8) holds.*

Equation (5.1) and the limiting equation $Rh = f$ are basic in random fields estimation theory, and the kernel $R(x, y)$ in this theory is a covariance function, so $R(x, y)$ is a non-negative definite kernel:

$$(R\varphi, \varphi) \geq 0\,, \quad \forall\varphi(x) \in C_0^\infty(R^n)\,.$$

If $p < q$, then the inequality $(R\varphi, \varphi) \geq C(\varphi, \varphi)$, $C = \text{const} > 0$, $\forall\varphi(x) \in C_0^\infty(R^n)$ does not hold.

In [Ramm and Shifrin (1991); Ramm and Shifrin (1993); Ramm and Shifrin (1995)] a method was developed for finding asymptotics of the solution to equation (5.1) with kernel $R(x, y)$ satisfying equation (5.5) with $Q(x, D_x)$ and $P(x, D_x)$ being differential operators with constant coefficients.

Our purpose is to generalise this theory to the case of operators with variable coefficients.

In Chapter 8 the limiting equation $Rh = f$ is studied for the above class of kernels. In Chapter 2 the class of kernels $R(x, y)$, which are kernels of positive rational functions of an arbitrary selfadjoint in $L^2(R^n)$ elliptic operator, was studied.

In Section 5.2 we prove some auxiliary results. In Section 5.3 the asymptotics of the solution to equation (5.1) is constructed in case $n = 1$, that is, for one-dimensional integral equations of class $\mathcal{R}$ defined below formula

(5.1). In Section 5.4 examples of applications of the proposed asymptotical solutions are given. In Section 5.5 the asymptotics of the solution to equation (5.1) is constructed in the case $n > 1$, and in Section 5.6 examples of applications are given.

5.2 Auxiliary results

Lemma 5.1 *Assume (5.4) and suppose that $G(\infty, y) = 0$. Then*

$$Q^*(y, D_y)G(x, y) = \delta(x - y). \tag{5.11}$$

Proof. Let $\varphi(x) \in C_0^\infty(R^n)$. Then

$$\begin{aligned}(Q^*(y, D_y)Q(x, D_x)G(x, y), \varphi(y)) &= (\delta(x - y), Q(y, D_y)\varphi(y)) \\ &= Q(x, D_x)\varphi(x). \end{aligned} \tag{5.12}$$

Also one has:

$$(Q^*(y, D_y)Q(x, D_x)G(x, y), \varphi(y)) = Q(x, D_x)(Q^*(y, D_y)G(x, y), \varphi(y)). \tag{5.13}$$

Therefore

$$Q(x, D_x)(Q^*(y, D_y)G(x, y), \varphi(y)) = Q(x, D_x)\varphi(x). \tag{5.14}$$

Because of (5.6) and of the condition $G(\infty, y) = 0$ this implies

$$(Q^*(y, D_y)G(x, y), \varphi(y)) = \varphi(x), \quad \forall\varphi(x) \in C_0^\infty(R^n), \tag{5.15}$$

so (5.11) follows. □

Consider now the case $n = 1$:

$$P(y, D_y) = \sum_{i=0}^{p} a_i(y)\frac{d^i}{dy^i}, \quad Q(x, D_x) = \sum_{j=0}^{q} b_j(x)\frac{d^j}{dx^j}, \quad x \in R^1, \ y \in R^1 \tag{5.16}$$

In this case $D = (c, d)$, $\overline{D} = [c, d]$.

Lemma 5.2 *If $g(y)$ is a smooth function in $\overline{D}$, then*

$$\int_c^d [P(y, D_y)G(x, y)]\, g(y)\, dy = \int_c^d G(x, y)\, [P^*(y, D_y)g(y)]\, dy + K_2 g(x), \tag{5.17}$$

where

$$K_2g(x) := \sum_{k=1}^{p}\sum_{j=1}^{k}(-1)^{j-1}\left[\frac{\partial^{k-j}G(x,y)}{\partial y^{k-j}}\,\frac{d^{j-1}(a_k(y)g(y))}{dy^{j-1}}\right]\Bigg|_c^d, \tag{5.18}$$

and

$$K_2 = 0 \ \text{ if } \ p = 0\,. \tag{5.19}$$

Proof. Use definition (5.16) of $P(y, D_y)$ in (5.17), integrate by parts, and get formulas (5.17) - (5.19). □

Lemma 5.3 *If $g(y)$ is a smooth function in $\overline{D}$, then*

$$\int_c^d G(x,y)\,[Q(y,D_y)g(y)]\,dy = \int_c^d [Q^*(y,D_y)G(x,y)]\,g(y)\,dy + K_1g(x)\,, \tag{5.20}$$

where

$$K_1g(x) := \sum_{m=1}^{q}\sum_{i=1}^{m}(-1)^{i-1}\left[\frac{d^{m-i}g(y)}{dy^{m-i}}\,\frac{\partial^{i-1}(b_m(y)G(x,y))}{\partial y^{i-1}}\right]\Bigg|_c^d. \tag{5.21}$$

Proof. Similarly to Lemma 5.2, integrations by parts yield the desired formulas. □

Consider the case $n > 1$.

Lemma 5.4 *If $P(y, D_y)$ is defined in (5.3) and $g(x)$ is a smooth function in $\overline{D}$, then*

$$\int_D [P(y,D_y)G(x,y)]\,g(y)\,dy = \int_D G(x,y)\,[P^*(y,D_y)g(y)]\,dy + M_2g(x)\,, \tag{5.22}$$

where

$$M_2g(x) := \sum_{1\le\alpha\le p}\sum_{k=1}^{n}\sum_{\gamma_k=1}^{\alpha_k}(-1)^{\gamma_k+\alpha_{k+1}+\alpha_{k+2}+\cdots+\alpha_n-1}$$

$$\times\int_{\partial D}\frac{\partial^{|\alpha|-\alpha_n-\alpha_{n-1}-\cdots-\alpha_{k+1}-\gamma_k}G(x,y)}{\partial y_1^{\alpha_1}\partial y_2^{\alpha_2}\cdots\partial y_{k-1}^{\alpha_{k-1}}\partial y_k^{\alpha_k-\gamma_k}}$$

$$\times \frac{\partial^{\alpha_{k+1}+\alpha_{k+2}+\cdots+\alpha_n+\gamma_k-1}(a_\alpha(y)g(y))}{\partial y_k^{\gamma_k-1}\partial y_{k+1}^{\alpha_{k+1}}\partial y_{k+2}^{\alpha_{k+2}}\cdots\partial y_n^{\alpha_n}} N_k(y)\,dS_y\,. \tag{5.23}$$

Here ∂D is the boundary of D, $y \in \partial D$, $N_k(y)$ is the k - th component of the unit normal N to ∂D at the point y, pointing into $D' := R^n \setminus D$, and if $\alpha_k = 0$ then the summation over γ_k should be dropped.

Proof. Apply Gauss' formula (i.e. integrate by parts). □

Lemma 5.5 *If $Q(x, D_x)$ is defined in (5.3) and $g(y)$ is a smooth function in $\overline{D}$, then*

$$\int_D G(x,y)\,[Q(y,D_y)g(y)]\,dy = \int_D [Q^*(y,D_y)G(x,y)]\,g(y)\,dy + M_1 g(x)\,, \tag{5.24}$$

where

$$M_1 g(x) := \sum_{1\le\beta\le q}\sum_{k=1}^{n}\sum_{\gamma_k=1}^{\beta_k}(-1)^{\gamma_k+\beta_{k+1}+\beta_{k+2}+\cdots+\beta_n-1}$$

$$\times \int_{\partial D} \frac{\partial^{\beta_{k+1}+\beta_{k+2}+\cdots+\beta_n+\gamma_k-1}(b_\beta(y)G(x,y))}{\partial y_k^{\gamma_k-1}\partial y_{k+1}^{\beta_{k+1}} y_{k+2}^{\beta_{k+2}}\cdots\partial y_n^{\beta_n}}$$

$$\times \frac{\partial^{|\beta|-\beta_n-\beta_{n-1}-\cdots-\beta_{k+1}-\gamma_k} g(y)}{\partial y_1^{\beta_1}\partial y_2^{\beta_2}\cdots\partial y_{k-1}^{\beta_{k-1}}\partial y_k^{\beta_k-\gamma_k}} N_k(y)\,dS_y \tag{5.25}$$

Here $y \in \partial D$, and if $\beta_k = 0$ then the summation over γ_k should be dropped.

Remark 5.1 *For any smooth in $\overline{D}$ function $g(x)$, one has*

$$Q(x,D_x)K_j g(x) = 0\,,\quad x\in(c,d)\,,\quad j=1,2\,, \tag{5.26}$$

and

$$Q(x,D_x)M_j g(x) = 0\,,\quad x\in D\,,\quad j=1,2\,. \tag{5.27}$$

Formulas (5.26) and (5.27) follow from the definitions of K_j and M_j and from equation (5.4).

5.3 Asymptotics in the case $n = 1$

To construct asymptotic solutions to equation (5.1) with $R(x,y) \in \mathcal{R}$ we reduce this equation to a differential equation with special, non - standard, boundary conditions.

Theorem 5.2 *Equation (5.1) is equivalent to the problem:*

$$\varepsilon Q(x,D_x)h(x,\varepsilon) + P^*(x,D_x)h(x,\varepsilon) = Q(x,D_x)f(x)\,, \quad x \in (c,d) \tag{5.28}$$

with the following conditions

$$\varepsilon K_1 h(x,\varepsilon) - K_2 h(x,\varepsilon) = K_1 f(x)\,. \tag{5.29}$$

Proof. If $h(x,\varepsilon)$ solves (5.1) and $R(x,y)$ satisfies (5.2), one gets

$$\varepsilon h(x,\varepsilon) + \int_c^d \left[P(y,D_y)G(x,y)\right] h(y,\varepsilon)\, dy = f(x)\,. \tag{5.30}$$

From (5.30) and (5.17) one gets:

$$\varepsilon h(x,\varepsilon) + \int_c^d G(x,y)\left[P^*(y,D_y)h(y,\varepsilon)\right] dy + K_2 h(x,\varepsilon) = f(x)\,. \tag{5.31}$$

Applying $Q(x,D_x)$ to (5.31) and using (5.4) and (5.26), yields (5.28). Let us check (5.29). From (5.28) and (5.31) one gets:

$$\varepsilon h(x,\varepsilon) + \int_c^d G(x,y)\, Q(y,D_y)\left[f(y) - \varepsilon h(y,\varepsilon)\right] dy + K_2 h(x,\varepsilon) = f(x)\,. \tag{5.32}$$

From (5.32) and (5.20) one obtains

$$\varepsilon h(x,\varepsilon) + \int_c^d \left[Q^*(y,D_y)G(x,y)\right]\left(f(y) - \varepsilon h(y,\varepsilon)\right) dy$$

$$+K_1(f - \varepsilon h)(x,\varepsilon) + K_2 h(x,\varepsilon) = f(x)\,. \tag{5.33}$$

From(5.33) and (5.11) one concludes:

$$\varepsilon h(x,\varepsilon) + f(x) - \varepsilon h(x,\varepsilon) + K_1 f(x) - \varepsilon K_1 h(x,\varepsilon) + K_2 h(x,\varepsilon) = f(x)$$

This relation yields (5.29).

Let us now assume (5.28) and (5.29) and prove that $h(x,\varepsilon)$ solves (5.1). Indeed, (5.2) and (5.17) imply

$$\varepsilon h(x,\varepsilon)+\int_c^d R(x,y)h(y,\varepsilon)\,dy=\varepsilon h(x,\varepsilon)+\int_c^d [P(y,D_y)G(x,y)]\,h(y,\varepsilon)\,dy$$

$$=\varepsilon h(x,\varepsilon)+\int_c^d G(x,y)\,[P^*(y,D_y)h(y,\varepsilon)]\,dy+K_2h(x,\varepsilon)\,. \tag{5.34}$$

From (5.34) and (5.28) one gets

$$\varepsilon h(x,\varepsilon)+Rh(x,\varepsilon)=\varepsilon h(x,\varepsilon)$$

$$+\int_c^d G(x,y)\,Q(y,D_y)(f(y)-\varepsilon h(y,\varepsilon))\,dy+K_2h(x,\varepsilon)\,. \tag{5.35}$$

From (5.35) and (5.20) one obtains:

$$\varepsilon h(x,\varepsilon)+Rh(x,\varepsilon)=\varepsilon h(x,\varepsilon)$$

$$+\int_c^d [Q^*(y,D_y)G(x,y)]\,(f(y)-\varepsilon h(y,\varepsilon))\,dy+K_1(f-\varepsilon h)(x,\varepsilon)+K_2h(x,\varepsilon)\,.$$

This relation and equation (5.11) yield:

$$\begin{aligned}\varepsilon h(x,\varepsilon)+Rh(x,\varepsilon)&=\varepsilon h(x,\varepsilon)+f(x)-\varepsilon h(x,\varepsilon)\\&\quad+K_1f(x)-\varepsilon K_1h(x,\varepsilon)+K_2h(x,\varepsilon)\,,\end{aligned}$$

and, using (5.29), one gets (5.1). Theorem 5.2 is proved. □

This theorem is used in our construction of the asymptotic solution to (5.1). Let us look for this asymptotics of the form:

$$h(x,\varepsilon)=\sum_{l=0}^{\infty}\varepsilon^l(u_l(x)+w_l(x,\varepsilon))=\sum_{l=0}^{\infty}\varepsilon^l h_l(x,\varepsilon)\,, \tag{5.36}$$

where the series in (5.36) is understood in the asymptotical sense as follows:

$$h(x,\varepsilon)=\sum_{l=0}^{L}\varepsilon^l(u_l(x)+w_l(x,\varepsilon))+O(\varepsilon^{L+1})\ \text{as}\ \varepsilon\to 0$$

where $O(\varepsilon^{L+1})$ is independent of x and $u_l(x)$ and $w_l(x,\varepsilon)$ are some functions.

Here $u_0(x)$ is an arbitrary solution to the equation

$$P^*(x,D_x)u_0(x) = Q(x,D_x)f(x)\,. \tag{5.37}$$

If $u_0(x)$ is chosen, the function $w_0(x,\varepsilon)$ is constructed as a unique solution to the equation:

$$\varepsilon Q(x,D_x)w_0(x,\varepsilon) + P^*(x,D_x)w_0(x,\varepsilon) = 0\,, \tag{5.38}$$

which satisfies the conditions

$$\varepsilon K_1 w_0(x,\varepsilon) - K_2 w_0(x,\varepsilon) = K_1 f(x) + K_2 u_0(x)\,. \tag{5.39}$$

Theorem 5.3 *The function $h_0(x,\varepsilon) = u_0(x) + w_0(x,\varepsilon)$ solves the equation*

$$\varepsilon h_0(x,\varepsilon) + Rh_0(x,\varepsilon) = f(x) + \varepsilon u_0(x)\,. \tag{5.40}$$

Proof. From (5.2) and (5.17) one gets:

$$\varepsilon h_0(x,\varepsilon) + Rh_0(x,\varepsilon) = \varepsilon h_0(x,\varepsilon) + \int_c^d [P(y,D_y)G(x,y)]\, h_0(y,\varepsilon)\,dy$$

$$= \varepsilon h_0(x,\varepsilon) + \int_c^d G(x,y)P^*(y,D_y)h_0(y,\varepsilon)\,dy + K_2 h_0(x,\varepsilon)\,. \tag{5.41}$$

From (5.37) and (5.38) it follows that

$$P^*(y,D_y)h_0(y,\varepsilon) = P^*(y,D_y)(u_0(y) + w_0(y,\varepsilon)) = P^*(y,D_y)u_0(y)$$

$$+P^*(y,D_y)w_0(y,\varepsilon) = Q(y,D_y)f(y) - \varepsilon Q(y,D_y)w_0(y,\varepsilon)$$

$$= Q(y,D_y)\,[f(y) - \varepsilon w_0(y,\varepsilon)]\,. \tag{5.42}$$

From (5.42) and from the definition of $h_0(x,\varepsilon)$ one derives:

$$P^*(y,D_y)h_0(y,\varepsilon) = Q(y,D_y)\,[f(y) - \varepsilon h_0(y,\varepsilon) + \varepsilon u_0(y)]\,. \tag{5.43}$$

From (5.43) and (5.41) one gets:

$$\varepsilon h_0(x,\varepsilon) + Rh_0(x,\varepsilon) = \varepsilon h_0(x,\varepsilon) + \int_c^d G(x,y)\, Q(y,D_y)[f(y)$$

$$-\varepsilon h_0(y,\varepsilon) + \varepsilon u_0(y)]\, dy + K_2 h_0(x,\varepsilon)\,. \tag{5.44}$$

Equations (5.44) and (5.20) yield:

$$\varepsilon h_0(x,\varepsilon) + Rh_0(x,\varepsilon) = \varepsilon h_0(x,\varepsilon)$$

$$+ \int_c^d [Q^*(y,D_y)G(x,y)]\,(f(y) - \varepsilon h_0(y,\varepsilon) + \varepsilon u_0(y))\, dy$$

$$+K_1(f(x) - \varepsilon h_0(x,\varepsilon) + \varepsilon u_0(x)) + K_2 h_0(x,\varepsilon)\,. \tag{5.45}$$

From (5.45) and (5.11) one derives:

$$\varepsilon h_0(x,\varepsilon) + Rh_0(x,\varepsilon) = \varepsilon h_0(x,\varepsilon) + f(x) - \varepsilon h_0(x,\varepsilon) + \varepsilon u_0(x)$$

$$+K_1 f(x) - \varepsilon K_1 h_0(x,\varepsilon) + \varepsilon K_1 u_0(x) + K_2 h_0(x,\varepsilon)\,. \tag{5.46}$$

This implies:

$$\varepsilon h_0(x,\varepsilon) + Rh_0(x,\varepsilon) = f(x) + \varepsilon u_0(x) + K_1 f(x)$$

$$-\varepsilon K_1 w_0(x,\varepsilon) + K_2 u_0(x) + K_2 w_0(x,\varepsilon)\,. \tag{5.47}$$

Equations (5.47) and (5.39) yield (5.40). Theorem 5.3. is proved. □

Let us construct *higher order approximations.* If $l \geq 1$ then $u_l(x)$ is chosen to be an *arbitrary particular solution to the equation*

$$P^*(x,D_x)u_l(x) = -Q(x,D_x)u_{l-1}(x)\,. \tag{5.48}$$

After $u_l(x)$ is fixed, the function $w_l(x,\varepsilon)$ is constructed as the unique solution to the equation

$$\varepsilon Q(x,D_x)w_l(x,\varepsilon) + P^*(x,D_x)w_l(x,\varepsilon) = 0\,, \tag{5.49}$$

satisfying the conditions

$$\varepsilon K_1 w_l(x,\varepsilon) - K_2 w_l(x,\varepsilon) = -K_1 u_{l-1}(x) + K_2 u_l(x)\,. \tag{5.50}$$

Theorem 5.4 *The function $h_l(x,\varepsilon) = u_l(x) + w_l(x,\varepsilon)$ solves the equation*

$$\varepsilon h_l(x,\varepsilon) + Rh_l(x,\varepsilon) = -u_{l-1}(x) + \varepsilon u_l(x)\,. \tag{5.51}$$

Proof. The proof is similar to that of Theorem 5.3 and is omitted. □

Define

$$H_L(x,\varepsilon) = \sum_{l=0}^{L} \varepsilon^l h_l(x,\varepsilon)\,. \tag{5.52}$$

Theorem 5.5 *The function $H_L(x,\varepsilon)$ solves the equation*

$$\varepsilon H_L(x,\varepsilon) + RH_L(x,\varepsilon) = f(x) + \varepsilon^{L+1} u_L(x)\,. \tag{5.53}$$

Proof. From (5.52) one gets

$$\varepsilon H_L(x,\varepsilon) + RH_L(x,\varepsilon) = \varepsilon \sum_{l=0}^{L} \varepsilon^l h_l(x,\varepsilon) + \sum_{l=0}^{L} \varepsilon^l Rh_l(x,\varepsilon)$$

$$= \sum_{l=0}^{L} \varepsilon^l \left[\varepsilon h_l(x,\varepsilon) + Rh_l(x,\varepsilon)\right]. \tag{5.54}$$

Using (5.40), (5.51) and (5.54) yield (5.54). Theorem 5.5 is proved. □

Theorem 5.6 *If the function $f(x)$ is sufficiently smooth in $\overline{D}$, then it is possible to choose a solution $u_0(x)$ to (5.37) and a solution $u_l(x)$ to (5.48) so that the following inequality holds*

$$\|H_L(x,\varepsilon) - h(x,\varepsilon)\|_- \leq C\varepsilon^{L+1}\,, \tag{5.55}$$

where $C =$ const > 0 *does not depend on ε, but it depends on $f(x)$.*

Proof. From (5.1) and (5.54) one obtains

$$\varepsilon(H_L(x,\varepsilon) - h(x,\varepsilon)) + R(H_L(x,\varepsilon) - h(x,\varepsilon)) = \varepsilon^{L+1} u_L(x)\,. \tag{5.56}$$

From (5.56) it follows that

$$\varepsilon(H_L(x,\varepsilon) - h(x,\varepsilon), H_L(x,\varepsilon) - h(x,\varepsilon))$$

$$+(R(H_L(x,\varepsilon) - h(x,\varepsilon)), H_L(x,\varepsilon) - h(x,\varepsilon)) = \varepsilon^{L+1}(u_L(x), H_L(x,\varepsilon) - h(x,\varepsilon))\,. \tag{5.57}$$

Using (5.8) one obtains

$$c_3\|H_L(x,\varepsilon) - h(x,\varepsilon)\|_-^2 \leq \varepsilon^{L+1}\|u_L(x)\|_+ \|H_L(x,\varepsilon) - h(x,\varepsilon)\|_-. \tag{5.58}$$

Inequality (5.55) follows from (5.58) if the norm $\|u_L(x)\|_+$ is finite.

Consider $L = 0$. If $f(x) \in H^{3(q-p)/2}(D)$ then it is possible to find a solution of (5.37) $u_0(x) \in H^{(q-p)/2}(D)$. Thus the norm $\|u_0(x)\|_+$ is finite.

For $L = 1$ suppose that $f(x) \in H^{5(q-p)/2}(D)$. Then there exist a solution to (5.37) $u_0(x) \in H^{3(q-p)/2}(D)$ and a solution to (5.38) $u_1(x) \in H^{(q-p)/2}(D) = H_+$ so that the norm $\|u_1(x)\|_+$ is finite.

If $f(x) \in C^\infty(\overline{D})$ then the approximation $H_L(x, \varepsilon)$ satisfying (5.55) can be constructed for an arbitrary large L. □

5.4 Examples of asymptotical solutions: case $n = 1$

Example 5.1 Let

$$\varepsilon h(x,\varepsilon) + \int_{-1}^{1} e^{-a|x-y|} r(y) h(y,\varepsilon)\, dy = f(x)\,, \tag{5.59}$$

where $r(y) \geq C^2 > 0$ is a given function.

In this example the operators $P(y, D_y)$ and $Q(x, D_x)$ act on an arbitrary, sufficiently smooth, function $g(x)$ according to the formulas:

$$P(y, D_y)g(y) = r(y)g(y)\,, \text{ i.e., } p = 0\,,$$

and

$$Q(x, D_x)g(x) = -\frac{1}{2a}\frac{d^2 g(x)}{dx^2} + \frac{a}{2} g(x)\,.$$

One has

$$Q(x, D_x)e^{-a|x-y|} = \delta(x-y)\,, \text{ so } G(x,y) = e^{-a|x-y|}\,.$$

Equation (5.37) yields

$$u_0(x) = \frac{-f''(x) + a^2 f(x)}{2ar(x)}\,, \tag{5.60}$$

and (5.38) takes the form

$$\frac{\varepsilon}{2a}(-w_0''(x,\varepsilon) + a^2 w_0(x,\varepsilon)) + r(x) w_0(x,\varepsilon) = 0\,. \tag{5.61}$$

If one looks for the main term of the asymptotics of $h(x,\varepsilon)$, then one can solve in place of (5.61) the following equation

$$-\frac{\varepsilon}{2a}\, w''_{0a}(x,\varepsilon) + r(x) w_{0a}(x,\varepsilon) = 0\,, \tag{5.62}$$

where $w_{0a}(x,\varepsilon)$ is the main term of the asymptotics of $w_0(x,\varepsilon)$.

We seek asymptotics of the bounded, as $\varepsilon \to 0$, solutions to (5.61) and (5.62). To construct the asymptotics, one may use the method, developed in [Vishik and Lusternik (1962)]. Namely, near the point $x = -1$ one sets $x = y - 1$, $y \geq 0$, and writes (5.62) as:

$$-\frac{\varepsilon}{2a}\, v''_a(y,\varepsilon) + r(y-1)\, v_a(y,\varepsilon) = 0, \tag{5.63}$$

where $v_a(y,\varepsilon) := w_{0a}(y-1,\varepsilon)$.

Put $y = t\sqrt{\varepsilon}$ and denote $\varphi_a(t,\varepsilon) := v_a(t\sqrt{\varepsilon},\varepsilon)$. Then

$$-\frac{1}{2a}\frac{d^2\varphi_a(t,\varepsilon)}{dt^2} + r(t\sqrt{\varepsilon}-1)\,\varphi_a(t,\varepsilon) = 0\,. \tag{5.64}$$

Neglecting the term $t\sqrt{\varepsilon}$ in the argument of r is possible if we are looking for the main term of the asymptotics of φ_a. Thus, consider the equation:

$$-\frac{1}{2a}\frac{d^2\varphi_a(t,\varepsilon)}{dt^2} + r(-1)\,\varphi_a(t,\varepsilon) = 0\,. \tag{5.65}$$

Its solution is

$$\varphi_a(t,\varepsilon) = C_1 e^{-\sqrt{2ar(-1)}\,t} + C_2 e^{\sqrt{2ar(-1)}\,t}\,.$$

Discarding the unbounded, as $t \to +\infty$, part of the solution, one gets

$$\varphi_a(t,\varepsilon) = C_1 e^{-\sqrt{2ar(-1)}\,t}\,.$$

Therefore, the main term of the asymptotics of $w_{0a}(x,\varepsilon)$ near the point $x = -1$ is:

$$w_{0a}(x,\varepsilon) = C_1 e^{-\sqrt{2ar(-1)/\varepsilon}\,(1+x)}\,, \quad C_1 = \text{const.} \tag{5.66}$$

Similarly one gets near the point $x = 1$

$$w_{0a}(x,\varepsilon) = D_1 e^{-\sqrt{2ar(1)/\varepsilon}\,(1-x)}\,, \quad D_1 = \text{const}\,. \tag{5.67}$$

From (5.66) and (5.67) one derives the main term of the asymptotics of the bounded, as $\varepsilon \to 0$, solution to equation (5.62):

$$w_{0a}(x,\varepsilon) = C_1 e^{-\sqrt{2ar(-1)/\varepsilon}\,(1+x)} + D_1 e^{-\sqrt{2ar(1)/\varepsilon}\,(1-x)}\,. \tag{5.68}$$

Now the problem is to find the constants C_1 and D_1 from condition (5.39). Since $p = 0$, formula (5.19) yields $K_2 = 0$, and (5.39) is:

$$\varepsilon K_1 w_0(x, \varepsilon) = K_1 f(x). \tag{5.69}$$

From (5.69) and (5.21) one gets

$$\varepsilon \left\{ w_{0a}(y, \varepsilon) \frac{\partial G(x,y)}{\partial y}\bigg|_{-1}^{1} - \frac{dw_{0a}(y,\varepsilon)}{dy} G(x,y)\bigg|_{-1}^{1} \right\}$$

$$= f(y) \frac{\partial G(x,y)}{\partial y}\bigg|_{-1}^{1} - f'(y)G(x,y)|_{-1}^{1}. \tag{5.70}$$

Note that $\dfrac{\partial G(x,y)}{\partial y} = -ae^{-a|x-y|} sgn(y-x)$, where $sgn(t) = t/|t|$, so

$$\frac{\partial G(x,1)}{\partial y} = -ae^{-a(1-x)}, \quad \frac{\partial G(x,-1)}{\partial y} = ae^{-a(1+x)}. \tag{5.71}$$

From (5.71) and (5.70) one obtains

$$\varepsilon\{w_{0a}(1,\varepsilon)(-a)e^{-a(1-x)} - w_{0a}(-1,\varepsilon)ae^{-a(1+x)}$$

$$-w'_{0a}(1,\varepsilon)e^{-a(1-x)} + w'_{0a}(-1,\varepsilon)e^{-a(1+x)}\} = -f(1)ae^{-a(1-x)}$$

$$-f(-1)ae^{-a(1+x)} - f'(1)e^{-a(1-x)} + f'(-1)e^{-a(1+x)}. \tag{5.72}$$

This implies:

$$\varepsilon\{aw_{0a}(1,\varepsilon) + w'_{0a}(1,\varepsilon)\} = af(1) + f'(1),$$

and

$$\varepsilon\{-aw_{0a}(-1,\varepsilon) + w'_{0a}(-1,\varepsilon)\} = -af(-1) + f'(-1). \tag{5.73}$$

Keeping the main terms in the braces, one gets:

$$\sqrt{\varepsilon 2ar(1)}\, D_1 = f'(1) + af(1),$$

and

$$-\sqrt{\varepsilon 2ar(-1)}\, C_1 = f'(-1) - af(-1).$$

Therefore

$$C_1 = \frac{-f'(-1) + af(-1)}{\sqrt{\varepsilon 2ar(-1)}}, \quad D_1 = \frac{f'(1) + af(1)}{\sqrt{\varepsilon 2ar(1)}}. \tag{5.74}$$

From (5.60), (5.68) and (5.74) one finds the main term of the asymptotics of the solution to (5.59):

$$h(x,\varepsilon) \approx \frac{-f''(x) + a^2 f(x)}{2ar(x)} + \frac{-f'(-1) + af(-1)}{\sqrt{\varepsilon 2ar(-1)}} e^{-\sqrt{2ar(-1)/\varepsilon}(1+x)}$$

$$+ \frac{f'(1) + af(1)}{\sqrt{\varepsilon 2ar(1)}} e^{-\sqrt{2ar(1)/\varepsilon}(1-x)}. \tag{5.75}$$

If $r(x) = \text{const}$, then (4.17) yields the asymptotic formula obtained in [Ramm and Shifrin (1991)].

Example 5.2 Consider the equation

$$\varepsilon h(x,\varepsilon) + \int_c^d G(x,y) h(y,\varepsilon)\, dy = f(x), \tag{5.76}$$

where $G(x,y)$ solves the problem

$$-\frac{\partial^2 G(x,y)}{\partial x^2} + a^2(x) G(x,y) = \delta(x-y), \quad G(\infty, y) = 0, \tag{5.77}$$

and $a^2(x) \geq \text{const} > 0$, $\forall x \in R^1$.

Here $P(y, D_y) = I$, $p = 0$, $Q(x, D_x) = -\frac{d^2}{dx^2} + a^2(x)$, $q = 2$.

One can write $G(x,y)$ as

$$G(x,y) = \begin{cases} \varphi_1(x)\varphi_2(y), & x < y, \\ \varphi_2(x)\varphi_1(y), & y < x, \end{cases} \tag{5.78}$$

where functions $\varphi_1(x)$ and $\varphi_2(x)$ are linearly independent solutions to the equation $Q(x, D_x)\varphi(x) = 0$, satisfying conditions $\varphi_1(-\infty) = 0$, $\varphi_2(+\infty) = 0$ and

$$\varphi_1'(x)\varphi_2(x) - \varphi_1(x)\varphi_2'(x) = 1. \tag{5.79}$$

By (5.37) one gets

$$u_0(x) = -f''(x) + a^2(x) f(x). \tag{5.80}$$

By (5.38) one obtains

$$\varepsilon(-w_0''(x,\varepsilon)+a^2(x)w_0(x,\varepsilon))+w_0(x,\varepsilon)=0\,.$$

The main term $w_{0a}(x,\varepsilon)$ of the asymptotics of $w_0(x,\varepsilon)$ solves the equation:

$$-\varepsilon w_{0a}''(x,\varepsilon)+w_{0a}(x,\varepsilon)=0\,.$$

Thus

$$w_{0a}(x,\varepsilon)=Ce^{-(x-c)/\sqrt{\varepsilon}}+De^{-(d-x)/\sqrt{\varepsilon}}\,. \tag{5.81}$$

Condition (5.39) takes the form (5.69). Using $w_{0a}(x,\varepsilon)$ in place of $w_0(x,\varepsilon)$ in (5.69), one gets, similarly to (5.70), the relation

$$\varepsilon\left\{w_{0a}(y,\varepsilon)\,\frac{\partial G(x,y)}{\partial y}\bigg|_c^d-w_{0a}'(y,\varepsilon)\,G(x,y)\big|_c^d\right\}$$
$$=f(y)\,\frac{\partial G(x,y)}{\partial y}\bigg|_c^d-f'(y)\,G(x,y)\big|_c^d\,.$$

Keeping the main terms, one gets

$$-\varepsilon w_{0a}'(y,\varepsilon)\,G(x,y)\big|_c^d=f(y)\,\frac{\partial G(x,y)}{\partial y}\bigg|_c^d-f'(y)\,G(x,y)\big|_c^d\,. \tag{5.82}$$

From (5.82) and (5.78) one obtains

$$\varepsilon\left\{-w_{0a}'(d,\varepsilon)\varphi_1(x)\varphi_2(d)+w_{0a}'(c,\varepsilon)\varphi_2(x)\varphi_1(c)\right\}$$

$$=f(d)\varphi_1(x)\varphi_2'(d)-f(c)\varphi_2(x)\varphi_1'(c)-f'(d)\varphi_1(x)\varphi_2(d)$$

$$+f'(c)\varphi_2(x)\varphi_1(c)\,. \tag{5.83}$$

Because $\varphi_1(x)$ and $\varphi_2(x)$ are linearly independent, it follows from (5.83)

$$-\varepsilon w_{0a}'(d,\varepsilon)\varphi_2(d)=f(d)\varphi_2'(d)-f'(d)\varphi_2(d)\,,$$

$$\varepsilon w_{0a}'(c,\varepsilon)\varphi_1(c)=-f(c)\varphi_1'(c)+f'(c)\varphi_1(c)\,. \tag{5.84}$$

Substitute (5.81) in (5.84) and keep the main terms, to get

$$-\varepsilon\,\frac{D}{\sqrt{\varepsilon}}\,\varphi_2(d)=f(d)\varphi_2'(d)-f'(d)\varphi_2(d)\,,$$

$$-\varepsilon \frac{C}{\sqrt{\varepsilon}} \varphi_1(c) = -f(c)\varphi_1'(c) + f'(c)\varphi_1(c) .$$

This yields the final formulas for the coefficients:

$$C = \frac{-f'(c)\varphi_1(c) + f(c)\varphi_1'(c)}{\sqrt{\varepsilon}\,\varphi_1(c)} , \quad D = \frac{f'(d)\varphi_2(d) - f(d)\varphi_2'(d)}{\sqrt{\varepsilon}\,\varphi_2(d)} . \tag{5.85}$$

From (5.80), (5.81) and (5.85) one gets the main term of the asymptotics of the solution to (5.76):

$$h(x,\varepsilon) \approx -f''(x) + a^2(x)f(x) + \frac{-f'(c)\varphi_1(c) + f(c)\varphi_1'(c)}{\sqrt{\varepsilon}\,\varphi_1(c)} e^{-(x-c)/\sqrt{\varepsilon}}$$

$$+ \frac{f'(d)\varphi_2(d) - f(d)\varphi_2'(d)}{\sqrt{\varepsilon}\,\varphi_2(d)} e^{-(d-x)/\sqrt{\varepsilon}} . \tag{5.86}$$

5.5 Asymptotics in the case $n > 1$

Consider equation (5.1) with $R(x,y) \in \mathcal{R}$. The method for construction of the asymptotics of the solution to (5.1) in the multidimensional case is parallel to the one developed in the case $n = 1$. The proofs are also parallel to the ones given for the case $n = 1$, and are omitted by this reason.

Let us state the basic results.

Theorem 5.7 *Equation (5.1) is equivalent to the problem*

$$\varepsilon Q(x, D_x)h(x,\varepsilon) + P^*(x, D_x)h(x,\varepsilon) = Q(x, D_x)f(x) , \tag{5.87}$$

$$\varepsilon M_1 h(x,\varepsilon) - M_2 h(x,\varepsilon) = M_1 f(x) . \tag{5.88}$$

Proof. One uses Lemmas 5.1, 5.4 and 5.5 and formula (5.27) to prove Theorem 5.7. □

To construct the asymptotics of the solution to equation (5.1), let us look for the asymptotics of the form:

$$h(x,\varepsilon) = \sum_{l=0}^{\infty} \varepsilon^l (u_l(x) + w_l(x,\varepsilon)) = \sum_{l=0}^{\infty} \varepsilon^l h_l(x,\varepsilon) , \tag{5.89}$$

where $u_0(x)$ is an arbitrary solution to the equation

$$P^*(x, D_x)u_0(x) = Q(x, D_x)f(x) , \tag{5.90}$$

and if some $u_0(x)$ is found, then $w_0(x,\varepsilon)$ is uniquely determined as the solution to the problem

$$\varepsilon Q(x,D_x)w_0(x,\varepsilon)+P^*(x,D_x)w_0(x,\varepsilon)=0\,, \tag{5.91}$$

$$\varepsilon M_1w_0(x,\varepsilon)-M_2w_0(x,\varepsilon)=M_1f(x)+M_2u_0(x)\,. \tag{5.92}$$

Theorem 5.8 *The function $h_0(x,\varepsilon)=u_0(x)+w_0(x,\varepsilon)$ solves the equation*

$$\varepsilon h_0(x,\varepsilon)+Rh_0(x,\varepsilon)=f(x)+\varepsilon u_0(x)\,. \tag{5.93}$$

Let us construct higher order terms of the asymptotics. Define $u_l(x)$ $(l\geq 1)$ as an arbitrary solution to the equation

$$P^*(x,D_x)u_l(x)=-Q(x,D_x)u_{l-1}(x)\,. \tag{5.94}$$

After finding $u_l(x)$, one finds $w_l(x,\varepsilon)$ as the unique solution to the problem

$$\varepsilon Q(x,D_x)w_l(x,\varepsilon)+P^*(x,D_x)w_l(x,\varepsilon)=0\,, \tag{5.95}$$

$$\varepsilon M_1w_l(x,\varepsilon)-M_2w_l(x,\varepsilon)=-M_1u_{l-1}(x)+M_2u_l(x)\,. \tag{5.96}$$

Theorem 5.9 *The function $h_l(x,\varepsilon)=u_l(x)+w_l(x,\varepsilon)$ solves the equation*

$$\varepsilon h_l(x,\varepsilon)+Rh_l(x,\varepsilon)=-u_{l-1}(x)+\varepsilon u_l(x)\,. \tag{5.97}$$

Define

$$H_L(x,\varepsilon)=\sum_{l=0}^{L}\varepsilon^l h_l(x,\varepsilon). \tag{5.98}$$

From Theorems 5.8 and 5.9 one derives

Theorem 5.10 *The function $H_L(x,\varepsilon)$ solves the equation*

$$\varepsilon H_L(x,\varepsilon)+RH_L(x,\varepsilon)=f(x)+\varepsilon^{L+1}u_L(x)\,. \tag{5.99}$$

Theorem 5.11 *If the function $f(x)$ is sufficiently smooth in $\overline{D}$, then it is possible to choose a solution $u_0(x)$ to (5.90) and a solution $u_l(x)$ to (5.94), so that the following inequality holds*

$$\|H_L(x,\varepsilon)-h(x,\varepsilon)\|_-\leq C\varepsilon^{L+1},$$

where $C=\text{const}>0$ does not depend on ε, but it depends on $f(x)$.

5.6 Examples of asymptotical solutions: case $n > 1$

Example 5.3

Consider the equation

$$\varepsilon h(x,\varepsilon) + \int_{S_1} G(x,y)s(|y|)h(y,\varepsilon)\,dy = 1\,, \tag{5.100}$$

where $x = (x_1, x_2)$, $y = (y_1, y_2)$, $|y| = \sqrt{y_1^2 + y_2^2}$, $s(|y|)$ is a known smooth positive function, $s(|y|) \geq C^2 > 0$, $G(x,y) = \frac{1}{2\pi} K_0(a|x-y|)$, $K_0(r)$ is the MacDonalds function, $(-\Delta_x + a^2)G(x,y) = \delta(x-y)$, S_1 is a unit disk centered at the origin.

In this example $P(y, D_y)g(y) = s(|y|)g(y)$, $p = 0$, $Q(x, D_x) = -\Delta_x + a^2$, $q = 2$.

Let us construct the main term of the asymptotics of the solution to (5.100). By (5.90) one gets

$$s(|x|)u_0(x) = (-\Delta_x + a^2)1 = a^2\,.$$

Thus

$$u_0(x) = \frac{a^2}{s(|x|)}. \tag{5.101}$$

Equation (5.91) yields:

$$\varepsilon(-\Delta_x + a^2)w_0(x,\varepsilon) + s(|x|)w_0(x,\varepsilon) = 0\,. \tag{5.102}$$

The main term $w_{0a}(x,\varepsilon)$ of the asymptotics of $w_0(x,\varepsilon)$ solves the equation

$$-\varepsilon\Delta_x w_{0a}(x,\varepsilon) + s(|x|)w_{0a}(x,\varepsilon) = 0\,. \tag{5.103}$$

In polar coordinates one gets

$$-\varepsilon\left(\frac{\partial^2 w_{0a}(r,\varphi,\varepsilon)}{\partial r^2} + \frac{1}{r}\frac{\partial w_{0a}(r,\varphi,\varepsilon)}{\partial r} + \frac{1}{r^2}\frac{\partial^2 w_{0a}(r,\varphi,\varepsilon)}{\partial \varphi^2}\right)$$

$$+ s(r)w_{0a}(r,\varphi,\varepsilon) = 0\,. \tag{5.104}$$

By radial symmetry $w_{0a}(r,\varphi,\varepsilon) = w_{0a}(r,\varepsilon)$, so

$$-\varepsilon\left(\frac{d^2 w_{0a}(r,\varepsilon)}{dr^2} + \frac{1}{r}\frac{dw_{0a}(r,\varepsilon)}{dr}\right) + s(r)w_{0a}(r,\varepsilon) = 0\,. \tag{5.105}$$

The asymptotics of the solution to (5.105) we construct using the method of [Vishik and Lusternik (1962)]. Let $r = 1 - \varrho$. Then

$$-\varepsilon\left(\frac{d^2 w_{0a}(\varrho,\varepsilon)}{d\varrho^2} - \frac{1}{1-\varrho}\frac{dw_{0a}(\varrho,\varepsilon)}{d\varrho}\right) + s(1-\varrho)\, w_{0a}(\varrho,\varepsilon) = 0\,.$$

Put $\varrho = t\sqrt{\varepsilon}$ and keep the main terms, to get

$$-\frac{d^2 w_{0a}(t)}{dt^2} + s(1) w_{0a}(t) = 0\,, \tag{5.106}$$

so

$$w_{0a}(t) = Ce^{-\sqrt{s(1)}\,t} + De^{\sqrt{s(1)}\,t}.$$

Keeping exponentially decaying, as $t \to +\infty$, solution one obtains:

$$w_{0a}(t) = Ce^{-\sqrt{s(1)}\,t}\,.$$

Therefore

$$w_{0a}(r,\varepsilon) = Ce^{-\sqrt{s(1)/\varepsilon}\,(1-r)}\,. \tag{5.107}$$

To find the constant C in (5.107) we use condition (5.92). Since $p = 0$, one concludes $M_2 = 0$, and (5.92) takes the form

$$\varepsilon M_1 w_{0a}(x,\varepsilon) = M_1 f(x) = M_1 1\,. \tag{5.108}$$

From (5.25) and (5.108) one gets:

$$\varepsilon \int\limits_{\partial S_1} \left[\frac{\partial G(x,y)}{\partial N_y}\, w_0(y,\varepsilon) - G(x,y)\,\frac{\partial w_0(y,\varepsilon)}{\partial N_y}\right] dl_y$$

$$= \int\limits_{\partial S_1} \left[\frac{\partial G(x,y)}{\partial N_y}\, 1 - G(x,y)\,\frac{\partial 1}{\partial N_y}\right] dl_y\,,$$

where dl_y is the element of the arclength of ∂S_1.

If one replaces $w_0(y,\varepsilon)$ by $w_{0a}(y,\varepsilon)$ in the above formula then one gets

$$\varepsilon \int\limits_{\partial S_1} \left[\frac{\partial G(x,y)}{\partial N_y}\, w_{0a}(y,\varepsilon) - G(x,y)\,\frac{\partial w_{0a}(y,\varepsilon)}{\partial N_y}\right] dl_y = \int\limits_{\partial S_1} \frac{\partial G(x,y)}{\partial N_y}\, dl_y\,. \tag{5.109}$$

The main term in (5.109) can be written as:

$$-\varepsilon \int_{\partial S_1} G(x,y) \frac{\partial w_{0a}(y,\varepsilon)}{\partial N_y} \, dl_y = \int_{\partial S_1} \frac{\partial G(x,y)}{\partial N_y} \, dl_y \,. \tag{5.110}$$

By (5.107) for $y \in \partial S_1$ one gets

$$\frac{\partial w_{0a}(y,\varepsilon)}{\partial N_y} = \sqrt{\frac{s(1)}{\varepsilon}} \, C \,. \tag{5.111}$$

From (5.110) and (5.111) one obtains

$$-\sqrt{\varepsilon s(1)} \, C \int_{\partial S_1} G(x,y) \, dl_y = \int_{\partial S_1} \frac{\partial G(x,y)}{\partial N_y} \, dl_y \,, \quad \forall x \in S_1 \,. \tag{5.112}$$

For $x = 0$ and $y \in \partial S_1$ one gets

$$G(0,y) = \frac{1}{2\pi} K_0(a) \,, \quad \frac{\partial G(0,y)}{\partial N_y} = \frac{1}{2\pi} \left. \frac{dK_0(ar)}{dr} \right|_{r=1}$$

$$= \frac{a}{2\pi} K_0'(ar) \Big|_{r=1} = -\frac{a}{2\pi} K_1(a) \,.$$

These relations and (5.112) imply:

$$-\sqrt{\varepsilon s(1)} \, C K_0(a) = -a K_1(a) \,.$$

Therefore

$$C = \frac{a K_1(a)}{\sqrt{\varepsilon s(1)} \, K_0(a)} \,. \tag{5.113}$$

From (5.101), (5.107) and (5.113) one finds the main term of the asymptotics of the solution to (5.100):

$$h(x,\varepsilon) \approx \frac{a^2}{s(|x|)} + \frac{a K_1(a)}{\sqrt{\varepsilon s(1)} \, K_0(a)} \, e^{-\sqrt{s(1)/\varepsilon} \, (1-|x|)} \,. \tag{5.114}$$

If $s(|x|) = 1$, then (5.114) agrees with the earlier result, obtained in [Ramm and Shifrin (1995)].

Example 5.4 Consider the equation

$$\varepsilon h(x,\varepsilon) + \int_{B_1} G(x,y) s(|y|) h(y,\varepsilon) \, dy = 1 \,, \tag{5.115}$$

where $x = (x_1, x_2, x_3)$, $y = (y_1, y_2, y_3)$, $s(|y|)$ is a smooth positive function, $s(|y|) \geq C^2 > 0$, $G(x, y) = \dfrac{e^{-a|x-y|}}{4\pi|x-y|}$, $P(y, D_y)g(y) = s(|y|)g(y)$, so $p = 0$, $(-\Delta_x + a^2)G(x, y) = \delta(x - y)$, so $Q(x, D_x) = -\Delta_x + a^2$, $q = 2$, B_1 is a unit ball centered at the origin.

The main term of the asymptotics is constructed by the method of Section 5. By (5.90) one gets

$$s(|x|)u_0(x) = (-\Delta_x + a^2)1 = a^2 .$$

Thus

$$u_0(x) = \frac{a^2}{s(|x|)} . \tag{5.116}$$

By (5.91)

$$\varepsilon(-\Delta_x + a^2)w_0(x, \varepsilon) + s(|x|)w_0(x, \varepsilon) = 0 .$$

Keeping the main terms $w_{0a}(x, \varepsilon)$ of the asymptotics of $w_0(x, \varepsilon)$, one gets

$$-\varepsilon\Delta_x w_{0a}(x, \varepsilon) + s(|x|)w_{0a}(x, \varepsilon) = 0 .$$

In spherical coordinates this equation for the spherically symmetric solution becomes:

$$-\varepsilon\left(\frac{d^2 w_{0a}(r, \varepsilon)}{dr^2} + \frac{2}{r}\frac{dw_{0a}(r, \varepsilon)}{dr}\right) + s(r)w_{0a}(r, \varepsilon) = 0 . \tag{5.117}$$

Let $r = 1 - \varrho$. Then (5.117) can be written as:

$$-\varepsilon\left(\frac{d^2 w_{0a}(\varrho, \varepsilon)}{d\varrho^2} - \frac{2}{1-\varrho}\frac{dw_{0a}(\varrho, \varepsilon)}{d\varrho}\right) + s(1-\varrho)\, w_{0a}(\varrho, \varepsilon) = 0 .$$

Put $\varrho = t\sqrt{\varepsilon}$ and keep the main terms in the above equation to get

$$-\frac{d^2 w_{0a}(t)}{dt^2} + s(1)w_{0a}(t) = 0 . \tag{5.118}$$

The exponentially decaying, as $t \to +\infty$, solution to (6.19) is:

$$w_{0a}(t) = Ce^{-\sqrt{s(1)}\, t} .$$

Therefore

$$w_{0a}(x, \varepsilon) = Ce^{-\sqrt{s(1)/\varepsilon}\,(1-|x|)} . \tag{5.119}$$

The constant C in (5.119) is determined from conditions (5.92), which in this example can be written as

$$\varepsilon M_1 w_0(x,\varepsilon) = M_1 f(x) = M_1 1\,. \tag{5.120}$$

Using formulas (5.25) and (5.120) one gets

$$\varepsilon \int\limits_{\partial B_1} \left[\frac{\partial G(x,y)}{\partial N_y} w_0(y,\varepsilon) - G(x,y)\frac{\partial w_0(y,\varepsilon)}{\partial N_y}\right] dS_y = \int\limits_{\partial B_1} \frac{\partial G(x,y)}{\partial N_y}\, dS_y\,.$$

Replacing $w_0(y,\varepsilon)$ by $w_{0a}(y,\varepsilon)$ and keeping the main terms, one obtains

$$-\varepsilon \int\limits_{\partial B_1} G(x,y)\frac{\partial w_{0a}(y,\varepsilon)}{\partial N_y}\, dS_y = \int\limits_{\partial B_1} \frac{\partial G(x,y)}{\partial N_y}\, dS_y\,. \tag{5.121}$$

From (5.119) for $y \in \partial B_1$ one derives

$$\frac{\partial w_{0a}(y,\varepsilon)}{\partial N_y} = \sqrt{\frac{s(1)}{\varepsilon}}\, C\,. \tag{5.122}$$

From (5.122) and (5.121) it follows that

$$-\sqrt{\varepsilon s(1)}\, C \int\limits_{\partial B_1} G(x,y)\, dS_y = \int\limits_{\partial B_1} \frac{\partial G(x,y)}{\partial N_y}\, dS_y\,. \tag{5.123}$$

Put $x = 0$ in (5.123). Let us compute the corresponding integrals:

$$\int\limits_{\partial B_1} G(0,y)\, dS_y = \frac{1}{4\pi}\int\limits_{\partial B_1} \frac{e^{-a|y|}}{|y|}\, dS_y = \frac{e^{-a}}{4\pi}\int\limits_{\partial B_1} dS_y = e^{-a}\,. \tag{5.124}$$

Note that:

$$\frac{\partial G(0,y)}{\partial N_y} = \frac{1}{4\pi}\frac{\partial}{\partial r}\left(\frac{e^{-ar}}{r}\right)\bigg|_{r=1} = -\frac{1}{4\pi}\left(ae^{-a} + e^{-a}\right).$$

Thus

$$\int\limits_{\partial B_1} \frac{\partial G(0,y)}{\partial N_y}\, dS_y = -\frac{1}{4\pi}\, e^{-a}(a+1)\int\limits_{\partial B_1} dS_y = -e^{-a}(a+1)\,. \tag{5.125}$$

From (5.123), (5.124) and (5.125)) one gets, setting $x = 0$, the relation

$$-\sqrt{\varepsilon s(1)}\, C e^{-a} = -e^{-a}(a+1)\,.$$

This yields

$$C = \frac{a+1}{\sqrt{\varepsilon s(1)}}. \tag{5.126}$$

From (5.116), (5.119) and (5.126) the main term of the asymptotics of the solution to equation (5.115) follows:

$$h(x,\varepsilon) \approx \frac{a^2}{s(|x|)} + \frac{a+1}{\sqrt{\varepsilon s(1)}} e^{-\sqrt{s(1)/\varepsilon}\,(1-|x|)}. \tag{5.127}$$

y If $s(x) = 1$, formula (5.127) yields a result obtained in [Ramm and Shifrin (1995)].

Let us summarize briefly our results. In this paper we constructed asymptotics of the solution to (5.1) as $\varepsilon \to +0$, and demonstrated how the L^2 - solution to (5.1) tends to a distributional solution of the limiting equation $Rh(x) = f(x)$.

Chapter 6

Estimation and Scattering Theory

In recent years a number of papers have appeared in which the three-dimensional ($3D$) inverse scattering problem is associated with the random fields estimation problem. In this Chapter we give a brief presentation of the direct and inverse scattering theory in the three-dimensional case and outline the connection between this theory and the estimation theory. This connection, however, is less natural and significant than in one-dimensional case, due to the lack of causality in the spacial variables. In Chapter 1 the direct scattering problem is studied, in Chapter 2 the inverse scattering problem is studied, in Chapter 3 the connection between the estimation theory and inverse scattering is discussed.

6.1 The direct scattering problem

6.1.1 *The direct scattering problem*

Consider the problem

$$\ell_q u - k^2 u := [-\nabla^2 + q(x) - k^2]u = 0 \quad \text{in} \quad R^3, \quad k > 0 \tag{6.1}$$

$$u = \exp(ik\theta \cdot x) + A(\theta', \theta, k)r^{-1}\exp(ikr) + o(r^{-1}), \quad r = |x| \to \infty, \quad \theta' = \frac{x}{r} \tag{6.2}$$

where $\theta, \theta' \in S^2$, S^2 is the unit sphere in R^3, and $o(r^{-1})$ in (6.2) is uniform in $\theta, \theta' \in S^2$.

The function u is called the scattering solution, the function $A(\theta', \theta, k)$ is called the scattering amplitude, the function $q(x)$ is the potential.

Let us assume that

$$q \in Q := \{q : q = \overline{q}, \quad |q| \leq c(1 + |x|)^{-a}, \quad a > 3\}. \tag{6.3}$$

The bar in this chapter stands for complex conjugate (and not for mean value). By c we denote various positive constants. By Q_M we denote the following class of q

$$Q_m := \left\{q : q^{(j)} \in Q, \quad 0 \leq |j| \leq m\right\}, \tag{6.4}$$

so that $Q_0 = Q$.

The scattering theory is developed for q which may have local singularities and are described by some integral norms, but this is not important for our presentation here. Our purpose is to give a brief outline of the theory for the problem (6.1)-(6.3) with minimum technicalities. The following questions are discussed:

1) selfadjointness of ℓ_q,
2) the nature of the spectrum of ℓ_q ,
3) existence and uniqueness of the solution to (6.1)-(6.3),
4) eigenfunction expansion in scattering solutions,
   and
5) properties of the scattering amplitude.

The operator ℓ_q defined by the differential expression (6.1) on $C_0^\infty(R^3)$ is symmetric and bounded from below. Let us denote by ℓ_q its closure in $H = L^2(R^3)$.

Lemma 6.1 *The operator ℓ_q is selfadjoint.*

Proof. This lemma is a particular case of Lemma 8.5 in Section 8.2.4.□

Lemma 6.2 *1) The negative spectrum of ℓ_q is discrete and finite. 2) The positive spectrum is absolutely continuous. 3) The point $\lambda = 0$ belongs to the continuous spectrum but may not belong to the absolutely continuous spectrum.*

Proof. Let us recall Glazman's lemma: □

Lemma 6.3 *Negative spectrum of a selfadjoint operator A is discrete and finite if and only if*

$$\sup \dim \mathcal{M} < \infty, \tag{6.5}$$

where the supremum is taken over the set of subspaces $\mathcal{M}$ such that $(Au, u) \leq 0$ for $u \in \mathcal{M}$.

A proof is, e.g., in [Ramm (1986), p. 330] or [Glazman (1965), §3].

Therefore the first statement of Lemma 6.2 is proved if one proves that

$$N_- := \sup \dim \mathcal{M} < \infty, \quad \mathcal{M} := \left\{ u : \int_{R^3} |\nabla u|^2 dx < \int_{R^3} q_-(x)|u|^2 dx \right\}, \tag{6.6}$$

where N_- is the number of negative eigenvalues of ℓ_q counting their multiplicities and $q_- = \max\{0, -q(x)\}$. One has

$$q = q_+(x) - q_-(x), \quad q_+ = \max\{q, 0\}. \tag{6.7}$$

Let us write

$$\mathcal{M} = \left\{ u : 1 < \int q_- |u|^2 dx \left(\int |\nabla u|^2 dx \right)^{-1} \right\}. \tag{6.8}$$

The ratio of the quadratic forms in (6.8) has a discrete spectrum and the corresponding eigenfunctions solve the problem

$$q_- u = -\lambda \nabla^2 u \quad \text{in} \quad R^3$$

which can be written as

$$g_0 q_- u = \lambda u, \quad g_0 f := \int (4\pi|x - y|)^{-1} f(y) dy. \tag{6.9}$$

Let $q_-^{1/2} := p$. Then (6.9) can be written as

$$A\phi := p g_0 p \phi = \lambda \phi, \quad \phi := pu. \tag{6.10}$$

The operator A, defined in (6.10), is compact, selfadjoint, and nonnegative-definite in $L^2(R^3)$ if $q \in Q$. Therefore the number of the eigenvalues λ_n of the problem (6.10) which satisfy the inequality $\lambda_n > 1$, is finite. This number is the dimension of $\mathcal{M}$ defined in (6.8). Thus $N_- < \infty$, where N_- is defined in (6.6). Statement 1) of Lemma 6.2 is proved.

Only a little extra work is needed to give an estimate of N_- from above. Namely,

$$A^2 \phi_j = \lambda_j^2 \phi_j, \quad Tr A^2 = \sum_{j=1}^{\infty} \lambda_j^2 \geq \sum_{\lambda_j > 1} \lambda_j^2 \geq \sum_{\lambda_j > 1} 1 = N_-. \tag{6.11}$$

Thus

$$N_{-} \leq Tr p g_0 q_{-} g_0 p = \iint \frac{q_{-}(x)q_{-}(y)dxdy}{(4\pi)^2|x-y|^2}. \tag{6.12}$$

The right-hand side of (6.12) is finite if $q_{-} \in Q$. Note that if $q \in Q$, then $q_{-} \in Q$ and $q_{+} \in Q$.

6.1.2 *Properties of the scattering solution*

Lemma 6.4 *The scattering solution exists and is unique.*

Proof. The scattering solution solves the integral equation

$$u = u_0 - gqu, \quad gf := \int \frac{\exp(ik|x-y|)}{4\pi|x-y|} f(y)dy, \tag{6.13}$$

where

$$u_0 := \exp(ik\theta \cdot x). \tag{6.14}$$

Conversely, the solution to (6.13) is the scattering solution with

$$A(\theta', \theta, k) = -\frac{1}{4\pi}\int \exp(-ik\theta' \cdot y)g(y)u(y,\theta,k)dy. \tag{6.15}$$

It is not difficult to check that if $q \in Q$ then the operator

$$T(k)u := gqu \tag{6.16}$$

is compact in $C(R^3)$. Therefore, the existence of the solution to (6.13) follows from the uniqueness of the solution to the homogeneous equation

$$u = -Tu, \quad u \in C(R^3) \tag{6.17}$$

by Fredholm's alternative.

If u solves (6.17) then u solves equation (6.1) and satisfies the radiation condition

$$\lim_{r\to\infty} \int_{|s|=r} \left| \frac{\partial u}{\partial r} - iku \right|^2 ds = 0. \tag{6.18}$$

Since $q = \overline{q}$, the function $\overline{u}$ solves equation (6.1) and Green's formula yields

$$\lim_{r\to\infty} \int_{|s|=r} \left(u\frac{\partial \overline{u}}{\partial r} - \overline{u}\frac{\partial u}{\partial r} \right) ds = 0. \tag{6.19}$$

From (6.18) and (6.19) it follows that

$$\lim_{r\to\infty}\int_{|s|=r}\left[\left|\frac{\partial u}{\partial r}\right|^2+k^2|u|^2\right]ds=0. \tag{6.20}$$

Any solution to (6.1) which satisfies condition (6.20) has to vanish identically according to a theorem of Kato [Kato (1959)]. Thus $u=0$ and Lemma 6.4 is proved. □

Lemma 6.5 *Let $f\in L^2(R^3)$ be arbitrary. Define*

$$\tilde f(\xi):=(2\pi)^{-3/2}\int f(x)\overline{u(x,\xi)}dx,\quad f_j:=(f,u_j),\quad 1\le j\le N_-, \tag{6.21}$$

where u_j are the orthonormalized eigenfunctions corresponding to the discrete spectrum of ℓ_q:

$$\ell_q u_j=\lambda_j u_j,\quad 1\le j\le N_-,\quad \lambda_j<0, \tag{6.22}$$

$$(u_j,u_m):=\int u_j(x)\overline{u_m(x)}dx=\delta_{jm}, \tag{6.23}$$

$$\xi\in R^3,\quad |\xi|=k,\quad \xi=k\theta,\quad \theta\in S^2. \tag{6.24}$$

Then

$$f(x)=(2\pi)^{-3/2}\int\tilde f(\xi)u(x,\xi)d\xi+\sum_{j=1}^{N_-}f_j u_j(x). \tag{6.25}$$

Formulas (6.21), (6.25) are analogous to the usual Fourier inversion formulas. They reduce to the latter if $q(x)=0$.

The proof of Lemma 6.5 requires some preparations. We follow the scheme used in [Ramm (1963); Ramm (1963b); Ramm (1965); Ramm (1968b); Ramm (1969d); Ramm (1970); Ramm (1971b); Ramm (1987); Ramm (1988b)] and in [Ramm (1986), p. 47].

Let $G(x,y,k)$ be the resolvent kernel of ℓ_q:

$$(\ell_q-k^2)G(x,y,k)=\delta(x-y)\quad\text{in}\quad R^3. \tag{6.26}$$

This kernel solves the equation

$$G(x,y,k)=g(x,y,k)-\int g(x,z,k)q(z)G(z,y,k)dz. \tag{6.27}$$

This equation is similar to (6.13) and can be written as

$$(I+T)G = g, \tag{6.28}$$

where T is defined in (6.16). Therefore, as in the proof of Lemma 6.4, the solution to the equation (6.27) exists and is unique in the space $C_y(R^3)$ of functions of the form $G = c|x-y|^{-1} + v(x,y)$ where $v(x,y)$ is continuous and $c = \text{const}$, $\| G \| := |c| + \max_{x\in R^3} |v(x,y)|$. The operator T is compact in $C_y(R^3)$ if $q \in Q$. The homogeneous equation (6.28) has only the trivial solution.

The operator $T = T(k)$ depends continuously on $k \in \mathbf{C}_+ := \{k : Imk \geq 0\}$. Therefore $[I+T(k)]^{-1}$ is a continuous function of k in the region $\mathbf{C}_+ \cap \Delta(k)$, where $\Delta(k)$ is a neighborhood of a point k_0, $\Delta(k) := \{k : |k-k_0| < \delta\}$ $\delta > 0$, and the operator $I + T(k_0)$ is invertible. Since for any $k > 0$ the operator $I + T(k)$ is invertible, it follows that $G(x,y,k)$ is continuous in k in the region $\mathbf{C}_+ \cap \Delta(0,\infty)$, that is in a neighborhood of the positive semiaxis in $\mathbf{C}_+$. The continuity holds for any x, y fixed, $x \neq y$, and also in the norm of C_y. This implies that the continuous spectrum of ℓ_q in the interval $(0,\infty)$ is absolutely continuous.

From the equation (6.27) it follows that

$$G(x,y,k) = g(r)u(y,-\theta,k)\left[1+o(1)\right], \quad r = |x| \to \infty, \quad \frac{x}{|x|} = \theta \tag{6.29}$$

where $g(r) := (4\pi r)^{-1} \exp(ikr)$ and $u(y,-\theta,k)$ is the scattering solution. In fact, $o(1) = 0\left(\frac{1}{|x|}\right)$ uniformly in $y \in D$, where $D \in R^3$ is an arbitrary fixed bounded domain. Indeed, it follows from (6.27) that

$$u(y,-\theta,k) = \exp(-ik\theta\cdot y) - \int \exp(-ik\theta\cdot z)q(z)G(z,y,k)dz. \tag{6.30}$$

The function (6.30) solves equation (6.1):

$$(\ell_q - k^2)u = q(y)\exp(-ik\theta\cdot y) - \int \exp(-ik\theta\cdot z)q(z)\left[\delta(z-y)\right]dz = 0,$$

and satisfies the condition (6.2) since the integral term in (6.30) satisfies the radiation condition. Therefore, the scattering solution can be defined by formula (6.29). This definition was introduced and used systematically in [Ramm (1987)].

The starting formula in the proof of the eigenfunction expansion theorem is the Cauchy formula

$$0 = \frac{1}{2\pi i} \int_{C_N} R_\lambda f d\lambda \tag{6.31}$$

where

$$R_\lambda := (A - \lambda I)^{-1}, \quad A = A^* = \ell_q. \tag{6.32}$$

C_N is a contour which consists of the circle $\gamma_N := \{\lambda : |\lambda| = N|\}$, of a finite number N_- of circles $\gamma_j := \{\lambda : |\lambda + \lambda_j| = \delta\}$ where $\lambda_j < 0$, $1 \le j \le N_-$, are negative eigenvalues of ℓ_q, $\delta > 0$ is a small number such that γ_j does not intersec with γ_m for $j \ne m$, and of a loop $\mathcal{L}_N$ which joins points $N - i0$ and $N + i0$ and goes from $N - i0$ to 0 and from 0 to $N + i0$. The circles γ_j, $1 \le j \le N_-$ are run clockwise and γ_N is run counterclockwise. The integral

$$\frac{1}{2\pi i} \int_{\gamma_j} R_\lambda f d\lambda = P_j f \tag{6.33}$$

where P_j is the orthoprojection in $H = L^2(R^3)$ onto the eigenspace of A corresponding to the eigenvalue λ_j. Note that there is no minus sign in front of the integral in (6.33) because γ_j is run clockwise and not counterclockwise.

One has:

$$\frac{1}{2\pi i} \int_{\mathcal{L}_N} R_\lambda f d\lambda = \frac{1}{\pi} \int_0^N Im R_{\lambda+i0} f d\lambda, \tag{6.34}$$

where we have used the relation

$$R_{\lambda-i0} f = \overline{R_{\lambda+i0} f}. \tag{6.35}$$

Formula (6.35) follows from the selfadjointness of A:

$$R_{\lambda-i0} = R^*_{\lambda+i0} \tag{6.36}$$

and from the symmetry of the kernel of the operator $R_\lambda(x, y)$.

Finally, for any selfadjoint A one has

$$\lim_{N\to\infty} -\frac{1}{2\pi i} \int_{\gamma_N} R_\lambda f d\lambda = f. \tag{6.37}$$

Indeed, if A is selfadjoint then

$$R_\lambda = \int_{-\infty}^{\infty} (t-\lambda)^{-1} dE_\lambda, \tag{6.38}$$

where E_λ is the resolution of the identity for A. Substitute (6.38) into (6.37) to get

$$\begin{aligned} \lim_{N\to\infty} -\frac{1}{2\pi i}\int_{\gamma_N}\int_{-\infty}^{\infty} \frac{dE_\lambda f}{t-\lambda} d\lambda &= \lim_{N\to\infty}\left\{\int_{-\infty}^{\infty} dE_\lambda f\left(-\frac{1}{2\pi i}\int_{\gamma_N}\frac{d\lambda}{t-\lambda}\right)\right\} \\ &= \lim_{N\to\infty}\int_{-N}^{N} dE_\lambda f = f. \end{aligned} \tag{6.39}$$

Here we have used the formula

$$-\frac{1}{2\pi i}\int_{\gamma_N}\frac{d\lambda}{t-\lambda} = \begin{cases} 1, & -N < t < N, \\ 0, & t > N \quad \text{or} \quad t < -N. \end{cases} \tag{6.40}$$

Using (6.31), (6.33), (6.34) and (6.37) one obtains

$$f = \sum_{j=1}^{N_-} f_j u_j(x) + \frac{1}{\pi}\int_0^{\infty} Im R_{\lambda+i0} f d\lambda, \tag{6.41}$$

where

$$f_j = (f, u_j), \quad 1 \le j \le N_- \tag{6.42}$$

and the sum in (6.41) is the term

$$\sum_j P_j f. \tag{6.43}$$

Let $\lambda = k^2$ in (6.41). Then

$$\mathcal{J} := \frac{1}{\pi}\int_0^{\infty} Im R_{\lambda+i0} f d\lambda = \frac{2}{\pi}\int_0^{\infty}\left(\int Im G(x,y,k) f(y) dy\right) k dk. \tag{6.44}$$

We wish to show that the term (6.44) is equal to the integral in (6.25). This can be done by expressing $ImG(x,y,k)$ via the scattering solutions. Green's formula yields:

$$G(x,y,k) - \overline{G}(x,y,k) = \int_{|s|=r}\left[\overline{G}(s,y)\frac{\partial G(x,s)}{\partial|s|} - G(s,y)\frac{\partial\overline{G}(x,s)}{\partial|s|}\right] ds. \tag{6.45}$$

Take $r \to \infty$ and use (6.29) to get

$$2iImG(x,y,k) = \lim_{r\to\infty} \int_{|s|=r} \overline{g}(r)\overline{u(y,-\theta,k)}ikgu(x,-\theta,k) + g(r)u(y,-\theta,k)ik\overline{g}u(x,-\theta,k) = \frac{2ik}{(4\pi)^2}\int_{S^2} u(x,\theta,k)\overline{u(y,\theta,k)}d\theta. \tag{6.46}$$

Thus

$$\frac{2}{\pi}ImG(x,y,k) = \frac{k}{(2\pi)^3}\int_{S^2} u(x,\theta,k)\overline{u(y,\theta,k)}d\theta. \tag{6.47}$$

Substitute (6.47) into (6.44) to get

$$\mathcal{J} = \frac{1}{(2\pi)^{3/2}}\int_0^\infty \int_{S^2} \tilde{f}(\xi)u(x,\theta,k)k^2dk = \frac{1}{(2\pi)^{3/2}}\int \tilde{f}(\xi)u(x,\xi)d\xi. \tag{6.48}$$

Here $\xi = k\theta$, $d\xi = k^2dkd\theta$, $\tilde{f}(\xi)$ is given by (6.21). From (6.41), (6.44), and (6.48) formula (6.25) follows. Lemma 6.5 is proved.

Remark 6.1 *Let us give a discussion of the passage from (6.44) and (6.47) to (6.48). First note that our argument yields Parseval's equality:*

$$(f,h) = (\tilde{f}(\xi),\tilde{h}(\xi)) + \sum_j f_j\overline{h}_j \tag{6.49}$$

and the formula for the kernel of the operator $\frac{dE_\lambda}{d\lambda}$, where the derivative is understood in the weak sense

$$\frac{dE_\lambda(x,y)}{d\lambda} = \frac{1}{\pi}ImG(x,y,\sqrt{\lambda}) = \frac{\sqrt{\lambda}}{16\pi^3}\int_{S^2} u(x,\theta,\sqrt{x})\overline{u(y,\theta,\sqrt{\lambda})}d\theta, \quad \lambda > 0. \tag{6.50}$$

To check (6.49) one writes

$$(f,h) = \sum_j f_j\overline{h}_j + \left(\int_0^\infty dE_\lambda f, \quad \int_0^\infty dE_\mu h\right) = \sum_j f_j\overline{h}_j + \int_0^\infty d(E_\lambda f,h)$$

where we have used the orthogonality of the spectral family: $E(\Delta)E(\Delta') = E(\Delta\cap\Delta')$. Furthermore, using (6.50) one obtains

$$\int_0^\infty d(E_\lambda f,g) = \int \tilde{f}(\xi)\tilde{h}(\xi)d\xi.$$

The last two formulas yield (6.49). The passage from (6.44) and (6.47) to (6.48) is clear if $f \in L^2(R^3) \cap L^1(R^3)$*: in this case the integral* $\tilde{f}(\xi) := \int f(y)\overline{u(y,\xi)}dy$ *converges absolutely. If* $f \in L^2(R^3)$ *then one can establish formula (6.25) by a limiting argument. Namely, let* $\mathcal{F}$ *be the operator of the Fourier transform,* $\mathcal{F}f = \left[\tilde{f}(\xi), \{f_j\}\right] := [\mathcal{F}_c f, \mathcal{F}_d f]$ *where the brackets indicate that the Fourier transform is a set of the coefficients* f_j*, corresponding to the discrete spectrum of* ℓ_q *and the function* $\tilde{f}(\xi)$ *corresponding to the continuous spectrum of* ℓ_q*. The operator* $\mathcal{F}$ *is isometric by (6.49), and if it is defined originally on a dense in* $L^2(R^3)$ *set* $L^2(R^3) \cap L^1(R^3)$ *it can be uniquely extended by continuity on all of* $L^2(R^3)$*. Formula (6.21) is therefore well defined for* $f \in L^2(R^3)$*. If formula (6.25) is proved for* $f \in L^2(R^3) \cap L^1(R^3)$*, it remains valid for any* $f \in L^2(R^3)$ *because the inverse of* $\mathcal{F}$ *is also an isometry from* $Ran\mathcal{F}$ *onto* $L^2(R^3)$*. Let us note finally that* $Ran\mathcal{F}_C = L^2(R^3)$*, where* $\mathcal{F}_C f := \tilde{f}(\xi)$*, and* $\mathcal{F}_C$ *is an isometry from* $L^2(R^3)$ *onto* $L^2(R^3)$*,* $\mathcal{F}_C^* \mathcal{F}_C f = f - E_0 f$*,* $\mathcal{F}_C \mathcal{F}_C^* \tilde{f} = \tilde{f}$*. Here* $E_0 f$ *is the projection of* f *onto the linear span of the eigenfunctions of* ℓ_q*,* $E_0 f = \sum_j P_j f$*. This follows from the formula* $N(\mathcal{F}_C^*) = \{0\}$*, the proof of which is the same as in [Ramm (1986), p. 51].*

6.1.3 *Properties of the scattering amplitude*

Let us now formulate some properties of the scattering amplitude $A(\theta', \theta, k)$.

Lemma 6.6 *If* $q \in Q$ *then the scattering amplitude has the properties*

$$A(\theta', \theta, -k) = \overline{A(\theta', \theta, k)}, \quad k > 0 \quad \textit{(reality)} \tag{6.51}$$

$$A(\theta', \theta, k) = A(-\theta, -\theta', k), \quad \textit{(reciprocity)} \tag{6.52}$$

$$\frac{A(\theta', \theta, k) - \overline{A(\theta, \theta', k)}}{2i} = \frac{k}{4\pi}\int_{S^2} A(\theta', \alpha, k)\overline{A(\theta, \alpha, k)}d\alpha \quad \textit{(unitarity)}. \tag{6.53}$$

In particular, if $\theta' = \theta$ *in (6.53) then one obtains the identity*

$$ImA(\theta, \theta, k) = \frac{k}{4\pi}\int_{S^2} |A(\theta, \alpha, k)|^2 d\alpha \quad \textit{(optical theorem)}. \tag{6.54}$$

Proof. 1) Equation (6.51) follows from the real-valuedness of $q(x)$. Indeed, $\overline{u(x, \theta, -k)}$ and $u(x, \theta, k)$, $k > 0$, solve the same integral equation

(6.13). Since this integral equation has at most one solution, it follows that

$$u(x, \theta, k) = \overline{u(x, \theta, -k)}, \quad k > 0. \tag{6.55}$$

Equation (6.51) follows from (55) immediately.

2) The proof of (6.52)-(6.54)is somewhat longer and since it can be found in [Ramm (1975), pp. 54-56] we refer the reader to this book. □

Let us define the S-matrix

$$S = I - \frac{k}{2\pi i} A \tag{6.56}$$

where $S : L^2(S^2) \to L^2(S^2)$ is considered to be an operator on $L^2(S^2)$ with the kernel

$$S(\theta', \theta, k) = \delta(\theta - \theta') - \frac{k}{2\pi i} A(\theta', \theta, k). \tag{6.57}$$

The unitarity of S

$$S^* S = I \tag{6.58}$$

implies

$$\frac{A - A^*}{2i} = \frac{k}{4\pi} A^* A \tag{6.59}$$

which is (6.53) in the operator notation.

6.1.4 *Analyticity in k of the scattering solution*

Define

$$\phi := \exp(-ik\theta \cdot x) u(x, \theta, k). \tag{6.60}$$

Then ϕ solves the equation

$$[I + T_\theta(k)] \, \phi = 1 \tag{6.61}$$

where

$$T_\theta(k)\phi := \int \frac{\exp\left[ik|x - y| - ik\theta \cdot (x - y)\right]}{4\pi |x - y|} q(y)\phi(y) dy. \tag{6.62}$$

The operator $T_\theta(k) : C(R^3) \to C(R^3)$ is compact and continuous (in the norm of operators) in the parameter $k \in \overline{\mathbf{C}_+} := \{k : Imk \geq 0\}$. If $q \in Q$, the operator $T_\theta(k)$ is analytic in k in $\mathbf{C}_+$ since $|x - y| - \theta \cdot (x - y) \geq 0$. The operator $I + T_\theta(k)$ is invertible for some $k \in \mathbf{C}_+$, for example, for $k \in \mathbf{C}_+$

sufficiently close to the positive real semiaxis, or for $k = a + ib$, where a and b are real numbers and $b > 0$ is sufficiently large. Indeed under the last assumption the norm of the operator $T_\theta(k)$ is less than one. Therefore by the well-known result, the analytic Fredholm's theorem, one concludes that $[I + T_\theta(k)]^{-1}$ is a meromorphic in $\mathbf{C}_+$ operator function on $C(R^3)$ (see [Ramm (1975), p. 57]). The poles of this function occur at the values k_j at which the operator $I + T_\theta(k_j)$ is not invertible. These values are

$$k_j = i\sqrt{|\lambda_j|} \tag{6.63}$$

where λ_j are the eigenvalues of the operator ℓ_q, and, possibly, the value $k = 0$. Indeed, if

$$[I + T_\theta(k_j)]\, v = 0, \quad v \in C(R^3), \quad k_j \in \mathbf{C}_+ \tag{6.64}$$

then the function

$$w := \exp(ik\theta \cdot x)v, \quad v \in C(R^3) \tag{6.65}$$

solves the equation

$$w = -Tw \tag{6.66}$$

where T is defined in (6.16). It follows from (6.65) and (6.66) that $w = O(|x|^{-1})$. This and equation (6.66) imply that

$$(\ell_q - k^2)w = 0 \tag{6.67}$$

and

$$w \in L^2(R^3). \tag{6.68}$$

Equation (6.67) follows from (6.66) immediately. Equation (6.68) can be easily checked if $k \in \mathbf{C}_+$, that is, if $k = a + ib$, $b > 0$, a is real. Indeed, use (6.66) , the assumption $q \in Q$, which implies that $q \in L^2(R^3) \cap L^1(R^3)$, and boundedness of w to get:

$$\begin{aligned} \| w \|^2_{L^2(R^3)} &\le \int dx \left| \int \frac{\exp(-b|x-y|)}{4\pi|x-y|} |q||w| dy \right|^2 \\ &\le c \int dx \int \frac{\exp(-2b|x-y|)}{|x-y|^2} |q(y)| dy \le c_1. \end{aligned} \tag{6.69}$$

Since the operator $\ell_q = -\Delta + q(x)$ is selfadjoint equations (6.67) and (6.68) imply $w = 0$ provided that k^2 is not real. Since $k \in \mathbf{C}_+$, the number k^2 is real if and only if $k = i\sqrt{|\lambda|}$, $k^2 = -|\lambda| < 0$. Equations (6.67) and

(6.68) with $k^2 = -|\lambda|$ imply that $\lambda = \lambda_j$ is an eigenvalue of ℓ_q. Therefore, the only points at which the operator $[I + T_\theta(k)]^{-1}$ has poles in $\mathbf{C}_+$ are the points (6.63) and, possibly, the point $k = 0$. One can prove [Ramm (1987)], [Ramm (1988b)] that if $q \in Q$ and $\ell_q \geq 0$ the number $\lambda = 0$ is not an eigenvalue of ℓ_q. However, even if $\ell_q \geq 0$ the point $\lambda = 0$ may be a resonance (half-bound state) for ℓ_q. This means that the equation

$$\Delta u = qu, \quad u \in C(R^3) \quad \text{and} \quad u \notin L^2(R^3) \tag{6.70}$$

may have a nontrivial solution which is not in $L^2(R^3)$. In this case the operator $I + T_\theta(0)$ is not invertible. Even if $q(x) \in C_0^\infty$ the operator ℓ_q may have a resonance at $\lambda = 0$. Even if $\ell_q \geq 0$ and q is compactly supported and locally integrable the operator ℓ_q may have a resonance at $\lambda = 0$.

Example 6.1 Let $B = \{x : |x| \leq 1, x \in R^3\}$. Let $u = |x|^{-1}$ for $|x| \geq 1$. Extend u inside B as a C^∞ real-valued function such that $u(x) \geq \delta > 0$ in B. This is possible since $u = 1$ on ∂B. Define

$$q(x) := \frac{\Delta u}{u}. \tag{6.71}$$

Then $q \in C_0^\infty$, $q = 0$ for $|x| \geq 1$, q is real-valued, $u \notin L^2(R^3)$, and the desired example is constructed. This argument does not necessarily lead to a nonnegative ℓ_q. In order to get $\ell_q \geq 0$ one needs an extra argument given in [Ramm (1987)]. Let us give a variant of this argument. The inequality $\ell_q \geq 0$ holds if and only if $(*)$ $\int \left[|\nabla\phi|^2 + q(x)|\phi|^2\right] dx \geq 0$ for all $\phi \in C_0^\infty(R^3)$. It is known that $\int |\nabla\phi|^2 dx \geq \int (4r^2)^{-1}|\phi|^2 dx$ for all $\phi \in C_0^\infty(R^3)$, $r := |x|$. Therefore $(*)$ holds if $(**)$ $(4r^2)^{-1} + q \geq 0$. Choose $u = r^{\gamma-1}(1 + \gamma - \gamma r)$, where $\gamma > 0$ is a sufficiently small number. Then q, defined by (71), satisfies $(**)$ as one can easily check. This q is integrable and $\ell_q \geq 0$. The function $u = r^{-1}$ for $r \geq 1$ and $u = r^{\gamma-1}(1+\gamma-\gamma r)$ solves the equation $\ell_q u = 0$ in R^3, $u \notin L^2(R^3)$, $\ell_q \geq 0$, $q = 0$ for $r \geq 1$, and q is locally integrable.

Exercise. Prove that the numbers (6.63) are simple poles of $[I + T_\theta(k)]^{-1}$.

6.1.5 *High-frequency behavior of the scattering solutions*

Assume now that $q \in Q_1$. Then the function ϕ defined in (6.60) can be written as

$$\phi = 1 + \frac{1}{2ik}\int_0^\infty q(x - r\theta)dr + o\left(\frac{1}{k}\right), \quad k \to +\infty. \tag{6.72}$$

If $q \in Q_m$, $m > 1$, more terms in the asymptotic expansion of ϕ as $k \to \infty$ can be written [Skriganov (1978)]. Formula (6.72) is well known and can be derived as follows.

Proof of Formula (6.72)

Step 1: Note that

$$\max_{\theta \in S^2} \| T_\theta^2(k) \| \to 0 \quad \text{as} \quad k \to +\infty. \tag{6.73}$$

This can be proved, as in [Ramm (1986), p. 390], by writing the kernel $B_\theta(x, y, k)$ of T_θ^2:

$$B_\theta(x, y, k) = \int \frac{\exp\left[ik(|x-z| + |z-y|)\right]}{16\pi^2 |x-z||z-y|} q(z) dz q(y) \exp\left[ik\theta \cdot (x-y)\right]. \tag{6.74}$$

Introduce the coordinates s, t, ψ defined by the formulas

$$z_1 = \ell st + \frac{x_1 + y_1}{2}, \quad z_2 = \ell\sqrt{(s^2-1)(1-t^2)} \cos\psi + \frac{x_2 + y_2}{2}$$

$$z_3 = \ell\sqrt{(s^2-1)(1-t^2)} \sin\psi + \frac{x_3 + y_3}{2}, \tag{6.75}$$

where

$$\ell = |x-y|/2, \quad |x-z|+|z-y| = 2\ell s, \quad |x-y|-|z-y| = 2\ell t, \quad \mathcal{J} = \ell^3(s^2-t^2), \tag{6.76}$$

and $\mathcal{J}$ is the Jacobian of the transformation

$$(z_1, z_2, z_3) \to (s, t, \psi), \quad 1 \le s < \infty, \quad -1 \le t \le 1, \quad 0 \le \psi < 2\pi.$$

In the new coordinates one obtains

$$|B_\theta(x, y, k)| \le \frac{1}{16\pi^2} |q(y)| \ell \left| \int_1^\infty \exp(2ik\ell s) p(s) ds \right|, \tag{6.77}$$

where

$$p(s) := \int_0^{2\pi} d\psi \int_{-1}^1 dt q_1(s, t, \psi)$$

and $q_1(s, t, \psi)$ is $q(z)$ in the new coordinates given by formula (6.75). One can choose a sufficiently large number $N > 0$ such tht

$$|x| > N \quad \text{or} \quad |y| > N \quad \text{implies} \quad \sup_{\theta \in S^2, k > 0} |B_\theta(x, y, k)| < \epsilon(N), \tag{6.78}$$

where $\epsilon(N) \to 0$ as $N \to \infty$.

If N is fixed, then for $|x| \le N$ and $|y| \le N$ it follows from (77) that

$$|B_\theta| \to 0 \quad \text{as} \quad k \to +\infty \tag{6.79}$$

since $p(s) \in L^1(1, \infty)$.

This proves (6.73). Note that in this argument it is sufficient to assume $q \in Q$.

Step 2: If (6.73) holds, one can write

$$\phi = 1 + \sum_{j=1}^{\infty} (-1)^j T_\theta^j(k) 1, \tag{6.80}$$

where the series in (6.80) converges in the norm of $C(R^3)$ if k is sufficiently large so that $\| T_\theta^2 \| < 1$. Note that if $\| T \| < 1$ then

$$(I + T)^{-1} = \sum_{j=0}^{\infty} (-1)^j T^j \tag{6.81}$$

and the series converges in the norm of operators. If $\| T^2 \| < 1$, formula (6.81) remains valid. Indeed

$$\begin{aligned} \sum_{j=0}^{\infty} (-1)^j T^j &= \sum_{j=0}^{\infty} (-1)^{2j} T^{2j} + T \sum_{j=0}^{\infty} (-1)^{2j+1} T^{2j} \\ &= (I - T^2)^{-1} - T(I - T^2)^{-1} \\ &= (I - T)(I - T^2)^{-1} = (I + T)^{-1}. \end{aligned} \tag{6.82}$$

In fact it is known that the series (6.81) converges and formula (6.81) holds if $\| T^m \| < 1$ for some integer $m \ge 1$.

As $k \to \infty$, each term in (6.80) has a higher order of smallness than the previous one. Therefore it is sufficient to consider the first term in the sum (6.80) and to check that

$$-T_\theta(k) 1 = \frac{1}{2ik} \int_0^\infty q(x - r\theta) dr + o\left(\frac{1}{k}\right), \quad k \to +\infty \tag{6.83}$$

in order to prove (6.72).

Step 3: Let us check (6.83). One has

$$T_\theta(k)1 = \int \frac{\exp\{ik[|x-y| - \theta\cdot(x-y)]\}}{4\pi|x-y|} q(y)dy$$
$$= \int_0^\infty \frac{drr^2\exp(ikr)}{4\pi r} \int_{S^2} \exp(ikr\theta\cdot\alpha)q(x+r\alpha)d\alpha, \quad (6.84)$$

where we set $y = x + z$, $z = r\alpha$, $\alpha \in S^2$. Use formula [Ramm (1986)], p. 55:

$$\int_{S^2} \exp(ikr\theta\cdot\alpha)f(\alpha)d\alpha$$
$$= 2\pi i\left[\tfrac{\exp(-ikr)}{kr}f(-\theta) - \tfrac{\exp(ikr)}{kr}f(\theta)\right] + o\left(\tfrac{1}{k}\right), \quad \text{as} \quad k\to\infty \quad (6.85)$$

which holds if $f \in C^1(S^2)$.

From (6.84) and (6.85) one obtains

$$T_\theta(k)1 = -\frac{1}{2ik}\int_0^\infty drq(x-r\theta) + o\left(\frac{1}{k}\right), \quad k\to+\infty \quad (6.86)$$

which is equivalent to (6.83). Formula (6.72) is proved. □

It follows from (6.72) that

$$u(x,\theta,k) = \exp(ik\theta\cdot x)\left[1 + \frac{1}{2ik}\int_0^\infty q(x-r\theta)\right] + o\left(\frac{1}{k}\right), \quad k\to+\infty \quad (6.87)$$

provided that $q \in Q_1$.

Another formula, which follows from (6.72), is

$$\theta\cdot\nabla_x \lim_{k\to+\infty}\{2ik[\phi(x,\theta,k)-1]\} = \theta\cdot\nabla_x\int_0^\infty q(x-r\theta)dr$$
$$= -\int_0^\infty \frac{\partial q}{\partial r}dr = q(x)$$

or

$$q(x) = \theta\cdot\nabla_x \lim_{k\to+\infty}\{2ik[\phi(x,\theta,k)-1]\}. \quad (6.88)$$

Note that the left side does not depend on θ, so that (6.88) is a compatibility condition on the function $\phi(x,\theta,k)$.

From (6.87) and (6.15) it follows that

$$A(\theta', \theta, k) = -\frac{1}{4\pi} \int \exp\left[ik(\theta - \theta') \cdot x\right] q(x)dx + 0\left(\frac{1}{k}\right), \quad k \to +\infty \tag{6.89}$$

provided that $q \in Q_1$. In particular,

$$A(\theta, \theta, k) = -\frac{1}{4\pi} \int q(x)dx + 0\left(\frac{1}{k}\right), \quad k \to +\infty. \tag{6.90}$$

6.1.6 *Fundamental relation between* u^+ *and* u^-

If one defines

$$u^+ := u(x, \theta, k), \quad u^- := u(x, -\theta, -k) \tag{6.91}$$

then one can prove that

$$u^+ = Su^- := u^- + \frac{ik}{2\pi} \int_{S^2} A(\theta', \theta, k) u^-(x, \theta', k) d\theta'. \tag{6.92}$$

Let us derive (6.92). We start with the equations

$$u^+ = u_0 - G^+ q u_0, \quad u_0 := \exp(ik\theta \cdot x), \tag{6.93}$$

$$u^- = u_0 - G^- q u_0, \tag{6.94}$$

where $G^+ = G$, where G is defined by the equation (6.27), and

$$G^- := \overline{G}. \tag{6.95}$$

Equations (6.93) and (6.94) one can easily check by applying the operator $\ell_q - k^2$ to these equations. Subtract (6.94) from (6.93) and use (6.95) to get

$$\begin{aligned} u^+ - u^- &= -2iImG^+ q u_0 \\ &= \frac{ik}{2\pi} \int_{S^2} d\theta' u^-(x, \theta', k) \left\{ -\frac{1}{4\pi} \int \overline{u^-(y, \theta', k)} q(y) u_0(y, \theta, k) \right\} \\ &= \frac{ik}{2\pi} \int_{S^2} A(\theta', \theta, k) u^-(x, \theta', k) d\theta'. \end{aligned} \tag{6.96}$$

The last equality in (6.96) follows from the definition (6.91), properties (6.51) and (6.52) of the scattering amplitude and the formula

$$ImG^+(x, y, k) = \frac{k}{16\pi^2} \int_{S^2} u^-(x, \theta', k) \overline{u^-(y, \theta', k)} d\theta' \tag{6.97}$$

which is similar to (6.46) and which follows from (6.46) and (6.91). Let us derive (6.97). Note that, by formulas (6.46) and (6.91), one has

$$ImG^+(x,y,k) = \overline{ImG^+(x,y,k)}, \quad u^{\pm}(x,\theta,-k) = \overline{u^{\pm}(x,\theta,k)}, \qquad (6.98)$$

$$\begin{aligned} ImG^+(x,y,k) &= \frac{k}{16\pi^2}\int_{S^2} u(x,\theta,k)\overline{u(y,\theta,k)}d\theta \\ &= \frac{k}{16\pi^2}\int_{S^2} u^-(x,-\theta,-k)\overline{u^-(y,-\theta,-k)}d\theta \\ &= \frac{k}{16\pi^2}\int_{S^2} \overline{u^-(x,\theta',k)}u^-(y,\theta',k)d\theta' \\ &= \frac{k}{16\pi^2}\int_{S^2} u^-(x,\theta',k)\overline{u^-(y,\theta',k)}d\theta'. \end{aligned} \qquad (6.99)$$

Here we used (6.98). Thus, formula (6.97) is obtained.

Note that

$$-4\pi A(\theta',\theta,k) = \int u_0(y,-\theta',k)q(y)u(y,\theta,k)dy \qquad (6.100)$$

and

$$\begin{aligned} -4\pi A(-\theta,-\theta',k) &= \int u_0(y,\theta,k)q(y)u(y,-\theta',k)dy \\ &= \int u_0(y,\theta,k)q(y)\overline{u^-(y,\theta',k)}. \end{aligned} \qquad (6.101)$$

From formula (6.52) it follows that the right sides of (6.100) and (6.101) are equal. This explains the last equation (6.96).

6.1.7 *Formula for* det $S(k)$ *and state the Levinson Theorem*

If $q(x) = \overline{q}(x)$ is decaying sufficiently fast, (for example, if $(1+|x|)q \in Q$, $x \in R^3$) then the operator $A : L^2(S^2) \to L^2(S^2)$ with kernel $A(\theta',\theta,k)$, $k > 0$, is in the trace class and

$$\det S(k) = \det\left(I + \frac{ik}{2\pi}A\right) = \exp\left[-\frac{ik}{2\pi}\int q(x)dx\right]\frac{d(-k)}{d(k)}, \quad k > 0 \qquad (6.102)$$

where

$$d(k) := \det_2(I + T(k)). \qquad (6.103)$$

The operator $T(k)$ in (6.103) is defined in (6.16) and the symbol $\det_2(I+T)$ is defined in Definition 8.7 on p. 300. If $k > 0$ and $q = \overline{q}$, then $d(-k) = \overline{d(k)}$, where the bar stands for complex conjugate. Therefore

$$\det S(k) = \exp\left[2i\delta(k)\right], \tag{6.104}$$

where

$$\delta(k) = -\frac{k}{4\pi}\int q(x)dx - \beta(k), \quad \beta(k) := arg d(k). \tag{6.105}$$

The Levinson Theorem says that

$$\delta(0) = \pi\left(m + \frac{\nu}{2}\right), \tag{6.106}$$

where m is the number of the bound states counting with their multiplicities, in other words m is the dimension of the subspace spanned by the eigenfunctions of ℓ_q corresponding to all of its negative eigenvalues, and $\nu = 1$ if $k = 0$ is a resonance and $\nu = 0$ otherwise, that is, if $I + T(0)$ is invertible. It is assumed that $\delta(k)$ is normalized in such a way that

$$\lim_{k\to\infty}\left[\delta(k) + \frac{k}{4\pi}\int q(x)dx\right] = 0 \tag{6.107}$$

or, according to (6.105), that

$$\lim_{k\to\infty}\beta(k) = 0. \tag{6.108}$$

Formula (6.106) follows from (6.105) and the argument principle applied to $d(k)$.

Formula (6.102) can be derived as follows:

$$\begin{aligned} d(-k) &:= \det_2\left(I + T(-k)\right) \\ &= \det_2\left\{[I + T(k)]\left[I + (I + T(k))^{-1}(T(-k) - T(k))\right]\right\} \\ &= \det_2\left[I + T(k)\right]\det\left[I + (I + T(k))^{-1}(T(-k) - T(k))\right] \times \\ &\quad \exp\left\{-Tr\left[T(-k) - T(k)\right]\right\}. \end{aligned} \tag{6.109}$$

Here we have used item 12) which precedes Definition 8.8 on p. 300. The operator $T(-k) - T(k)$ has the kernel $-\frac{2i\sin(k|x-y|)}{4\pi|x-y|}q(y)$, so that its trace is

$$Tr\left[T(-k) - T(k)\right] = \int -\frac{2ik}{4\pi}q(y)dy = -\frac{ik}{2\pi}\int q(y)dy. \tag{6.110}$$

Therefore formula (6.109) can be written as

$$\frac{d(-k)}{d(k)} \exp\left\{-\frac{ik}{2\pi}\int q(x)dx\right\} = \det\left[I + (I+T(k))^{-1}(T(-k)-T(k))\right]. \tag{6.111}$$

Finally one proves that

$$\det\left[I + (I+T(k))^{-1}(T(-k)-T(k))\right] = \det\left(I + \frac{ik}{2\pi}A\right). \tag{6.112}$$

Formulas (6.111) and (6.112) imply (6.102). Let us prove (6.112). Let $(I+T(k))^{-1} := B$. Then

$$\tau := (I+T(k))^{-1}(T(-k)-T(k)) = B\left[g(-k)-g(k)\right]q. \tag{6.113}$$

One has

$$g(-k)-g(k) = -\frac{2ik}{4\pi}\frac{\sin(k|x-y|)}{k|x-y|} = -\frac{ik}{2\pi}\frac{1}{4\pi}\int_{S^2}\exp\left\{ik\theta\cdot(x-y)\right\}d\theta. \tag{6.114}$$

Furthermore, if $u_0 := \exp(ik\theta\cdot x)$ then

$$Bu_0 = u(x,\theta,k), \tag{6.115}$$

where $u(x,\theta,k)$ is the scattering solution (6.1)-(6.2). Therefore the right-hand side of (6.113) is the operator in $L^2(R^3)$ with the kernel

$$\tau(x,y) := -\frac{ik}{8\pi^2}\int_{S^2} d\theta \exp(-ik\theta\cdot y)u(x,\theta,k)q(y). \tag{6.116}$$

Note that

$$Tr\tau = \int \tau(x,x)dx = \frac{ik}{2\pi}\int_{S^2} A(\theta,\theta,k)d\theta = Tr\left(\frac{ik}{2\pi}A\right). \tag{6.117}$$

From item (8) of Section 8.3.3, p. 299, it follows that (6.112) is valid provided that

$$Tr\tau^j = Tr\left(\frac{ik}{2\pi}A\right)^j, \quad j = 1,2,3\dots. \tag{6.118}$$

One can check (6.118) as we checked (6.117). Thus, formula (6.102) is derived.

6.1.8 *Completeness properties of the scattering solutions*

Theorem 6.1 *Let $h(\theta) \in L^2(S^2)$ and assume that*

$$\int_{S^2} h(\theta)u(x,\theta,k)d\theta = 0 \quad \forall x \in \Omega_R := \{x : |x| > R\}, \tag{6.119}$$

where $k > 0$ is fixed, $x \in R^3$. It is assumed that $q \in Q$. Then $h(\theta) = 0$. The same conclusion holds if one replaces $u(x,\theta,k)$ by $u(x,-\theta,-k)$ in (119) and if $x \in R^r$, $r \geq 2$.

Proof. The proof consists of two steps.

Step 1. The conclusion of Theorem 6.1 holds if $u(x,\theta,k)$ is replaced by $u_0(x,\theta,k) := \exp(ik\theta \cdot x)$ in (6.119). Indeed, if

$$\int_{S^2} h(\theta)\exp(ik\theta \cdot x)d\theta = 0 \quad \forall x \in \Omega_R \tag{6.120}$$

and a fixed $k > 0$, then the Fourier transform of the distribution $h(\theta)\delta_{S^2}$ vanishes for all sufficiently large x. The distribution $h(\theta)\delta_{S^2}$ is defined by the formula

$$\int \phi(y)\delta_{S^2}dy = \int_{S^2} \phi(\theta)h(\theta)d\theta \quad \forall \phi \in C_0^\infty(R^3). \tag{6.121}$$

Since $h(\theta)\delta_{S^2}$ has compact support, its Fourier transform is an entire function of x. If this entire function vanishes for all sufficiently large $x \in {}^3$, it vanishes identically. Therefore $h(\theta) = 0$.

Step 2. If (6.119) holds then (6.120) holds. Therefore, by Step 1, $h(\theta) = 0$. In order to prove that (6.119) implies (6.120) let us note that

$$u_0(x,\theta,k) = (I + T(k))\,u, \tag{6.122}$$

where $T(k)$ is defined by (6.16). For every $k > 0$ the operator $I + T(k)$ is an isomorphism of $C(R^3)$ onto $C(R^3)$. Applying the operator $I + T(k)$ to (6.119) and using (6.122) one obtains (6.120). Note that the operator $I + T(k)$ acts on $u(x,\theta,k)$ which is considered as a function of x while θ and k are parameters. Theorem 6.1 is proved. □

Theorem 6.1 is used in [Ramm (1978e)] for a characterization of the scattering data which we give in Section 6.2.5.

Another completeness property of the scattering solution can be formulated. Let

$$N_D(\ell_q) := \left\{w : w \in H^2(D), \ell_q w = 0 \quad \text{in} \quad D\right\}, \tag{6.123}$$

where $D \subset R^3$ is a bounded domain with a sufficiently smooth boundary Γ, for example $\Gamma \in C^{1,\alpha}$, $\alpha > 0$, suffices.

Theorem 6.2 *Let $q \in Q$, where Q is defined in (6.3). The closure in $L^2(D)$ (and in $H^1(D)$) of the linear span of the scattering solutions $\{u(x,\theta,k)\}$ $\forall\theta \in S^2$ and any fixed $k > 0$ contains $N_D(\ell_q - k^2)$.*

Proof. We first prove the statement concerning the $L^2(D)$ closure. Let $f \in N_D(\ell_q - k^2)$ and assume that

$$\int_D f u(x,\theta,k)dx = 0 \quad \forall\theta \in S^2. \tag{6.124}$$

Define

$$v(x) := \int_D G(x,y,k)f(y)dy, \tag{6.125}$$

where G is uniquely defined by equation (6.27). Use (6.29) and (6.123) to conclude that

$$v(x) = O\left(|x|^{-2}\right) \quad \text{as} \quad |x| \to \infty. \tag{6.126}$$

Since

$$(\ell_q - k^2)v = 0 \quad \text{in} \quad \Omega := R^3 \setminus D \tag{6.127}$$

and (6.126) holds, one concludes applying Kato's theorem (see [Kato (1959)]) that $v = 0$ in Ω (see the end of the proof of Lemma 6.4 in Section 6.1.2). In particular,

$$v = v_N = 0 \quad \text{on} \quad \Gamma, \tag{6.128}$$

where v_N is the normal derivative of v on Γ. It follows from (6.124) that

$$(\ell_q - k^2)v = -f \quad \text{in} \quad D. \tag{6.129}$$

Since

$$(\ell_q - k^2)f = 0 \quad \text{in} \quad D \tag{6.130}$$

by the assumption, one can multiply (6.128) by $\overline{f}$, integrate over D and get

$$\begin{aligned} -\int_D |f|^2 dy &= \int_D \overline{f}(\ell_q - k^2) v dy \\ &= \int_D (\ell_q - k^2)\overline{f} v dx + \int_\Gamma (\overline{f} v_N - \overline{f}_N v) ds \\ &= \int_D \overline{(\ell_q - k^2 f)} v dx = 0. \end{aligned} \tag{6.131}$$

Here we have used (6.128) and the real-valuedness of the potential. It follows from (6.131) that $f = 0$. The first statement of Theorem 6.2 is proved.

In order to prove the second statement which deals with completeness in $H^1(D)$, one assumes

$$\int_D (fu + \nabla f \cdot \nabla u) dx = 0 \quad \forall \theta \in S^2 \tag{6.132}$$

and some $f \in N_D(\ell_q - k^2)$. Integrate (6.131) by parts to get

$$\int_D (-\Delta f + f) u dx + \int_\Gamma f_N u ds = 0, \quad \forall \theta \in S^2. \tag{6.133}$$

Define

$$v := \int_D (-\Delta f + f) G(x, y, k) dy + \int_\Gamma G(x, s, k) f_N ds. \tag{6.134}$$

Argue as above to conclude that $v = 0$ in Ω and

$$v = v_N^- = 0 \quad \text{on} \quad \Gamma, \tag{6.135}$$

where v_N^- is the limit value of v_N on Γ from Ω. By the jump formula for the normal derivative of the single-layer potential (see, e.g., [Ramm (1986), p. 14]) one has

$$v_N^+ - v_N^- = f_N. \tag{6.136}$$

Since $v_N^- = 0$ it follows that $v_N^+ = f_N$ on Γ. Thus

$$v = 0, \quad v_N^+ = f_N \quad \text{on} \quad \Gamma, \tag{6.137}$$

and

$$(\ell_q - k^2) v = \Delta f - f \quad \text{in} \quad D. \tag{6.138}$$

Multiply (6.138) by $\overline{f}$, integrate over D and then by parts to get

$$-\int \left[|\nabla f|^2 + |f|^2\right] dx + \int_\Gamma \overline{f} f_N ds = \int_D (\ell_q - k^2)\overline{f} v dx + \int_\Gamma (\overline{f} v_N^+ - \overline{f}_N v)\, ds. \tag{6.139}$$

From (6.130), (6.137) and (6.139) it follows that

$$\int_D (|\nabla f|^2 + |f|^2)\, dx = 0. \tag{6.140}$$

Thus, $f = 0$. Theorem 6.2 is proved. □

6.2 Inverse scattering problems

6.2.1 *Inverse scattering problems*

The inverse scattering problem consists of finding $q(x)$ given $A(\theta', \theta, k)$. One should specify for which values of θ', θ and k the scattering amplitude is given.

Problem 1 $A(\theta', \theta, k)$ is given for all $\theta', \theta \in S^2$ and all $k > 0$. Find $q(x)$.

Problem 2 $A(\theta', \theta, k)$ is given for all $\theta', \theta \in S^2$ and a fixed $k > 0$.

Problem 3 $A(\theta', \theta, k)$ is given for a fixed $\theta \in S^2$ and all $\theta' \in S^2$ and all $k > 0$.

Problem 1 has been studied much. We will mention some of the results relevant to estimation theory.

Problem 2 has been solved recently [Ramm (2005)] but we do not describe the results since they are not connected with the estimation theory

Problem 3 is open, but a partial result is given in [Ramm (1989)].

6.2.2 *Uniqueness theorem for the inverse scattering problem*

The uniqueness of the solution to Problem 1 follows immediately from formula (1.89). Indeed, if $A(\theta', \theta, k)$ is known for all $\theta', \theta \in S^2$ and all $k > 0$, then take an arbitrary $\xi \in R^3$, an arbitrary sequence $k_n \to +\infty$, and find a sequence $\theta_n, \theta'_n \in S^2$ such that

$$\lim_{n\to\infty} (\theta_n - \theta'_n)k_n = \xi, \quad k_n \to +\infty. \tag{6.141}$$

This is clearly possible. Pass to the limit $k_n \to \infty$ in (6.89) to get

$$-4\pi \lim_{k_n\to\infty} k_n(\theta_n - \theta'_n) = \xi A(\theta'_n, \theta_n, k_n) = \int \exp(i\xi \cdot x)q(x)dx. \tag{6.142}$$

Therefore the Fourier transform of q is uniquely determined. Thus q is uniquely determined. We have proved

Lemma 6.7 *If $q \in Q_1$ then the knowledge of $A(\theta', \theta, k)$ on $S^2 \times S^2 \times R_+$, $R_+ := (0, \infty)$, determines $q(x)$ uniquely.*

In fact, our proof shows that it suffices to have the knowledge of A for an arbitrary sequence $k_n \to \infty$ and for some θ' and θ such that for any $\xi \in R^3$ one can choose θ'_n and θ_n such that (6.1) holds.

The reconstruction of $q(x)$ from the scattering data via formula (6.142) requires the knowledge of the high frequency data. These data are not easy to collect in the quantum mechanics problems, and for very high energies the Schrödinger equation is no longer a good model for the physical processes.

Therefore much effort was spent in order to find a solution to Problem 1 which uses all of the scattering data; to find necessary and sufficient conditions for a function $A(\theta', \theta, k)$ to be the scattering amplitude for a potential q from a certain class, e.g. for $q \in Q_m$, this is called a characterization problem; and to give a stable reconstruction of q given noisy data.

6.2.3 *Necessary conditions for a function to be a scatterng amplitude*

A number of necessary conditions for $A(\theta', \theta, k)$ to be the scattering amplitude corresponding to $q \in Q_1$ follow from the results of Section 6.1 of this scatteringtheory. Let us list some of these **necessary conditions**:

1) reality, reciprocity and unitarity: that is, formulas (6.51)-(6.54)
2) high-frequency behavior: formulas (6.89), (6.90), (6.142).

Other necessary conditions will be mentioned later (see formulas (6.158) and (6.159) below). Some necessary and sufficient conditions for $A(\theta', \theta, k)$ to be the scattering amplitude for a $q \in Q_1$ are given first in [Ramm

(1987e)], see also [Ramm (2005)]. These conditions can not be checked algorithmically: they are formulated in terms of the properties of the solutions to certain integral equations whose kernel is the given function $A(\theta', \theta, k)$.

6.2.4 *A Marchenko equation (M equation)*

Define

$$\eta(x, \theta, \alpha) := \frac{1}{2\pi} \int_{-\infty}^{\infty} [\phi(x, \theta, k) - 1] \exp(-ik\alpha) dk, \tag{6.143}$$

where ϕ is defined by (6.60) and has property (6.72) as $k \to \infty$, provided that $q \in Q_1$.

For simplicity we assume that ℓ_q has no bound states. (6.144)

Under this assumption ϕ is analytic in $\mathbf{C}_+$ and continuous in $\overline{\mathbf{C}}_+ \setminus 0$. Let us assume that $k = 0$ is not an exceptional point, that is, ϕ is continuous in $\overline{\mathbf{C}}_+$.

Start with equation (6.92) which we rewrite as

$$\phi(x, \theta, k) = \phi(x, -\theta, -k) + \frac{ik}{2\pi} \times \int_{S^2} A(\theta', \theta, k) \exp\left[ik(\theta' - \theta) \cdot x\right] \phi(x, -\theta', -k) d\theta \tag{6.145}$$

or

$$\begin{aligned} \phi(x, \theta, k) - 1 = \phi(x, -\theta, -k) - 1 \\ + \frac{ik}{2\pi} \int_{S^2} A(\theta', \theta, k) \exp\left[ik(\theta' - \theta) \cdot x\right] \left[\phi(x, -\theta', -k) - 1\right] d\theta' \\ + \frac{ik}{2\pi} \int_{S^2} A(\theta', \theta, k) \exp\left[ik(\theta' - \theta) \cdot x\right] d\theta'. \end{aligned} \tag{6.146}$$

Take the Fourier transform of (6.146) and use (6.143) to get

$$\eta(x, \theta, \alpha) = \eta(x, -\theta, -\alpha) + \int_{-\infty}^{\infty} \int_{S^2} B(\alpha - \beta) \eta(x, -\theta', -\beta) d\theta' d\beta + \eta_0. \tag{6.147}$$

Here

$$\eta_0 := \int_{-\infty}^{\infty} \exp(ik\alpha) \left\{ \frac{ik}{2\pi} \int_{S^2} A(\theta', \theta, k) \exp\left[ik(\theta' - \theta) \cdot x\right] d\theta' \right\} dk \tag{6.148}$$

and the integral term in (6.147) is

$$\frac{1}{2\pi}\int_{-\infty}^{\infty} dk \exp(-ik\alpha)\frac{ik}{2\pi}\int_{S^2} A(\theta',\theta,k)\exp\left[ik(\theta'-\theta)\cdot x\right]\times$$
$$[\phi(x,-\theta',-k)-1]\,d\theta'$$
$$=\int_{-\infty}^{\infty} d\beta\int_{S^2} d\theta' B(\alpha-\beta,\theta',\theta,x)\eta(x,-\theta',-\beta)d\theta', \qquad (6.149)$$

where

$$B(\alpha,\theta',\theta,x) := \frac{1}{2\pi}\int_{-\infty}^{\infty} dk\exp(-ik\alpha)\frac{ik}{2\pi}A(\theta',\theta,k)\exp\left[ik(\theta'-\theta)\cdot x\right]. \qquad (6.150)$$

The Fourier transform in (6.150) is understood in the sense of distributions.

Under the assumption (4), the analyticity of $\phi(x,\theta,k)$ in k in the region $\mathbf{C}_+$ and the decay of ϕ as $|k|\to\infty$, $k\in\mathbf{C}_+$, which follows from (6.72) imply that

$$\eta(x,\theta,\alpha)=0 \quad \text{for} \quad \alpha<0. \qquad (6.151)$$

Therefore the right hand side of (6.149) can be written as

$$\int_0^{\infty} d\beta\int_{S^2} d\theta' B(\alpha+\beta,\theta',\theta,x)\eta(x,-\theta',\beta) := \int_0^{\infty}\mathcal{B}(\alpha+\beta)\eta(\beta)d\beta. \qquad (6.152)$$

Equation (6.147) now takes the form of the Marchenko equation

$$\eta(x,\theta,\alpha)=\int_0^{\infty}\mathcal{B}(\alpha+\beta)\eta(\beta)d\beta+\eta_0, \quad \alpha>0, \qquad (6.153)$$

where we took into account that

$$\eta(x,-\theta,-\alpha)=0 \quad \text{for} \quad \alpha>0 \qquad (6.154)$$

according to (6.151). The function η_0 in (6.153) is defined in (6.148), and the integral operator in (6.153) is defined in (6.152). The kernel of the operator in (6.153) is defined by (6.150) and is known if the scattering amplitude is known. If $A(\theta',\theta,k)$ is the scattering amplitude corresponding to a $q\in Q_1^0$, where Q_1^0 is the subset of Q_1 which consists of the potentials with no bound states, then equation (6.153) has a solution η with the following properties: if one defines η for $\alpha<0$ by formula (6.151) then the function

$$\phi(x,\theta,k) := 1+\int_0^{\infty} d\alpha\exp(ik\alpha)\eta(x,\theta,\alpha) \qquad (6.155)$$

solves the equation

$$\nabla_x^2 \phi + 2ik\theta \cdot \nabla_x \phi - q(x)\phi = 0 \tag{6.156}$$

and the function $u := \exp(ik\theta \cdot x)\phi$ solves the Schrödinger equation

$$\ell_q u = 0. \tag{6.157}$$

In particular, the function

$$\frac{(\nabla^2 + k^2)u}{u} := q(x) \tag{6.158}$$

does not depend on θ (this is a compatibility condition).

Another compatibility condition gives formula (6.88). This formula can be written as

$$q(x) = -2\theta \cdot \nabla_x \eta(x, \theta, +0). \tag{6.159}$$

Indeed, it follows from (6.155) that

$$\lim_{k \to \infty} \{2ik(\phi - 1)\} = -2\eta(x, \theta, +0). \tag{6.160}$$

Formula (6.159) follows from (6.88) and (6.160).

The compatibility condition (6.159) and the Marchenko equation (6.153) appeared in [Newton (1982)] where condition (6.159) was called the "miracle" condition since the left side of (6.159) does not depend on θ). The above derivation is from [Ramm (1992)].

6.2.5 *Characterization of the scattering data in the 3D inverse scattering problem*

Let us write $A \in \mathcal{A}_Q$ if $A := A(\theta', \theta, k)$ is the scattering amplitude corresponding to a potential $q \in Q$. Assuming that $q \in Q$, we have proved in Section 6.1 that equation (6.92), which we rewrite as

$$\begin{aligned} v(x, \theta, k) &= v(x, -\theta, -k) + \frac{ik}{2\pi} \int_{S^2} A(\theta', \theta, k) v(x, -\theta', -k) d\theta' \\ &+ \frac{ik}{2\pi} \int_{S^2} A(\theta', \theta, k) \exp(ik\theta' \cdot x) d\theta' \end{aligned} \tag{6.161}$$

has a solution v for all $x \in R^3$ and all $k > 0$, where

$$v := u(x, \theta, k) - \exp(ik\theta \cdot x) := u - u_0. \tag{6.162}$$

This v has the following properties:

$$v = A(\theta', \theta, k)g(r) + o(r^{-1}), \quad r = |x| \to \infty, \quad \theta' = xr^{-1} \tag{6.163}$$

which is equation (6.2), and

$$\frac{(\Delta + k^2)(u_0 + v)}{u_0 + v} = q(x) \in Q \tag{6.164}$$

which is equation (6.158).

These properties are necessary for $A \in \mathcal{A}_Q$. It turns out that they are also sufficient for $A \in \mathcal{A}_Q$. Let us formulate the basic result (see [Ramm (1992)]).

Theorem 6.3 *For $A \in \mathcal{A}_Q$ it is necessary and sufficient that equation (6.161) has a solution v such that (6.164) holds and*

$$v = A_q(\theta', \theta, k)g(r) + o(r^{-1}), \quad r = |x| \to \infty, \quad xr^{-1} = \theta'. \tag{6.165}$$

The function A_q defined by (6.165) is equal to the function $A(\theta', \theta, k)$ which is the given function, the kernel of equation (6.161), and it is equal to the scattering amplitude corresponding to the function $q(x)$ defined by (6.164). There is at most one solution to equation (6.161) with properties (6.163), (6.164) and (6.165).

Proof. We have already proved the necessity part. Let us prove the sufficiency part. Let $A(\theta'\theta, k)$ be a given function such that equation (6.161) has a solution with properties (6.164) and (6.165). First, it follows that u defined by the formula

$$u := \exp(ik\theta \cdot x) + v \tag{6.166}$$

is the scattering solution for the potential $q(x)$ defined by formula (6.164). Since the scattering solution is uniquely determined (see Lemma 6.4 in Section 6.1.2) one concludes that the function $A_q(\theta', \theta, k)$ defined by formula (6.165) is the scattering amplitude corresponding to the potential $q(x)$ defined by formula (6.164).

Secondly, let us prove that

$$A_q(\theta', \theta, k) = A(\theta', \theta, k) \tag{6.167}$$

where $A(\theta', \theta, k)$ is the given function, the kernel of equation (6.161). Note that we have proved in 6.1 that v satisfies the equation

$$v(x,\theta,k) = v(x,-\theta,-k) + \frac{ik}{2\pi}\int_{S^2} A_1(\theta',\theta,k)v(x,-\theta',-k)d\theta' + \frac{ik}{2\pi}\int_{S^2} A_q(\theta',\theta,k)\exp(ik\theta'\cdot x)d\theta'. \quad (6.168)$$

This is equation (6.92) written in terms of v. Subtract (6.168) from (6.161) to get

$$0 = \int_{S^2} [A(\theta',\theta,k) - A_q(\theta',\theta,k)]\, u(x,-\theta',-k)d\theta', \quad \forall x \in R^3. \quad (6.169)$$

Equation (6.169) and Theorem 6.1 from Section 6.1.8, imply (6.167).

The last statement of Theorem 6.3 can be proved as follows. Suppose there are two (or more) solutions v_j, $j = 1,2$, to equation (6.161) with properties (6.164) and (6.165). Let $q_j(x)$ and $A_j(\theta',\theta,k)$, $j = 1,2$ be the corresponding potentials and scattering amplitudes. If $q_1 = q_2$ then $v_1 = v_2$ by the uniqueness of the scattering solution (Lemma 6.4, p. 114). If $q_1 \not\equiv q_2$ then $w := v_1 - v_2 \not\equiv 0$. The function w solves the equation

$$w(x,\theta,k) = w(x,-\theta,-k) + \frac{ik}{2\pi}\int_{S^2} A(\theta'',\theta,k)w(x,-\theta'',-k)d\theta'', \quad \forall x \in \mathbf{R}^3. \quad (6.170)$$

Note that

$$w(x,\theta,k) = [A_1(\theta',\theta,k) - A_2(\theta',\theta,k)]\, g(r) + o(r^{-1}), \quad r \to \infty, \quad xr^{-1} = \theta' \quad (6.171)$$

and

$$w(x,-\theta,-k) = [A_1(\theta',-\theta,-k) - A_2(\theta',-\theta,-k)]\, \overline{g}(r) + o(r^{-1}), \quad r \to \infty, \quad xr^{-1} = \theta' \quad (6.172)$$

where $g(r) := r^{-1}\exp(ikr)$.

From (6.170), (6.171) and (6.172) it follows that

$$[A_1(\theta',\theta,k) - A_2(\theta',\theta,k)]\, g(r) = B(\theta',\theta,k)\overline{g}(r) + o(r^{-1}), \quad r \to \infty \quad (6.173)$$

where the expression for $B(\theta',\theta,k)$ is not important for our argument. It follows from (6.173) that

$$A_1(\theta',\theta,k) = A_2(\theta',\theta,k) \quad (6.174)$$

so that $B(\theta', \theta, k) = 0$). By Lemma 6.7, it follows that $q_1 = q_2$. Theorem 6.3 is proved. □

Exercise. Prove that if $ag(r) = b\overline{g}(r) + o(r^{-1})$, $r \to \infty$, $k > 0$, where a and b do not depend on r then $a = b = 0$.

Hint: Write $a \exp(ikr) = b \exp(-ikr) + o(1)$; choose $r_n = \frac{n\pi}{k}$, $n \to \infty$. Derive that $a = b$. Then choose $r_n' = \frac{n\pi + \frac{\pi}{2}}{k}$, $n \to \infty$. Derive $-a = b$. Thus $a = b = 0$.

Another characterization of the class of scattering amplitudes is given in [Ramm (1992)]. A characterization of the class of scattering amplitudes at a fixed $k > 0$ is given in [Ramm (1988)].

6.2.6 *The Born inversion*

The scattering amplitude in the Born approximation is defined to be

$$A_B(\theta', \theta, k) := -\frac{1}{4\pi} \int \exp\{ik(\theta - \theta') \cdot x\} q(x) dx \tag{6.175}$$

which is formula (6.15) with $u(y, \theta, k)$ substituted by $u_0(y, \theta, k) := \exp(ik\theta \cdot y)$. The Born inversion is the inversion for $q(x)$ of the equation

$$\int \exp\{ik(\theta - \theta') \cdot x\} q(x) dx = -4\pi A(\theta', \theta, k) \tag{6.176}$$

which comes from setting

$$A(\theta', \theta, k) = A_B(\theta', \theta, k). \tag{6.177}$$

The first question is: does a $q(x) \in Q$ exist such that (6.177) holds for all $\theta', \theta \in S^2$ and all $k > 0$?

The answer is no, unless $q(x) = 0$ so that $A_B(\theta', \theta, k) = A(\theta', \theta, k) = 0$.

Theorem 6.4 *Assume that $q \in Q$. If (6.177) holds for all $\theta', \theta \in S^2$ and all $k > 0$ then $q(x) = 0$.*

Proof. Since $q = \overline{q}$, it follows from (6.175) that

$$A_B(\theta, \theta, k) - \overline{A_B(\theta, \theta, k)} = 0. \tag{6.178}$$

From (6.178), (6.177) and (1.54) one concludes that

$$\int_{S^2} |A_B(\theta, \alpha, k)|^2 d\alpha = 0 \quad \forall\theta \in S^2 \quad \text{and all} \quad k > 0. \tag{6.179}$$

Thus $A_B(\theta, \alpha, k) = 0$ for all $\theta, \alpha \in S^2$ and $k > 0$. This and (6.175) imply that $q(x) = 0$. Theorem 6.4 is proved. □

Remark 6.2 *If $q \in Q$ is compactly supported and (6.177) holds for all $\theta', \theta \in S^2$ and a fixed $k > 0$ then $q(x) = 0$. This follows from the uniqueness theorem proved in [Ramm (1992)].*

It follows from Theorem 6.4 that the scattering amplitude $A(\theta', \theta, k)$ cannot be a function of $p := k(\theta - \theta')$ only. The Born inversion in practice reduces to choosing a $p \in \mathbf{R}^3$, finding $\theta, \theta' \in S^2$ and $k > 0$ such that

$$p = k(\theta - \theta'), \tag{6.180}$$

writing equation (6.176) as

$$\tilde{q}(p) := \int \exp(ip \cdot x) q(x) dx = -4\pi \left[A(p) + \eta\right] \tag{6.181}$$

where

$$A(p) := -\frac{1}{4\pi} \int q(x) \exp(ip \cdot x) dx = -\frac{1}{4\pi} \tilde{q}(p) \tag{6.182}$$

and η is defined as

$$\eta := A(\theta', \theta, k)|_{k(\theta - \theta') = p} - A(p). \tag{6.183}$$

One then wishes to neglect η and compute $q(x)$ by the formula

$$q(x) = \frac{-4\pi}{(2\pi)^3} \int A(p) \exp(-ip \cdot x) dp. \tag{6.184}$$

However, the data are the values $A(p) + \eta$ or, if the measurements are noisy, the values

$$A(p) + \eta + \eta_1 := B_\delta(p) \tag{6.185}$$

where η_1 is noise and $\delta > 0$ is defined by formula (6.186) below.

The question is: assuming that $\delta > 0$ is known such that

$$|\eta + \eta_1| < \delta \tag{6.186}$$

how does one compute $q_\delta(x)$, such that

$$|q_\delta(x) - q(x)| \le \epsilon(\delta) \to 0 \quad \text{as} \quad \delta \to 0. \tag{6.187}$$

In other words, how does one compute a stable approximation of $q(x)$ given noisy values of $A(p)$ as in (6.185)? This question is answered in [Ramm (1992)]. We present the answer here. Define

$$q_\delta(x) := -4\pi(4\pi)^{-3} \int_{|p|\le R(\delta)} B_\delta(p) \exp(-ip \cdot x) dp, \quad R(\delta) = c_0 \delta^{-\frac{1}{2b}} \tag{6.188}$$

where the constants $c_0 > 0$ and $b > \frac{3}{2}$ will be specified below (see formulas (6.191) and (6.190)).

Theorem 6.5 *The following stability estimate holds:*

$$|q_\delta(x) - q(x)| \le c_1 \delta^{1-\frac{3}{2b}} \tag{6.189}$$

provided that

$$|\tilde{q}(p)| \le c_2 (1 + |p|^2)^{-b}, \quad b > \frac{3}{2}. \tag{6.190}$$

The constants c_0 and c_1 are given by the formulas

$$c_0 = \left(\frac{c_2}{4\pi}\right)^{\frac{1}{2b}} \tag{6.191}$$

$$c_1 = \left[\frac{2}{3\pi}\left(\frac{1}{4\pi}\right)^{\frac{3}{2b}} + \frac{1}{2\pi^2(2b-3)(4\pi)^{\frac{3-2b}{2b}}}\right] c_2^{\frac{3}{2b}}. \tag{6.192}$$

Proof. Using (6.182) and (6.190), one obtains

$$\begin{aligned}
|q_\delta(x) - q(x)| &\le 4\pi(2\pi)^{-3} \times \\
&\quad \left| \int_{|p|<R} B_\delta(p) \exp(-ip \cdot x) dp - \int \exp(-ip \cdot x) A(p) dp \right| \\
&\le (2\pi^2)^{-1} \left\{ \frac{4}{3}\pi R^3 \delta + c_2 \int_R^\infty \frac{r^2 dr}{(1+r^2)^b} \right\} \\
&\le \frac{2}{3\pi} \delta R^3 + \frac{c_2}{2\pi^2} \frac{R^{3-2b}}{2b-3} \\
&:= \phi(\delta, R).
\end{aligned} \tag{6.193}$$

For a fixed $\delta > 0$, minimize $\phi(\delta, R)$ in R to get

$$\phi_{\min} = c_1 \delta^{1-\frac{3}{2b}} \quad \text{if} \quad R(\delta) = \left(\frac{c_2}{4\pi\delta}\right)^{\frac{1}{2b}} \tag{6.194}$$

where c_1 is given by (52). Theorem 6.5 is proved. □

The practical conclusions, which follow from Theorem 6.5, are:

1) The Born inversion needs a regularization. One way to use a regularization is given by formula (6.188). If one would take the integral in (6.188) over all of R^3, or over too large a ball, the error of the Born inversion might have been unlimited.
2) Even if the error η of the Born approximation for solving the direct scattering problem is small, it does not imply that the error of the Born inversion (that is the Born approximation for solving the inverse scattering problem) is small.

The second conclusion can be obtained in a different way, a more general one. Let $\mathcal{B}(q) = A(*)$, where $\mathcal{B}$ is a nonlinear map which sends a potential $q \in Q$ into a scattering amplitude A. The Born approximation is a linearization of $(*)$. Let us write it as

$$\mathcal{B}'(q_0)(q - q_0) = A - \mathcal{B}(q_0). \tag{6.195}$$

The inverse of the operator $\mathcal{B}'(q_0)$ is unbounded on the space of functions with the sup norm. Therefore small in absolute value errors in the data may lead to large errors in the solution $q - q_0$. The Born approximation is a linearization around $q_0 = 0$. The distorted wave Born approximation is a linearization around the reference potential q_0. In both cases the basic conclusion is the same: without regularization the Born inversion may lead to large errors even if the perturbation $q - q_0$ is small (in which case Born's approximation is accurate for solving the direct problem).

Let us discuss another way to recover $q(x)$ from $A(\theta', \theta, k)$ given for large k. This way has a computational advantage of the following nature. One does not need to find θ', θ, and k such that (6.180) holds and one integrates over $S^2 \times S^2$ instead of R^3 in order to recover $q(x)$ stably from the given noisy data $A_\delta(\theta', \theta, k)$:

$$|A_\delta(\theta', \theta, k) - A(\theta', \theta, k)| \le \delta. \tag{6.196}$$

We start with the following known formula ([Saito (1982)])

$$\lim_{k\to\infty} k^2 \int_{S^2}\int_{S^2} A(\theta',\theta,k)\exp\{ik(\theta'-\theta)\cdot x\}\,d\theta d\theta' = -2\pi\int |x-y|^{-2}q(y)dy. \tag{6.197}$$

To estimate the rate of convergence in (6.197) one substitutes for $A(\theta',\theta,k)$ its expression (6.15) to get

$$\begin{aligned}\mathcal{J} := \frac{k^2}{-4\pi}\int dyq(y)\int_{S^2}\exp\{ik\theta'\cdot(x-y)\}\,d\theta' \\ \left\{\int_{S^2}\exp\{-ik\theta\cdot(x-y)\}\,d\theta + \int_{S^2}\exp(-ik\theta\cdot x)(\phi-1)d\theta\right\}\end{aligned} \tag{6.198}$$

where ϕ is defined by (6.60). A simple calculation yields

$$\int_{S^2}\exp\{ik\theta\cdot(x-y)\}\,d\theta = 4\pi\frac{\sin(k|x-y|)}{k|x-y|}. \tag{6.199}$$

Thus

$$\begin{aligned}\mathcal{J} &= -4\pi\int dyq(y)\left\{\frac{\sin^2(k|x-y|)}{|x-y|^2} + \frac{\sin(k|x-y|)}{4\pi|x-y|}\times\right. \\ &\left.\int_{S^2}\exp(-ik\theta\cdot x)\left[\phi(y,\theta,k)-1\right]d\theta\right\} \\ &:= \mathcal{J}_1+\mathcal{J}_2.\end{aligned} \tag{6.200}$$

One has

$$\begin{aligned}\mathcal{J}_1 &= -4\pi\int dyq(y)\frac{1-\cos(2k|x-y|)}{2|x-y|^2} \\ &= -2\pi\int q(y)|x-y|^{-2}dy - 2\pi\int_0^\infty dr\cos(2kr)Q(x,r)\end{aligned} \tag{6.201}$$

where

$$Q(x,r) := \int_{S^2} q(x+r\alpha)d\alpha, \quad r=|y-x|. \tag{6.202}$$

Let us assume that

$$q\in Q_2 := \left\{q : q=\overline{q}, |q|+|Dq|+|D^2q|\in Q\right\}. \tag{6.203}$$

Then, integrating by parts, one obtains

$$\begin{aligned}\left|\int_0^\infty dr \cos 2krQ(x,r)\right| &= \left|\frac{-1}{2k}\int_0^\infty dr \sin(2kr)\frac{\partial Q}{\partial r}\right| \\ &= \left|\frac{1}{4k^2}\left[\cos(2kr)\frac{\partial Q}{\partial r}\Big|_0^\infty - \int_0^\infty dr \cos(2kr)\frac{\partial^2 Q}{\partial r^2}\right]\right| \\ &\le \frac{c}{k^2}, \quad k > 1. \end{aligned} \tag{6.204}$$

Here $c = \text{const} > 0$ does not depend on x:

$$c = \max_{\substack{x\in R^3 \\ r\ge 0}} \int_{S^2} |\nabla q(x + r\alpha)| d\alpha + \max_{x\in R^3} \int_0^\infty dr \left|\frac{\partial^2 Q}{\partial r^2}\right|, \tag{6.205}$$

$$\left|\frac{\partial^2 Q}{\partial r^2}\right| \le c_1(1 + |x - r|)^{-a}, \quad a > 3 \tag{6.206}$$

and

$$\begin{aligned}\int_0^\infty dr(1 + |x - r|)^{-a} &\le \int_0^\infty dr\,(1 + ||x| - r|)^{-a} \le \int_{-\infty}^\infty dr\,(1 + ||x| - r|)^{-a} \\ &\le 2\int_0^\infty dr(1 + r)^{-a} \le c_2 \end{aligned} \tag{6.207}$$

so that the right-hand side of (6.205) is bounded uniformly in $x \in R^3$. Thus

$$\mathcal{J}_1 = -2\pi \int |x - y|^2 q(y) dy + O(k^{-2}), \quad k \to \infty \tag{6.208}$$

provided (6.203) holds.

Note that if $q \in L^2_{\text{loc}}(R^3)$ and no a priori information about its smoothness is known, then one obtains only $o(1)$ in place of $O(k^{-2})$ in (6.208).

From (1.72) it follows that

$$\mathcal{J}_2 \le ck^{-1}, \quad k > 1. \tag{6.209}$$

Thus, assuming (6.203),

$$\mathcal{J} = -2\pi \int |x - y|^{-2} q(y) dy + O(k^{-1}), \quad k \to \infty \tag{6.210}$$

where $O(k^{-1})$ is uniform in x.

Therefore (6.197) can be written as

$$k^2 \int_{S^2} \int_{S^2} A(\theta', \theta, k) \exp\{ik(\theta' - \theta) \cdot x\} \, d\theta d\theta'$$
$$= -2\pi \int |x - y|^{-2} q(y) dy + O(k^{-1}). \tag{6.211}$$

The equation

$$-2\pi \int |x - y|^{-2} q(y) dy = f(x), \quad x \in R^3 \tag{6.212}$$

is solvable analytically. Take the Fourier transform of (6.212) to get

$$\tilde{q}(p) = -\frac{1}{4\pi^3} |p| \tilde{f}(p). \tag{6.213}$$

Here we have used the formula

$$\widetilde{|x|^2} := \int |x|^{-2} \exp(ip \cdot x) dx = 2\pi \int_0^\infty dr \int_0^\pi \exp(i|p| r \cos\theta) \sin\theta d\theta$$
$$= 4\pi \int_0^\infty dr \frac{\sin |p| r}{|p| r} = \frac{2\pi^2}{|p|}. \tag{6.214}$$

thus

$$q(x) = \frac{-1}{32\pi^6} \int \exp(-ip \cdot x) |p| \tilde{f}(p) dp. \tag{6.215}$$

Assume for a moment that the term $O(k^{-1})$ in (71) is absent. Then, applying formula (6.215), taking $f(x)$ to be the left-hand side of (6.211), and taking the Fourier transform of $\tilde{f}(p)$, one would obtain

$$q(x) = -\frac{1}{32\pi^6} \int dp |p| \exp(-ip \cdot x) \times$$
$$\left\{ k^2 \int_{S^2} \int_{S^2} A(\theta', \theta, k) (2\pi)^3 \delta\left[p + k(\theta' - \theta)\right] d\theta d\theta' \right\}$$
$$= -\frac{k^3}{4\pi^3} \int_{S^2} \int_{S^2} d\theta d\theta' A(\theta', \theta, k) |\theta' - \theta| \exp\{ik(\theta' - \theta) \cdot x\}. \tag{6.216}$$

This formula appeared in [Somersalo, E. *et al.* (1988)]. If A_δ is known in place of A, and (6.196) holds, then formula (6.216), with A_δ in place of A, gives

$$q_\delta(x) := -\frac{k^3}{4\pi^3} \int_{S^2} \int_{S^2} d\theta d\theta' A_\delta(\theta', \theta, k) |\theta' - \theta| \exp\{ik(\theta' - \theta) \cdot x\}. \tag{6.217}$$

Neglecting the term $O(k^{-1})$ in (6.211), one obtains

$$|q - q_\delta| \le \delta k^3 \int_{S^2}\int_{S^2} d\theta d\theta' |\theta - \theta'| (4\pi^3)^{-1} = \delta k^3 \frac{16}{3\pi}, \tag{6.218}$$

where we have used the formula:

$$\begin{aligned}\int_{S^2}\int_{S^2} |\theta - \theta'| d\theta d\theta' &= \int_{S^2} d\theta \int_{S^2} |\theta - \theta'| d\theta' = 4\pi \int_{S^2} |\theta - \theta'| d\theta' \\ &= 8\pi^2 \int_0^\pi \sqrt{2 - 2\cos\gamma}\sin\gamma d\gamma = \frac{64\pi^2}{3}.\end{aligned} \tag{6.219}$$

It is now possible to take into account the term $O(k^{-1})$ in (6.211). Note that, as follows from (6.72) and (6.200),

$$\mathcal{J}_2 \le \frac{1}{2k} \int dy |q(y)| \frac{1}{|x-y|} \int_{S^2} d\theta \int_0^\infty |q(x - r\theta)| dr + o(k^{-1}). \tag{6.220}$$

One has

$$c \int dy |q(y)| |x-y|^{-1} \int_0^\infty \int_{S^2} d\theta \left(1 + |x - r\theta|^2\right)^{-a/2} dr, \quad a > 3 \tag{6.221}$$

and

$$\begin{aligned}\int_{S^2} \left(1 + |x - r\theta|^2\right)^{-a/2} d\theta &= 2\pi \int_0^\pi \frac{\sin\gamma d\gamma}{(1 + |x|^2 + |r|^2 - 2|x| r\cos\gamma)^{a/2}} \\ &= 2\pi \int_{-1}^1 \frac{dt}{(1 + |x|^2 + |r|^2 - 2|x| r t)^{a/2}}\end{aligned}$$

$$= 2\pi \frac{\left[1 + \left||x| - r\right|^2\right]^{-a/2+1} - \left[1 + (|x| + r)^2\right]^{-a/2+1}}{2|x||r|\left(\frac{a}{2} - 1\right)}. \tag{6.222}$$

Moreover

$$\int_0^\infty dr \frac{\left[1 + \left||x| - r\right|^2\right]^{-a/2+1} - \left[1 + (|x| + r)^2\right]^{-a/2+1}}{r} \le \frac{c}{|x|}, \quad |x| \ge 1 \tag{6.223}$$

where $c > 0$ is a constant. Therefore

$$\mathcal{J}_2 \le \frac{ck^{-1}}{1 + |x|^2}. \tag{6.224}$$

This means that the $L^2(R^3)$ norm of $\mathcal{J}_2$ as a function of x is $O(k^{-1})$ as $k \to \infty$. Therefore, if one takes into account the $O(k^{-1})$ term is (71), one obtains in place of (6.218) the following estimate

$$\| q - q_\delta \|_{L^2(R^3)} \leq c(\delta k^3 + k^{-1}), \quad c = \text{const} > 0. \tag{6.225}$$

Minimization of the right-hand side of (6.225) in k yields

$$\| q - q_\delta \|_{L^2(R^3)} \leq c_1 \delta^{-1/4} \quad \text{for} \quad k_m = (3\delta)^{-1/4} \tag{6.226}$$

where $k_m = k_m(\delta)$ is the minimizer of the right-hand side of (6.225) on the interval $k > 0$.

Therefore, if the data $A_\delta(\theta', \theta, k)$ are noisy, so that (6.196) holds, one should not take k in formula (6.217) too large. The quasioptimal k is given in (6.226), and formula (6.217) with $k = k_m$ gives a stable approximation of $q(x)$.

Let us finally discuss (6.225). The term ck^{-1} has already been discussed. The first term $c\delta k^3$ has been discussed for the estimate in sup norm. In the case of L^2 norm one has to estimate the $L^2(R^3)$ norm of the function

$$h(x) := \int_{S^2} \int_{S^2} a(\theta', \theta) \exp\{ik(\theta' - \theta) \cdot x\} \, d\theta' d\theta, \quad a := A - A_\delta \tag{6.227}$$

given that

$$|a| \leq \delta, \quad |\nabla_\theta a| + |\nabla_{\theta'} a| \leq m_1 \tag{6.228}$$

where ∇_θ, $\nabla_{\theta'}$ are the first derivatives in θ and θ', and $m_1 = \text{const} > 0$. In order to estimate h we use the following formula [Ramm (1986), p.54]

$$\int_{S^2} \exp(ikr\theta \cdot \alpha) f(\theta) d\theta = \frac{2\pi i}{k} \left[\frac{\exp(-ikr)}{r} f(-\alpha) - \frac{\exp(ikr)}{r} f(\alpha) \right] + o\left(\frac{1}{r}\right), \quad r \to +\infty, \quad k > 0, \quad \alpha \in S^2. \tag{6.229}$$

This formula is proved under the assumption $f \in C^1(S^2)$. From (6.227)-(6.229) one obtains

$$|h(x)| \leq \frac{c\delta}{1 + |x|^2}. \tag{6.230}$$

Thus

$$\| h \|_{L^2(R^3)} \leq c\delta. \tag{6.231}$$

By c we denote various positive constants. From (6.231) one obtains the first term in (6.225) for the case of the estimate in $L^2(R^3)$ norm.

6.3 Estimation theory and inverse scattering in R^3

Consider for simplicity the filtering problem which is formulated in Chapter 2. Let

$$\mathcal{U} = s + n(x), \quad x \in R^3 \tag{6.232}$$

where the useful signal $s(x)$ has the properties

$$\overline{s(x)} = 0, \quad \overline{s^*(x)s(y)} = R_s(x,y) \tag{6.233}$$

$$\overline{s(x)n(y)} = 0 \tag{6.234}$$

and the noise is white

$$\overline{n(x)} = 0, \quad \overline{n^*(x)n(y)} = \delta(x-y). \tag{6.235}$$

In this section the star stands for complex conjugate and the bar stands for the mean value.

The optimal linear estimate of $s(x)$ is given by

$$\hat{s}(x) = \int_D h(x,y)\mathcal{U}(y)dy. \tag{6.236}$$

Here the optimality is understood in the sense of minimum of variance of the error of the estimate, as in Chapter 2. Other notations are also the same as in Chapter 2. In particular, $D \subset R^3$ is a bounded domain in which $\mathcal{U}$ is observed.

It is proved in Chapter 2 that the optimal $h(x,y)$ solves equation 2.11 which in the present case is of the form:

$$Rh := h(x,z) + \int_D R_s(z,y)h(x,y)dy = R_s(z,x), \quad x,y \in \overline{D} \tag{6.237}$$

or, if one changes $z \to y$ and $y \to z$, this equation takes the form

$$h(x,y) + \int_D R_s(y,z)h(x,z)dz = R_s(y,x), \quad x,y \in \overline{D}. \tag{6.238}$$

Note that under the assumptions (6.233)-(4) one has

$$R(x,y) = R_s(x,y) + \delta(x-y), \quad f(x,y) = R_s(x,y) \tag{6.239}$$

where $R(x,y)$ and $f(x,y)$ are defined in Chapter 2 (see formulas (2.3) and (2.10)). The basic equation (6.238) is a Fredholm's second kind equation with positive definite in $L^2(D)$ operator R, $R \geq I$. It is uniquely solvable in $L^2(D)$ (that is, it has a solution in $L^2(D)$ and the solution is unique).

There are many methods to solve (6.238) numerically. In particular an iterative method can easily be constructed for solving (6.238). This method converges as a geometrical series (see Section 3.2, Lemma 3.1).

Projection methods can be easily constructed and applied to (6.238) (see Section 3.2, Lemma 3.3). In [Levy and Tsitsiklis (1985)] and [Yagle (1988)] attempts are made to use for the numerical solution of the equation (6.238) some analogues of Levinson's recursion which was used in the one-dimensional problems when $R_s(x,y) = R_s(x-y)$.

In the one-dimensional problems causality plays important role. In the three-dimensional problems causality plays no role: the space is isotropic in contrast with time.

Therefore, in order to use the ideas similar to Levinson's recursion one needs to assume that the domain D is parametrized by one parameter. In [Levy and Tsitsiklis (1985)] the authors assume D to be a disc (so that the domain is determined by the radius of the disc; this radius is the parameter mentioned above). Of course, one has to impose severe restrictions on the correlation function $R_s(x,y)$. In [Levy and Tsitsiklis (1985)] it is assumed that

$$R_s(x,y) = R_s(|x-y|). \tag{6.240}$$

This means that $s(x)$ is an isotropic random field.

In [Yag] the case is considered when $D \subset R^3$ is a ball and $R_s(x,y)$ solves the equation

$$\Delta_x R_s(x,y) = \Delta_y R_s(x,y), \tag{6.241}$$

where Δ_x is the Laplacian. If $R_s(x,y) = R_s(x-y)$ then (6.241) holds.

Let us derive a differential equation for the optimal filter $h(x,y)$. The derivation is similar to the one given in Section 8.4.8 for Kalman filters. Let us apply the operator $\Delta_x - \Delta_y$ to (6.238) assuming (6.241) and taking

$D = \{z : |z| \le |x|\}$:

$$0 = (\Delta_x - \Delta_y)h(x,y) + \Delta_x \int_{|z|\le|x|} h(x,z)R_s(y,z)dz$$
$$- \int_{|z|\le|x|} h(x,z)\Delta_z R_s(y,z)dz$$
$$= (\Delta_x - \Delta_y)h(x,y) + \int_{|z|\le|x|} (\Delta_x - \Delta_z)h(x,z)R_s(y,z)dz - \mathcal{J}, \tag{6.242}$$

where, as we will prove,

$$\mathcal{J} := r^2 \int_{S^2} R_s(y, r\beta)q(x, r\beta)d\beta, \quad r = |x|. \tag{6.243}$$

Here S^2 is the unit sphere in R^3 and

$$q = q(r,\alpha,\beta) := q(x, r\beta) := -2\frac{d}{dr}h(r\alpha, r\beta) - \frac{4}{r}h(r\alpha, r\beta)$$
$$= -\frac{2}{r^2}\frac{d}{dr^2}\left[r^2 h(r\alpha, r\beta)\right], \quad x = r\alpha. \tag{6.244}$$

Let us prove (6.243) and (6.244). Integrate by parts the second integral in (6.242) to get

$$\int_{|z|\le|x|} h(x,z)\Delta_z R_s(y,z)dz = \int_{|z|\le|x|} \Delta_z h(x,z)R_s(y,z)dz$$
$$+ \int_{|z|=r} \left[h(x,z)\frac{\partial}{\partial|z|}R_s(y,z) - \frac{\partial h(x,z)}{\partial|z|}R_s(y,z)\right] dz. \tag{6.245}$$

The first integral in (6.242) can be written as

$$\mathcal{J}_1 := \left(\frac{d^2}{dr^2} + \frac{2}{r}\frac{d}{dr} + \frac{\Delta^*}{r^2}\right)\int_0^r d\rho\rho^2 \int_{S^2} h(r\alpha, \rho\beta)R_s(y, \rho\beta)d\beta, \tag{6.246}$$

where Δ^* is the angular part of the Laplacian and $z = \rho\beta$, where $\rho = |z|$

and $\beta \in S^2$. One has

$$\mathcal{J}_1 = \int_0^r d\rho\rho^2 \int_{S^2} \frac{1}{r^2}\Delta^* h(r\alpha,\rho\beta)R_s(y,\rho\beta)d\beta + 2r\int_{S^2} h(r\alpha,r\beta)R_s(y,r\beta)d\beta$$
$$+ \int_0^r d\rho\rho^2 \int_{S^2} \frac{2}{r}\frac{d}{dr}h(r\alpha,\rho\beta)R_s(y,\rho\beta)d\beta$$
$$+ \frac{d}{dr}\left[r^2\int_{S^2} h(r\alpha,r\beta)R_s(y,r\beta)d\beta\right.$$
$$\left.+ \int_0^r d\rho\rho^2 \int_{S^2} \frac{d}{dr}h(r\alpha,\rho\beta)R_s(y,\rho\beta)d\beta\right]$$
$$= \int_{|z|\le r} \Delta_x h(x,z)R_s(y,z)dz + 4r\int_{S^2} h(r\alpha,r\beta)R_s(y,r\beta)d\beta$$
$$+ r^2\int_{S^2} \frac{d}{dr}h(r\alpha,r\beta)R_s(y,r\beta)d\beta + r^2\int_{S^2} h(r\beta,r\beta)\frac{d}{dr}R_s(y,r\beta)d\beta$$
$$+ r^2\int_{S^2} \frac{\partial}{\partial r}h(r\alpha,\rho\beta)\bigg|_{\rho=r} R_s(y,r\beta)d\beta \tag{6.247}$$

From (6.245) and (6.247) one obtains

$$-\mathcal{J} = r^2\int_{S^2}\left[-h(r\alpha,r\beta)\frac{\partial}{\partial r}R_s(y,r\beta) + \frac{\partial h(r\alpha,\rho\beta)}{\partial\rho}\bigg|_{\rho=r} R_s(y,r\beta)\right]d\beta$$
$$+ *4r\int_{S^2} h(r\alpha,r\beta)R_s(y,r\beta)d\beta$$
$$+ r^2\int_{S^2}\left[\frac{d}{dr}h(r\alpha,r\beta)R_s(y,r\beta)d\beta + h(r\alpha,r\beta)\frac{d}{dr}R_s(y,r\beta)\right.$$
$$\left.+ \frac{\partial}{\partial r}h(r\alpha,\rho\beta)\bigg|_{\rho=r} R_s(y,r\beta)\right]d\beta$$
$$= 2r^2\int_{S^2}\frac{d}{dr}h(r\alpha,r\beta)R_s(y,r\beta)d\beta + 4r\int_{S^2} h(r\alpha,r\beta)R_s(y,r\beta)d\beta$$
$$= r^2\int_{S^2} R_s(y,r\beta)q(x,r\beta)d\beta, \tag{6.248}$$

where q is given by (6.244).

In order to derive a differential equation for h let us assume that

$$R_s(x,y) = R_s(y,x), \tag{6.249}$$

write equation (6.238) for $D = \{z : |z| \le |x|\}$ as

$$\mathcal{H}(x,y) + \int_{|z|\le|x|} R_s(y,z)\mathcal{H}(x,z)dz = R_s(y,x), \quad |y| \le |x|. \tag{6.250}$$

Note the restiction $|y| \le |x|$ in (6.250). Multiply (6.250) by $q(r\alpha, r\beta)$, set in (6.250) $x = r\beta$, $r = |x|$, integrate over S^2 in β and then multiply by r^2 to get

$$r^2 \int_{S^2} \mathcal{H}(r\beta, y)q(r\alpha, r\beta)d\beta + \int_{|z|\le|x|} R_s(y,z)r^2 \int_{S^2} \mathcal{H}(r\beta, z)q(r\alpha, r\beta)d\beta$$
$$= r^2 \int_{S^2} R_s(y, r\beta)q(r\alpha, r\beta)d\beta. \tag{6.251}$$

Define

$$(\Delta_x - \Delta_y)\mathcal{H}(x,y) := \phi(x,y) \tag{6.252}$$

and set $x = r\alpha$ in (6.252). Write equation (6.242) as

$$\phi(r\alpha, y) + \int_{|z|\le|x|} \phi(r\alpha, z)R_s(y,z)dz = r^2 \int_{S^2} R_s(y, r\beta)q(r\alpha, r\beta)d\beta \tag{6.253}$$

or, in the operator form

$$(I + R_s)\phi = \psi, \tag{6.254}$$

where ψ is the right-hand side of (6.253).

Equation (6.251) is of the form

$$(I + R_s)\gamma = \psi \tag{6.255}$$

where

$$\gamma := r^2 \int_{S^2} \mathcal{H}(r\beta, y)q(r\alpha, r\beta)d\beta. \tag{6.256}$$

Since the operator $I + R_s \ge I$ is injective, it follows from (6.254) and (6.255) that $\phi = \gamma$. Thus

$$(\Delta_x - \Delta_y)\mathcal{H}(x,y) = r^2 \int_{S^2} \mathcal{H}(r\beta, y)q(r\alpha, r\beta)d\beta, \quad |y| \le |x| = r, \quad x = r\alpha, \tag{6.257}$$

where $\alpha, \beta \in S^2$ and q is given by (6.244) with $h(r\alpha, r\beta) = \mathcal{H}(r\alpha, r\beta)$. Let us formulate the result:

Lemma 6.8 *If $\mathcal{H}(x,y)$ solves equation (6.250) and the assumptions (6.241) and (6.249) hold, then $\mathcal{H}$ solves equation (6.257) with $q(r\alpha, r\beta)$ given by (6.244) with $h(r\alpha, r\beta) = \mathcal{H}(r\alpha, r\beta)$, $\alpha, \beta \in S^2$, $|x| = r$, $x = r\alpha$.*

If one defines

$$\mathcal{H}(x,y) = 0 \quad \text{for} \quad |y| > |x| \tag{6.258}$$

and put

$$\tilde{\mathcal{H}}(x,\xi) = (2\pi)^{-\frac{3}{2}} \int \mathcal{H}(x,y) \exp(i\xi \cdot y) dy, \quad \int := \int_{R^3}, \tag{6.259}$$

where the integral in (6.259) is taken actually over the ball $|y| \le |x|$ because of (6.258), then

$$\mathcal{H}(x,y) = (2\pi)^{-\frac{3}{2}} \int \tilde{\mathcal{H}}(x,\xi) \exp(-i\xi \cdot y) d\xi. \tag{6.260}$$

Substitute (6.260) into (6.257) (or, which is the same, Fourier transform (6.257) in the variable y) to get

$$(\Delta_x + \xi^2)\tilde{\mathcal{H}}(x,\xi) = r^2 \int_{S^2} \tilde{\mathcal{H}}(r\beta,\xi) q(r\alpha, r\beta) d\beta, \quad x = r\alpha. \tag{6.261}$$

Equation (6.261) is a Schrödinger equation with a non-local potential

$$Q\tilde{\mathcal{H}} := r^2 \int_{S^2} \tilde{\mathcal{H}}(r\beta,\xi) q(r\alpha, r\beta) d\beta. \tag{6.262}$$

Suppose that $\mathcal{H}(x,y)$ is computed for $|y| \le |x| \le a$. *Given this $\mathcal{H}(x,y)$, how does one compute the solution $h(x,y)$ to the equation (6.238) with $D = B_a = \{x : |x| \le a\}$?*

Write equation (6.238) for $D = B_a$ and $z = \rho\beta$ as

$$h(x,y,a) + \int_0^a \int_{S^2} R_s(y,\rho\beta) h(x,\rho\beta,a) d\beta \rho^2 d\rho = R_s(y,x). \tag{6.263}$$

Differentiate (6.263) in a to get

$$\frac{\partial h}{\partial a} + \int_0^a \int_{S^2} R_s(y,\rho\beta) \frac{\partial h(x,\rho\beta,a)}{\partial a} d\beta \rho^2 d\rho = -a^2 \int_{S^2} R(y,a\beta) h(x,a\beta,a) d\beta. \tag{6.264}$$

Let $x = a\alpha$ in (6.263). Multiply (6.263) by $-a^2 h(z, a\alpha, a)$ and integrate over S^2 to get:

$$\begin{aligned} &-a^2 \int_{S^2} h(a\alpha, y, a)h(z, a\alpha, a)d\alpha \\ &+ \int_0^a \int_{S^2} R_s(y, \rho\beta) \left\{-a^2 \int_{S^2} h(a\alpha, \rho\beta, a)h(z, a\alpha, a)\right\} \\ &= -a^2 \int_{S^2} R_s(y, a\alpha)h(z, a\alpha, a)d\alpha. \end{aligned} \tag{6.265}$$

The operator $I + R_s$ is injective. Therefore equations (6.264) and (6.265) have the same solution since their right-hand sides are the same. Set $z = x$ and $\alpha = \beta$ in (6.265), compare (6.264) and (6.265) and get

$$\frac{\partial h(x, y, a)}{\partial a} = -a^2 \int_{S^2} h(a\beta, y, a)h(x, a\beta, a)d\beta. \tag{6.266}$$

Note that

$$h(x, y) = \mathcal{H}(x, y) \quad \text{for} \quad |y| \le |x| \tag{6.267}$$

according to equation (6.250). Therefore (6.266) can be written as

$$\frac{\partial h(x, y, a)}{\partial a} = -a^2 \int_{S^2} \mathcal{H}(a\beta, y, a)h(x, a\beta, a)d\beta. \tag{6.268}$$

Equation (6.268) can be used for computing the function $h(x, y, a)$ for all $x, y \in B_a$, given $\mathcal{H}(x, y)$ for $|y| \le |x| \le a$. The value $h(x, a\beta, a)$ can be computed from equation (6.263):

$$h(x, a\beta, a) + \int_0^a \int_{S^2} R_s(a\beta, \rho\theta)h(x, \rho\theta, a)d\theta\rho^2 d\rho = R_s(a\beta, x). \tag{6.269}$$

The function $R_s(a\beta, z)$ is known for all z, β and the function $h(x, \rho\theta, a)$, $\rho < a$ is assumed to be computed recursively, as a grows, by equation (6.268).

Namely, let us assume that $\mathcal{H}(x, y)$ is computed for all values $|y| \le |x| \le A$, and one wants to compute $h(x, y)$ for all $x, y \in B_A := \{x : |x| \le A\}$. From (6.268) one has

$$\begin{aligned} h(x, y, (m+1)\tau) = h(x, y, m\tau) - \tau(m\tau)^2 \times& \\ \int_{S^2} \mathcal{H}(m\tau\beta, y, m\tau)h(x, m\tau\beta, m\tau)d\beta.& \end{aligned} \tag{6.270}$$

Here $m = 0, 1, 2, \dots, \tau > 0$ is a small number, the step of the increment of a. It follows from (6.263) that

$$h(x, y, 0) = R_s(y, x), \tag{6.271}$$

so that

$$h(x, y, \tau) = R_s(y, x), \tag{6.272}$$

$$h(x, y, 2\tau) = R_s(y, x) - \tau^3 \int_{S^2} \mathcal{H}(\tau\beta, y, \tau) h(x, \tau\beta, \tau) d\beta, \tag{6.273}$$

and so on. One can assume that $|y| > |x|$ because for $|y| \leq |x|$ one can use (6.267).

A formal connection of the estimation problem with scattering theory can be outlined as follows.

Let us *assume* that there exists a function $\mathcal{H}(x, y)$, $\mathcal{H}(x, y) = 0$ for $|y| > |x|$, such that the function $\phi(x, \theta, k)$ defined by the formula

$$\phi(x, \theta, k) := \exp(ik\theta \cdot x) - \int_{|y| \leq |x|} \exp(ik\theta \cdot y) \mathcal{H}(x, y) dy \tag{6.274}$$

is a solution to the Schrödinger equation

$$\left[\Delta + k^2 - q(x)\right] \phi = 0, \tag{6.275}$$

where $\Delta = \nabla^2$ is the Laplacian. This assumption is not justified presently, so that our argument is formal.

Taking the inverse Fourier transform of (6.274) in the variable $k\theta$, one obtains

$$-\frac{1}{(2\pi)^3} \int_0^\infty dk k^2 \int_{S^2} \left[\phi(x, \theta, k) - \exp(ik\theta \cdot x)\right] \exp(-ik\theta \cdot y) d\theta = \mathcal{H}(x, y). \tag{6.276}$$

Compute $(\Delta_x - \Delta_y)\mathcal{H}$ formally taking the derivatives under the integral signs in the left-hand side of (6.276) and using (6.275). The result is

$$(\Delta_x - \Delta_y)\mathcal{H}(x, y) = q(x)\mathcal{H}(x, y). \tag{6.277}$$

One is interested in the solution of (6.277) with the property $\mathcal{H}(x, y) = 0$ for $|y| > |x|$. Define $\tilde{\mathcal{H}}(x, \xi)$ by formula (6.259). Substitute (6.260) into (6.277) and differentiate in y formally under the sign of the integral to get

$$\left[\Delta_x + \xi^2 - q(x)\right] \tilde{\mathcal{H}}(x, \xi) = 0. \tag{6.278}$$

Therefore, if one compares (6.278) and (6.261) one can conclude that the right-hand side of (6.261) reduces to $q(x)\tilde{\mathcal{H}}(x, \xi)$. This means that

$$q(r\alpha, r\beta) = \frac{1}{r^2}\delta(\alpha - \beta) q(x), \quad x = r\alpha \tag{6.279}$$

where $\delta(\alpha - \beta)$ is the delta-function. If (6.279) holds then the non-local potential Q defined by (6.262) reduces to the local potential $q(x)$.

Equations (6.244) and (6.279) imply

$$-2\frac{d}{dr}\left[r^2\mathcal{H}(r\alpha, r\beta)\right] = \delta(\alpha - \beta)q(r\alpha). \tag{6.280}$$

Note that $h(r\alpha, r\beta) = \mathcal{H}(r\alpha, r\beta)$. From (6.280) one obtains

$$\mathcal{H}(r\alpha, r\beta) = -\frac{\delta(\alpha - \beta)}{2r^2}\int_0^r q(\rho\alpha)d\rho + R_s(0,0), \tag{6.281}$$

where we have used the equation

$$\mathcal{H}(0,0) = R_s(0,0) \tag{6.282}$$

which follows from (6.250).

Let us summarize the basic points of this section:

1) the solution to equation (6.250) solves equation (6.257) provided that the assumptions (6.241) and (6.249) hold;
2) the solution to equation (6.238) with $D = B_a$ is related to the solution to equation (6.250) for $|x| \leq a$ by the formulas (6.267) and (6.268);
3) if the solution to equation (6.250) is found then one can compute the solution to equation (6.238) with $D = B_a$ recursively using (6.270);
4) the solution to equation (6.250) solves the differential equation (6.257), and its Fourier transform solves the Schrödinger equation (6.261) with a non-local potential.

Chapter 7

Applications

In this chapter a number of questions arising in applications are discussed. All sections of this chapter are independent and can be read separately.

7.1 What is the optimal size of the domain on which the data are to be collected?

Suppose that one observes the signal

$$\mathcal{U}(x) = s(x) + n(x), \quad x \in D \subset R^r \tag{7.1}$$

in a bounded domain D which contains a point x_0. Assume for simplicity that D is a ball B_ℓ with radius ℓ centered at x_0. The problem is to estimate $s(x_0)$. As always in this book, $s(x)$ is a useful signal and $n(x)$ is noise. It is clear that if the radius ℓ, which characterizes the size of the domain of observation, is too large then the time and effort will be wasted in collecting data which do not improve the quality of the estimate significantly. On the other hand, if ℓ is too small then one can improve the estimate using more data. What is the optimal ℓ? This question is of interest in geophysics and many other applications.

Let us answer this question using the estimation theory developed in Chapter 2. We assume that the optimal estimate is linear and the optimization criterion is minimum of variance.

We also assume that the data are the covariance functions (1.3), that condition (1.2) holds, and that $R(x,y) \in \mathcal{R}$.

The optimal estimate is given by formula (2.15). Let us assume for simplicity that $P(\lambda) = 1$, and

$$|R(x,y)| \le c\exp(-a|x-y|) \quad c > 0, \quad |x-y| \ge \epsilon > 0 \tag{7.2}$$

where the last inequality allows a growth of $R(x,y)$ as $x \to y$, for example, $R(x,y) = (4\pi|x-y|)^{-1}\exp(-a|x-y|)$. Here c and a are positive constants, a^{-1} is the so-called correlation radius, and the function $f(x,y)$ defined by formula (1.3) is smooth. Concerning $f(x,y)$ we assume the same estimate as for $R(x,y)$:

$$|f(x,y)| \le c\exp(-a|x-y|), \quad c > 0, \quad |x-y| \ge \epsilon_0 > 0. \tag{7.3}$$

Under these assumptions the optimal filter is of the form

$$h = Q(\mathcal{L})f + h_s = h_0 + h_s \tag{7.4}$$

where h_s is the singular part of h which contains terms of the type $\{b(s)\delta_\Gamma\}^{(j)}$ (see Section 3.3), and h_0 is the regular part of h.

The optimal estimate is of the form

$$\hat{s}(x_0) = \int_{B_\ell} h_0(x_0,y)\mathcal{U}(y)dy + \int_{B_\ell} h_s(x_0,y)\mathcal{U}(y)dy. \tag{7.5}$$

The optimal size ℓ of the domain B_ℓ of observation is the size for which the second term in (7.5) is negligible compared with the first.

Example 7.1 Suppose that $r = 3$, $x_0 = 0$, $R(x,y) = (4\pi|x-y|)^{-1}\exp(-a|x-y|)$, and $|\partial^j f| \le M$, $0 \le |j| \le 2$, where j is a multi-index. Then by formula (2.85) one has

$$h(y) = (-\Delta + a^2)f(y) + \left(\frac{\partial f}{\partial|y|} - \frac{\partial u}{\partial|y|}\right)\delta_\Gamma, \tag{7.6}$$

where $h(y) = h(0,y)$, $f(y) = f(0,y)$, $\Gamma = \{x : |x| = \ell\}$. Therefore the optimal estimate is

$$\hat{s}(0) = \int_{B_\ell} h(y)\mathcal{U}(y)dy = \int_{B_\ell}(-\Delta + a^2)f\mathcal{U}dy + \int_\Gamma \left(\frac{\partial f}{\partial|y|} - \frac{\partial u}{\partial|y|}\right)\mathcal{U}(s)ds \tag{7.7}$$

and u is uniquely determined by f as the solution to the Dirichlet problem (2.22-2.23) which in our case is

$$(-\Delta + a^2)u = 0 \quad \text{if} \quad |x| \ge \ell, \quad u = f \quad \text{if} \quad |x| = \ell, \quad u(\infty) = 0. \tag{7.8}$$

The solution to problem (7.8) can be calculated analytically. One gets

$$u(r,\theta) = \sum_{n=0}^{\infty} f_n Y_n(\theta)\frac{h_n(iar)}{h_n(ia\ell)}, \tag{7.9}$$

where $\theta = (\vartheta, \phi)$ is a point on the unit sphere S^2 in R^3, a unit vector, $\{Y_n(\theta)\}$ is the system of spherical harmonics orthonormalized in $L^2(S^2)$, $f_n := \int_{S^2} f(\ell, \theta) Y_n^*(\theta) d\theta$, and $h_n(r)$ is the spherical Hankel function, $h_n(r) := \left(\frac{\pi}{2r}\right)^{1/2} H^{(1)}_{n+(1/2)}(r)$, where $H^{(1)}_n(r)$ is the Hankel function. The solution (7.9) is a three dimensional analogue of the solution (2.91).

The second integral on the right-hand side of (7.7) is of order of magnitude $O(\exp(-a\ell))$, while the first integral is of order of magnitude $O(1)$. Therefore, if we wish to be able to neglect the effects of the boundary term on the estimate with accuracy about 5 percent then we should choose $\ell = 3/a$. A practical recipe for choosing ℓ so that the magnitude of the boundary term in (7.7) is about γ percents of the magnitude of the volume term is

$$\ell = \frac{1}{a} \ln \frac{100}{\gamma}. \tag{7.10}$$

7.2 Discrimination of random fields against noisy background

Suppose that one observes a random field $\mathcal{U}(x)$ which can be of one of the forms

$$\mathcal{U}(x) = s_p(x) + n(x), \quad p = 0, 1. \tag{7.11}$$

Here $s_p(x)$, $p = 0, 1$, are deterministic signals and $n(x)$ is Gaussian random field with zero mean value

$$\overline{n} = 0. \tag{7.12}$$

In particular, if $s_0 = 0$ then the problem is to decide if the observed signal $\mathcal{U}(x)$ contains the signal $s_1(x)$ or is it just noise. In order to formulate the discrimination problem analytically we take as the optimality criterion the principle of maximum likelihood. Other optimal decision rules such as Neyman-Pearson or Bayes rules could be considered similarly.

Note that we assume in this section that noise in Gaussian. This is done because under such an assumption one can calculate the likelihood ratio analytically.

Let us first develop the basic tools for solving the discrimination problem. Let

$$R\phi_j := \int_D R(x, y) \phi_j(y) dy = \lambda_j \phi_j \quad \text{in} \quad D. \tag{7.13}$$

Here

$$R(x,y) := \overline{n^*(x)n(y)}, \tag{7.14}$$

$$\lambda_1 \geq \lambda_2 \geq \cdots > 0, \tag{7.15}$$

λ_j are the eigenvalues of the operator R counted according their multiplicities, and the ϕ_j are the corresponding normalized in $L^2(D)$ eigenfunctions. Let us define random variables n_j by the formula:

$$n_j = \lambda_j^{-1/2} \int_D n(x)\phi_j^*(x)dx. \tag{7.16}$$

From (7.12) it follows that

$$\overline{n}_j = 0. \tag{7.17}$$

Moreover, since $R(x,y) = R^*(y,x)$, one has

$$\begin{aligned} \overline{n_i n_j^*} &= (\lambda_i\lambda_j)^{-1/2} \int_D \int_D R(y,x)\phi_j(x)\phi_i^*(y)dxdy \\ &= (\lambda_i\lambda_j)^{-1/2} \int_D dx\phi_j(x)\lambda_i\phi_i^*(x) = \delta_{ij} = \begin{cases} 1 & i=j \\ 0 & i \neq j. \end{cases} \end{aligned} \tag{7.18}$$

The random variables n_j are called noncorrelated coordinates of the random field $n(x)$. One has

$$n(x) = \sum_{j=1}^{\infty} \lambda_j^{1/2} n_j \phi_j(x). \tag{7.19}$$

The series in (7.19) converges in the mean. The random variables n_j are Gaussian since the random field $n(x)$ is Gaussian. Define

$$s_{pj} := \lambda_j^{-1/2} \int_D s_p(x)\phi_j^*(x)dx, \quad p = 0, 1. \tag{7.20}$$

Then

$$s_p(x) + n(x) = \sum_{j=1}^{\infty} \lambda_j^{1/2} [s_{pj} + n_j]\phi_j(x). \tag{7.21}$$

Let

$$\mathcal{U}_{pj} := s_{pj} + n_j. \tag{7.22}$$

Then

$$\overline{\mathcal{U}_{pj}} = s_{pj}, \quad \overline{|\mathcal{U}_{pj} - s_{pj}|^2} = 1, \tag{7.23}$$

and $\mathcal{U}_{pj}$ are Gaussian.

Let $\mathbf{H}_p$ denote the hypothesis that the observed random field is $s_p(x) + n(x)$, $p = 0, 1$, and $f(u_1, \ldots, u_n \mid \mathbf{H}_p)$ is the probability density for the random variables $\mathcal{U}_{pj}$ under the assumption that the hypothesis $\mathbf{H}_p$ occured:

$$f(u_1, \ldots, u_n \mid \mathbf{H}_p) = (2\pi)^{-n} \exp\left\{ -\frac{1}{2} \sum_{j=1}^{n} |u_j - s_{pj}|^2 \right\}. \tag{7.24}$$

Here we used equations (7.23). Since $\mathcal{U}_{pj}$ are complex valued we took $(2\pi)^{-n}$ rather than $(2\pi)^{-n/2}$ as the normalizing constant in (7.24).

The likelihood ratio is defined as

$$\ell(u_1, \ldots, u_n) = \frac{f(u_1, \ldots, u_n \mid \mathbf{H}_1)}{f(u_1, \ldots, u_n \mid \mathbf{H}_0)}. \tag{7.25}$$

Therefore

$$\begin{aligned} \ln \ell(u_1, \ldots, u_n) &= -\frac{1}{2} \sum_{j=1}^{n} \left[|u_j - s_{1j}|^2 - |u_j - s_{0j}|^2 \right] \\ &= \frac{1}{2} \sum_{j=1}^{n} \left(|s_{0j}|^2 - |s_{1j}|^2 \right) + Re \sum_{j=1}^{n} u_j (s_{1j}^* - s_{0j}^*). \end{aligned} \tag{7.26}$$

We wish to compute the limit of the function (7.26) as $n \to \infty$. If this is done one can formulate the decision rule based on the maximum likelihood principle.

Note first that the system of eigenfunctions $\{\phi_j\}$ of the operator R is complete in $L^2(D)$ since we have assumed that the selfadjoint operator $R : L^2(D) \to L^2(D)$ is positive, that is $(R\phi, \phi) = 0$ implies $\phi = 0$ (see (7.15)). Indeed since

$$L^2(D) = c\ell\{\text{Ran} R\} \oplus N(R) \tag{7.27}$$

where $c\ell\{\text{Ran} R\}$ is the closure of the range of R, and $N(R)$ is the null space of R, and since $N(R) = \{0\}$ by assumption (7.15), one concludes that the closure of the range of R is the whole space $L^2(D)$. Thus the closure of the linear span of the eigenfunctions $\{\phi_j\}$ is $L^2(D)$ as claimed.

Let us assume that

$$R(x,y) \in \mathcal{R}. \tag{7.28}$$

Define the function $V(x)$ as the solution of minimal order of singularity of the equation

$$RV := \int_D R(x,y)V(y)dy = s_1(x) - s_0(x), x \in \overline{D}. \tag{7.29}$$

Thus

$$V(x) = R^{-1}(s_1 - s_0). \tag{7.30}$$

Using Parseval's equality one gets

$$\sum_{j=1}^{\infty} \lambda_j^{-1} c_j b_j^* = \int_D c(y) \left\{R^{-1}b(y)\right\}^* dy \tag{7.31}$$

where $c(y)$ and $b(y)$ are some functions for which the integral (7.31) converges,

$$c_j = \int_D c(y)\phi_j^* dy, \quad b_j = \int_D b(y)\phi_j^*(y)dy, \tag{7.32}$$

$$\lambda_j^{-1} b_j = \int_D R^{-1}b(y)\phi_j^* dy. \tag{7.33}$$

Therefore using formulas (7.16), (7.19), (7.20), (7.22), (7.30) and (7.31) one obtains

$$\sum_{j=1}^{\infty} \mathcal{U}_j (s_{1j}^* - s_{0j}^*) = \int_D \mathcal{U}(y)V^*(y)dy. \tag{7.34}$$

Let us assume that

$$\int_D s_0(R^{-1}s_0)^* dx < \infty, \quad \int_D s_1(R^{-1}s_1)^* dx < \infty. \tag{7.35}$$

Then, using selfadjointness of R^{-1}, one gets

$$\begin{aligned}\sum_{j=1}^{\infty}\left(|s_{0j}|^2-|s_{1j}|^2\right) &= \int_D\left[s_0(R^{-1}s_0)^*-s_1(R^{-1}s_1)^*\right]dx\\ &= -\int_D s_0V^*dx+\int_D s_0(R^{-1}s_1)^*\\ &\quad-\int_D R^{-1}(s_1-s_0)s_1^*dx-\int_D R^{-1}s_0s_1^*dx\\ &= -\int_D s_0V^*dx-\int_D s_1^*V dx. \qquad (7.36)\end{aligned}$$

Combining (7.36), (7.34) and (7.26) one obtains

Lemma 7.1 *There exists the limit*

$$\begin{aligned}\ln\ell\,(\mathcal{U}(x)) &= \lim_{n\to\infty}\ln\ell(u_1,\ldots,u_n)\\ &= Re\int_D\mathcal{U}(x)V^*(x)dx-\frac{1}{2}\int_D s_0V^*dx-\frac{1}{2}\int_D s_1^*(x)Vdx.\end{aligned} \qquad (7.37)$$

If the signals s_p, $p=0,1$, and the kernel $R(x,y)$ are real valued then (7.36) reduces to

$$\ln\ell\,(\mathcal{U}(x))=\int_D\left[\mathcal{U}(x)-\frac{s_0(x)+s_1(x)}{2}\right]V(x)dx. \qquad (7.38)$$

Suppose that the quantity on the right hand side of equation (7.37) (or (7.38)) has been calculated, so that the quantity $\ln\ell\,(\mathcal{U}(x))$ is known. Then we use

The maximum likelihood criterion: if $\ln\ell\,(\mathcal{U}(x))\geq 0$ then the decision is that hypothesis $\mathbf{H}_1$ occured, otherwise hypothesis $\mathbf{H}_0$ occured.

Therefore if

$$Re\int_D\mathcal{U}(x)V^*(x)dx\geq\frac{1}{2}\int_D s_0(x)V^*(x)dx+\frac{1}{2}\int_D s_1^*(x)V(x)dx \qquad (7.39)$$

then $\mathbf{H}_1$ occured. Here V is given by formula (7.30). If the opposite inequality holds in (7.39) then $\mathbf{H}_0$ occured.

If $s_p(x)$, $p=0,1$, $\mathcal{U}(x)$ and $R(x,y)$ are real valued then the inequality (7.39) reduces to

$$\int_D\mathcal{U}(x)V(x)dx\geq\frac{1}{2}\int_D\left[s_0(x)+s_1(x)\right]V(x)dx. \qquad (7.40)$$

The decision rule is: if (7.40) holds then $\mathbf{H}_1$ occured, otherwise $\mathbf{H}_0$ occured.

If one uses some other threshold criterion (such as Bayes, Neyman-Pearson, etc.) then one formulates the decision rule based on the inequality

$$Re \int_D \mathcal{U}(x)V^*(x)dx \geq \ln \kappa + \frac{1}{2}\int_D \left[s_0(x)V^*(x) + s_1^*(x)V(x)\right] dx, \quad (7.41)$$

where $\kappa > 0$ is a constant which is determined by the threshold. (See Section 8.4 for more details.)

The decision rule: *Practically, the decision rule based on the inequality (7.39) (or (7.41)) can be formulated as follows:*

1) *Given $s_p(x)$, $p = 0, 1$, solve equation (7.29) for $V(x)$ by formulas given in Theorem 2.1.*
2) *If $V(x)$ is found and $\mathcal{U}(x)$ is measured, then compute the integrals in formula (7.39) and check if the inequality (7.39) holds.*
3) *If yes, then the decision is that the observed signal is*

$$\mathcal{U} = s_1(x) + n(x). \quad (7.42)$$

Otherwise

$$\mathcal{U} = s_0(x) + n(x). \quad (7.43)$$

Example 7.2 Consider the problem of detection of signals against the background of white Gaussian noise. In this case $s_0(x) = 0$, $R(x, y) = \sigma^2\delta(x - y)$, we assume that the variance of the noise is σ^2. The solution to equation (7.29) is therefore

$$V = \sigma^{-2}s_1(x). \quad (7.44)$$

The inequality (7.39) reduces to

$$Re \int_D \mathcal{U}(x)s_1(x)dx \geq \frac{1}{2}\int_D |s_1^*(x)|^2 dx. \quad (7.45)$$

If (7.45) holds then the decision is that the observed signal $\mathcal{U}(x)$ is of the form

$$\mathcal{U}(x) = s_1(x) + n(x).$$

Otherwise one decides that

$$\mathcal{U}(x) = n(x).$$

The problem of discrimination between two signals $s_1(x)$ and $s_0(x)$ against the white Gaussian noise background is solved similarly. If

$$Re \int_D \mathcal{U}(x)(s_1 - s_0)^* dx \geq \frac{1}{2} \int_D \left(|s_1|^2 - |s_0|^2\right) dx \tag{7.46}$$

then the decision is that equation (7.42) holds. Otherwise one decides that (7.43) holds.

If all the signals are real-valued then inequality (36) can be written as

$$\int_D (\mathcal{U} - s_0)^2 dx \geq \int_D (\mathcal{U} - s_1)^2 dx. \tag{7.47}$$

The decision rule now has a geometrical meaning: if the observed signal $\mathcal{U}$ is closer to S_1 in $L^2(D)$ metric then the decision is that (7.42) holds. Otherwise one decides that (7.43) holds.

We have chosen a very simple case in order to demonstrate the decision rule for the problem for which all the calculations can be carried through in an elementary way. But the technique is the same for the general kernels $R \in \mathcal{R}$.

Example 7.3 Consider the problem of detection of a signal with unknown amplitude. Assume that the observed signal is either of the form

$$\mathcal{U}(x) = \gamma s(x) + n(x) \tag{7.48}$$

or

$$\mathcal{U}(x) = n(x). \tag{7.49}$$

Parameter γ is unknown, function $s(x)$ is known, $n(x)$ is a Gaussian noise with covariance function $R(x, y) \in \mathcal{R}$. Given the observed signal $\mathcal{U}(x)$ one wants to decide if the hypothesis $\mathbf{H}_1$ that (7.48) holds is true, or the hypothesis $\mathbf{H}_0$ that (7.49) holds is true. Moreover, one wants to estimate the value of γ. In formula (7.37) take

$$s_1 = \gamma s(x), \quad s_0 = 0. \tag{7.50}$$

Then, using the equation $Re\mathcal{U}^* V = Re\mathcal{U} V^*$, write

$$\ln \ell\left(\mathcal{U}(x)\right) = Re \int_D \mathcal{U}^*(x) V dx - \frac{\gamma^*}{2} \int_D s^* V dx \tag{7.51}$$

where $V(x)$ solves the equation

$$\int_D R(x, y) V(y) dy = \gamma s(x). \tag{7.52}$$

One should find the estimate of γ by the maximum likelihood principle from the equations

$$\frac{\partial \ln \ell}{\partial \gamma} = 0, \quad \frac{\partial \ln \ell}{\partial \gamma^*} = 0. \tag{7.53}$$

If, again for simplicity, one assumes that the noise is white with variance $\sigma^2 = 1$, so that $R(x, y) = \delta(x - y)$, then the solution to (7.52) is

$$V(x) = \gamma s(x), \tag{7.54}$$

and formula (7.51) reduces to

$$\ln \ell = Re\gamma \int_D \mathcal{U}^* s dx - \frac{|\gamma|^2}{2} \int_D |s|^2 dx. \tag{7.55}$$

Therefore equations (7.53) yield

$$\int_D \mathcal{U}^* s dx = \gamma^* \int_D |s|^2 dx, \quad \int_D \mathcal{U} s^* dx = \gamma \int_D |s|^2 dx, \tag{7.56}$$

so that the estimate $\hat{\gamma}$ of γ is

$$\hat{\gamma} = \int_D \mathcal{U} s^* dx \left(\int_D |s|^2 dx \right)^{-1}. \tag{7.57}$$

Exercise. Check that the estimate $\hat{\gamma}$ is unbiased, that is

$$\overline{\hat{\gamma}} = \gamma. \tag{7.58}$$

Hint: Use the equation $\overline{\mathcal{U}} = \gamma S(x)$.

Exercise. Calculate the variance:

$$\overline{(\hat{\gamma} - \overline{\hat{\gamma}})^2} = \frac{\sigma^2}{E}, \quad E := \int_D |s|^2 dx. \tag{7.59}$$

Estimate (7.59) shows that the variance of the estimate of γ decreases as the energy E of the signal $S(x)$ grows, which is intuitively obvious.

Assume that hypothesis $\mathbf{H}_0$ occured. Then the quantity $\hat{\gamma}$ defined by formula (7.57) is Gaussian with zero mean value and its variance equals $\frac{\sigma^2}{E}$ by formula (7.59). Therefore

$$\text{Prob}(|\hat{\gamma}| > b) = 2\text{erf}(bE^{1/2}/\sigma) \tag{7.60}$$

where

$$\text{erf}(x) := (2\pi)^{-1/2} \int_x^\infty \exp(-t^2/2)dt. \tag{7.61}$$

If one takes confidence level $\epsilon = 0,95$ and decides that hypothesis $\mathbf{H}_0$ occured if

$$2\text{erf}(\hat{\gamma} E^{1/2}/\sigma) > \epsilon, \tag{7.62}$$

then the decision rule for detection of a known signal with an unknown amplitude against the Gaussian white noise background is as follows:

1) given the observed signal $\mathcal{U}(x)$ calculate $\hat{\gamma}$ by formula (7.57),
2) calculate the left hand side in the formula (7.62); if the inequality (7.62) holds then the decision is that (7.49) holds; otherwise the decision is that (7.48) holds.

7.3 Quasioptimal estimates of derivatives of random functions

7.3.1 *Introduction*

Suppose that the observed signal in a domain $D \subset R^r$ is

$$\mathcal{U}(x) = s(x) + n(x), \quad x \in D \subset R^r, \tag{7.63}$$

where $s(x)$ is a useful signal and $n(x)$ is noise, $\overline{s} = \overline{n} = 0$. If one wishes to estimate $\partial^j s(x_0)$ optimally by the criterion of minimum of variance then one has a particular case of the problem studied in Chapter 2 with $As = \partial^j s$ (see formula (1.5) of Chapter 1). This estimation problem can be solved by the theory developed in Chapter 2. However, the basic integral equation for the optimal filter may be difficult to solve, the optimal filter may be difficult to implement, and the calculation of the optimal filter depends on the analytical details of the behavior of the spectral density of the covariance kernel $R(x,y) = \overline{u^*(x)u(y)}$, $R(x,y) \in \mathcal{R}$. That is, if one changes $\tilde{R}(\lambda)$ locally a little it ceases to be a rational function, for example.

One can avoid the above difficulties by constructing a quasioptimal estimate of the derivative which is easy to calculate under a general assumption about the spectral density, which is stable towards small local perturbations of the spectral density and depends basically on the asymptotic behavior of this density as $|\lambda| \to \infty$, and which is easy to implement practically.

The notion of quasioptimality will be specified later and it will be shown that the quasioptimal estimate is nearly as good as the optimal one. The basic ideas are taken from [Ramm (1968); Ramm (1972); Ramm (1981); Ramm (1984); Ramm (1985b)].

7.3.2 *Estimates of the derivatives*

Consider first the one-dimensional case: $\mathcal{U}(t) = s(t) + n(t)$. Assume that

$$|n(t)| \leq \delta, \tag{7.64}$$

$$|s''(t)| \leq M. \tag{7.65}$$

Let us assume for simplicity that $n(t)$ and $s(t)$ are defined on all of R^1, that $s(t)$ is an unknown deterministic function which satisfies (7.65), and the noise $n(t)$ is an arbitrary random function which satisfies (7.64).

Let A denote the set of all operators $T : C(R^1) \to C(R^1)$, linear and nonlinear, where $C(R^1)$ is the Banach space of continuous functions on R^1 with the norm $\| f \| = \max_{t\in R^1} |f(t)|$. Let

$$\Delta_h \mathcal{U} := (2h)^{-1}[\mathcal{U}(t+h) - \mathcal{U}(t-h)], \tag{7.66}$$

$$h(\delta) := (2\delta/M)^{1/2}, \quad \epsilon(\delta) := (2M\delta)^{1/2}. \tag{7.67}$$

First, let us consider the following problem: given $\mathcal{U}(t)$ and the numbers $\delta > 0$ and $M > 0$ such that (7.64) and (7.65) hold, find an estimate $\hat{\mathcal{U}}$ of $s'(t)$ such that

$$\| \hat{\mathcal{U}} - s'(t) \| \to 0 \quad \text{as} \quad \delta \to 0 \tag{7.68}$$

and such that this estimate is the best possible in the sense

$$\| \hat{\mathcal{U}} - s'(t) \| = \inf_{T\in A} \sup_{\substack{|s''|\leq M \\ |n|\leq\delta}} \| T\mathcal{U} - s' \| . \tag{7.69}$$

This means that among all estimates $T\mathcal{U}$ the estimate $\hat{\mathcal{U}}$ is the best one for the class of the data given by inequalities (7.64), (7.65). It turns out that this optimal estimate is the estimate (7.70) in the following theorem.

Theorem 7.1 *The estimate*

$$\hat{\mathcal{U}} := \Delta_{h(\delta)}\mathcal{U} \tag{7.70}$$

has the properties

$$\| \hat{\mathcal{U}} - s' \| \leq \epsilon(\delta) \tag{7.71}$$

and

$$\inf_{T \in A} \sup_{\substack{|s''| \leq M \\ |n| \leq \delta}} \| \hat{\mathcal{U}} - s' \| = \epsilon(\delta), \tag{7.72}$$

where $\epsilon(\delta)$ *and* $h(\delta)$ *are defined by (7.67) and* $\Delta_{h(\delta)}\mathcal{U}$ *is defined by (7.66).*

Proof. One has

$$|\Delta_h \mathcal{U} - s'| \leq |\Delta_h(\mathcal{U} - s)| + |\Delta_h s - s'| \leq \frac{\delta}{h} + \frac{Mh}{2}. \tag{7.73}$$

Indeed

$$|\Delta_h(\mathcal{U} - s)| \leq \frac{|n(t+h)| + |n(t-h)|}{2h} \leq \frac{\delta}{h},$$

and

$$\begin{aligned} &\left| \frac{s(t+h) - s(t-h)}{2h} - s' \right| \\ &= \left| \frac{s(t) + s'(t)h + s''(\xi_+)\frac{h^2}{2} - s(t) + s'(t)h - s''(\xi_-)\frac{h^2}{2}}{2h} \right| \leq \frac{Mh^2}{2h} \\ &= \frac{Mh}{2}, \end{aligned} \tag{7.74}$$

where $\xi_\pm$ are the points in the remainder in the Taylor formula and estimates (7.64), (7.65) were used.

For fixed $\delta > 0$ and $M > 0$, minimize the right side of (7.73) in $h > 0$ to get

$$\min_{h>0} \left(\frac{\delta}{h} + \frac{Mh}{2} \right) = \frac{\delta}{h(\delta)} + \frac{Mh(\delta)}{2} = \epsilon(\delta), \tag{7.75}$$

where $h(\delta)$ are $\epsilon(\delta)$ are defined in (7.67). This proves inequality (7.71). To prove (7.72), take

$$s_1 = -\frac{M}{2} t[t - 2h(\delta)] \quad 0 \leq t \leq 2h(\delta), \tag{7.76}$$

and extend it on R^1 so that

$$|s_1''(t)| \leq M, \quad |s_1(t)| \leq \delta, \quad \forall t \in R^1. \tag{7.77}$$

Here $h(\delta)$ is given by (7.67). The extension of s_1 with properties (7.77) is possible since on the interval $[0, 2h(\delta)]$ conditions (7.77) hold. Let

$$s_2(t) = -s_1(t). \tag{7.78}$$

One has

$$|s_p''| \le M, \quad |s_p| \le \delta, \quad p = 1, 2. \tag{7.79}$$

Take $\mathcal{U}(t) = 0$, $t \in R^1$. Then

$$|\mathcal{U}(t) - s_p(t)| \le \delta, \quad p = 1, 2. \tag{7.80}$$

Therefore one can consider $\mathcal{U}(t)$ as the observed value of both $s_1(t)$ and $s_2(t)$. Let $T \in A$ be an arbitrary operator on $C(R^1)$. Denote

$$T\mathcal{U}(t)\,\Big|_{t=0} = a. \tag{7.81}$$

One has

$$\begin{aligned}\sup_{\substack{|s''|\le M\\ |n|\le\delta}} \| T\mathcal{U} - s' \| &\ge \sup_{\substack{|s''|\le M\\ |n|\le\delta}} |T\mathcal{U}(0) - s'(0)| \ge \max\{|a - s_1'(0)|, |a - s_2'(0)|\} \\ &\ge \frac{1}{2}|s_1'(0) - s_2'(0)| = \epsilon(\delta),\end{aligned} \tag{7.82}$$

where $\epsilon(\delta)$ is given by (7.67). Taking infimum in $T \in A$ of both sides of (7.82) one obtains

$$\inf_{T\in A} \sup_{\substack{|s''|\le M\\ |n|\le\delta}} \| T\mathcal{U} - s' \| \ge \epsilon(\delta). \tag{7.83}$$

From (7.83) and (7.71) the desired inequality (7.72) follows. Theorem 7.1 is proved. □

7.3.3 *Derivatives of random functions*

Assume now that $s(t)$ is a random function, that $s(t)$ and $n(t)$ are uncorrelated, $\overline{n} = 0$, and $n = \sigma v$, where the variance of v, denoted by $\mathcal{D}[v]$, is 1, so that

$$\mathcal{D}[v] = 1, \mathcal{D}[n] = \sigma^2. \tag{7.84}$$

The problem is to find a linear estimate Lu such that

$$\mathcal{D}[L\mathcal{U} - s'] = \min \tag{7.85}$$

given the observed signal $\mathcal{U}(t) = s(t) + n(t)$. As was explained in section 7.3.1, we wish to find a quasioptimal linear estimate of s' such that this estimate is easy to compute, easy to implement, and is nearly as good as the optimal estimate.

Let us assume that

$$\mathcal{D}[s^{(m)}(t)] \leq M_m^2, \tag{7.86}$$

where $s^{(m)}(t)$ is the m-th derivative of $s(t)$. Let us seek the quasioptimal estimate among the estimates of the form

$$\Delta_h^{(Q)} s := h^{-1} \sum_{k=-Q}^{Q} A_K^{(Q)} s\left(t + \frac{kh}{Q}\right). \tag{7.87}$$

If $m = 2q$ or $m = 2q+1$ let us take $Q = q$. If one expands the expression on the right hand side of (7.87) in powers of h and requires that the order of the smallness as $h \to 0$ of the function $\Delta_h^{(Q)} s - s'$ be maximal, one obtains the following system for the coefficients $A_k^{(Q)}$:

$$\sum_{k=-Q}^{Q} \left(\frac{k}{Q}\right)^j A_k^{(Q)} = \delta_{1j}, \quad 0 \leq j \leq 2Q, \tag{7.88}$$

where

$$\delta_{1j} = \begin{cases} 0, & j \neq 1 \\ 1, & j = 1. \end{cases}$$

The system (7.88) is uniquely solvable since its determinant does not vanish: it is a Vandermonde determinant. One can find by solving system (7.88) that

$$A_0^{(1)} = 0, \quad A_{\pm 1}^{(1)} = \pm\frac{1}{2} \tag{7.89}$$

$$A_0^{(2)} = 0, \quad A_{\pm 1}^{(2)} = \pm\frac{4}{3}, \quad A_{\pm 2}^{(2)} = \mp\frac{1}{6} \tag{7.90}$$

$$A_0^{(3)} = 0, \quad A_{\pm 1}^{(3)} = \pm\frac{9}{4}, \quad A_{\pm 2}^{(3)} = \mp\frac{9}{20}, \quad A_{\pm 3}^{(3)} = \pm\frac{1}{20} \tag{7.91}$$

$$A_0^{(4)} = 0, \quad A_{\pm 1}^{(4)} = \pm\frac{16}{5}, \quad A_{\pm 2}^{(4)} = \mp\frac{4}{5}, \quad A_{\pm 3}^{(4)} = \pm\frac{16}{105}$$

$$A_{\pm 4}^{(4)} = \mp \frac{1}{70}. \tag{7.92}$$

We will need

Lemma 7.2 *Let $m = 2q+1$. Assume that the coefficients $A_k^{(q)}$ in (7.87) satisfy (7.88) with $0 \le j \le 2q$, and let*

$$c_m := \frac{m}{(m!)^2 q^{2m}} \sum_{k=-q}^{q} \left| A_k^{(q)} \right|^2 k^{2m}, \quad m = 2q+1. \tag{7.93}$$

Then

$$\mathcal{D}\left[\Delta_h^{(q)} s - s'\right] \le \gamma_m h^{2m-2}, \quad \gamma_m := c_m M_m^2, \tag{7.94}$$

where $\mathcal{D}$ is the symbol of variance.

In order to prove this lemma, one needs a simple

Lemma 7.3 *Let g_j be random variables and a_j be constants. If*

$$\mathcal{D}[g_j] \le M, \quad 1 \le j \le n \tag{7.95}$$

then

$$\mathcal{D}\left[\sum_{j=1}^{n} a_j g_j\right] \le nM \sum_{j=1}^{n} |a_j|^2. \tag{7.96}$$

Proof of Lemma 7.3 Note that

$$\left(\sum_{k=1}^{n} |b_k|\right)^2 \le n \sum_{k=1}^{n} |b_k|^2 \tag{7.97}$$

by Cauchy's inequality.

Let $f(x_1, \dots, x_n)$ be the probability density of the joint distribution of the random variables $g_1, g_2, \dots, g_n$. Let us assume without loss of generality that $\overline{g}_k = 0$. Denote $dx = dx_1 \dots dx_n$, $\int_{R^n} = \int$, $x = (x_1, \dots, x_n)$. Then

$$\begin{aligned}\mathcal{D}\left[\sum_{j=1}^{n} a_j g_j\right] &= \int \left|\sum_{j=1}^{n} a_j x_j\right|^2 f(x)dx \le \int \sum_{j=1}^{n} |a_j|^2 \sum_{j=1}^{n} |x_j|^2 f dx \\ &= \sum_{j=1}^{n} |a_j|^2 \sum_{j=1}^{n} \int x_j^2 f(x)dx = \sum_{j=1}^{n} |a_j|^2 \sum_{j=1}^{n} \mathcal{D}[g_j] \\ &\le nM \sum_{j=1}^{n} |a_j|^2. \end{aligned} \tag{7.98}$$

Lemma 7.3 is proved. □

Proof of Lemma 7.2 One has

$$\Delta_h^{(q)} s - s' = h^{-1} \sum_{k=-q}^{q} A_k^{(q)} \frac{h^m k^m}{m! q^m} s^{(m)}(t_k), \tag{7.99}$$

where t_K are the points in the remainder of Taylor's formula. Apply Lemma 7.3 to equation (7.99) and take into account the assumption (7.86) to get

$$\mathcal{D}\left[\Delta_h^q s - s'\right] \le \frac{h^{2m-2} M_m^2}{(m!)^2 q^{2m}} (2q+1) \sum_{k=-q}^{q} |A_k^{(q)}|^2 k^{2m} \tag{7.100}$$

which is equivalent to (7.34).

Lemma 7.2 is proved. □

Lemma 7.4 *One has*

$$\mathcal{D}\left[\Delta_h^{(q)} \mathcal{U} - s'\right] \le \phi(h), \tag{7.101}$$

where

$$\phi(h) := \gamma_m h^{2m-2} + \sigma^2 h^{-2} \sum_{k,j=-q}^{q} A_k^{(q)} A_j^{(q)} R\left(\frac{k-j}{q} h\right). \tag{7.102}$$

Here $m = 2q+1$,

$$R(t-\tau) := \overline{v^*(t) v(\tau)} \tag{7.103}$$

is the covariance function of $v(t)$*, and conditions (7.84) hold.*

Proof. By assumption s and v are uncorrelated. Therefore

$$\mathcal{D}\left[\Delta_h^{(q)}\mathcal{U} - s'\right] = \mathcal{D}\left[\Delta_h^{(q)} s - s'\right] + \sigma^2 \mathcal{D}\left[\Delta_h^{(q)} v\right] \tag{7.104}$$

$$\leq \gamma_m h^{2m-2} + \sigma^2 h^{-2} \sum_{k,j=-q}^{q} A_k^{(q)} A_j^{(q)} R\left(\frac{k-j}{q} h\right), \tag{7.105}$$

where we took into account that the coefficients $A_j^{(q)}$ are real numbers. Lemma 7.4 is proved. □

Definition 7.1 The estimate

$$Lu := \Delta_h^{(q)} \mathcal{U} \tag{7.106}$$

is called quasioptimal if h minimizes the function $\phi(h)$ defined by (7.102).

Thus, the quasioptimal estimate minimizes a natural majorant $\phi(h)$ of the variance of the estimate $\Delta_h^{(q)}\mathcal{U}$ among all estimates (7.106) with different h. The majorant $\phi(h)$ is natural because the equality sign can be attained in (7.101) (for example, if $s = 0$ then the equality sign is attained in (7.101)).

The quasioptimal filter is easy to calculate: it is sufficient to find minimizer of $\phi(h)$, $h > 0$. This filter is easy to implement: one needs only some multiplications, additions and time shift elements. We will compare the error estimates for optimal and quasioptimal filter shortly, but first consider an example.

Example 7.4 Let $m = 2$, $q = 1$,

$$\Delta_h^{(1)}\mathcal{U} = \frac{\mathcal{U}(t+h) - \mathcal{U}(t-h)}{2h}. \tag{7.107}$$

By formulas (7.93) and (7.89) for $m = 2$ and $q = 1$ one calculates

$$c_2 = \frac{1}{4}. \tag{7.108}$$

Let us assume that the constant $M_2^2 := M$ in the estimate (7.86) is known, the variance σ^2 of noise is known (see (7.84)) and the covariance function of $v(t)$ is

$$R(t) = \exp(-|t|). \tag{7.109}$$

Then formula (7.102) yields

$$\phi(h) = \frac{M}{4}h^2 + \frac{\sigma^2 h^{-2}}{2}\left[1 - \exp(-2h)\right] = \frac{Mh^2}{4} + \frac{\sigma^2 h^{-2}}{2}\left[R(0) - R(2h)\right]. \tag{7.110}$$

If $\sigma \ll 1$ then the minimizer of the function (7.110) should be small. Assuming $h \ll 1$ and using $1 - \exp(-h) \approx h$ for $h \ll 1$, one obtains

$$\phi(h) \approx \frac{Mh^2}{4} + \frac{\sigma^2}{h}, \quad h > 0. \tag{7.111}$$

It is easy to check that the function (7.111) attains its minimum at

$$h_{\min} = \left(\frac{2\sigma^2}{M}\right)^{1/3} \tag{7.112}$$

and

$$\min \phi \approx \sigma^{4/3} M^{1/3} \left(2^{2/3} + 2^{-1/3}\right) = 2.381\sigma^{4/3} M^{1/3}. \tag{7.113}$$

Note that if the minimizer is small, so that $h_{\min} \ll 1$, then the behavior of the covariance function is important only in a neighborhood of $t = 0$. But this behavior is determined only by the asymptotic behavior of the spectral density $\tilde{R}(\lambda)$ as $|\lambda| \to \infty$.

Let us now compare briefly optimal and quasioptimal estimates.

Let us define the spectral density $\tilde{R}_s(\lambda)$ of $s(t)$:

$$\tilde{R}_s(\lambda) := \int_{-\infty}^{\infty} \exp(-i\lambda t) R_s(t) dt, \tag{7.114}$$

where

$$R_s(t - \tau) := \overline{s^*(t)s(\tau)}. \tag{7.115}$$

We assume that

$$0 < \tilde{R}_s(\lambda) \le \frac{A}{(1+\lambda^2)^a}, \quad a \ge \frac{5}{2} \tag{7.116}$$

and that the spectral density of noise

$$0 < \tilde{R}(\lambda) \le \frac{B}{(1+\lambda^2)^b}, \quad b > 1. \tag{7.117}$$

One has for the quasioptimal estimate (7.107) formula (7.110), and

$$M = \mathcal{D}[s''(t)] = R_{s''}(0) := \overline{[s''(t)]^* s''(\tau)} \,\Big|_{t=\tau} = \left\{ \frac{\partial^4}{\partial t^2 \partial \tau^2} R_s(t-\tau) \right\} \Big|_{t=\tau}$$
$$= \frac{1}{2\pi} \int_{-\infty}^{\infty} \lambda^4 \tilde{R}_s(\lambda) d\lambda \le \frac{A}{2\pi} \int_{-\infty}^{\infty} \lambda^4 \frac{d\lambda}{(1+\lambda^2)^a} \le \text{const} A. \tag{7.118}$$

The term $R(0) - R(2h)$ in (7.110) can be written as

$$R(0) - R(2h) = \frac{1}{2\pi} \int_{-\infty}^{\infty} \tilde{R}(\lambda)[1 - \exp(2i\lambda h)]d\lambda. \tag{7.119}$$

Therefore the function $\phi(h)$ in (7.110) can be expressed entirely in terms of spectral densities. One has

$$\begin{aligned} |R(0) - R(2h)| &\leq \frac{B}{2\pi} \int_{-\infty}^{\infty} \frac{d\lambda}{(1+\lambda^2)^b} |1 - \exp(2i\lambda h)| \\ &= \frac{B}{\pi} \int_{-\infty}^{\infty} \frac{d\lambda}{(1+\lambda^2)^b} |\sin \lambda h| \leq \frac{Bh}{\pi} \int_{-\infty}^{\infty} \left| \frac{\sin \lambda h}{\lambda h} \right| \frac{\lambda d\lambda}{(1+\lambda^2)^b} \\ &\leq \text{const} \times Bh. \end{aligned} \tag{7.120}$$

It follows from (7.110), (7.118) and (7.120) that

$$\phi(h) \leq \text{const} \left(Ah^2 + \frac{B\sigma^2}{h} \right), \tag{7.121}$$

where const does not depend on A, B, σ and h but depends on a and b. If $a \geq \frac{5}{2}$ and $b > 1$ one can take an absolute constant in (7.121). This shows that estimate (7.121) does not depend much on the details of the behavior of the spectral densities. One can see from (7.121) that

$$\phi_{\min} \leq \text{const}\sigma^{4/3} \left(\frac{A}{B} \right)^{1/3}, \quad h_{\min} = \text{const} \left(\frac{B\sigma^2}{A} \right)^{1/3}. \tag{7.122}$$

This estimate shows how the variance of noise and the ratio $\frac{A}{B}$ influence the behavior of the error as $\sigma \to 0$. Note that

$$1 = \mathcal{D}[v] = R(0) = \int_{-\infty}^{\infty} \tilde{R}(\lambda)d\lambda \leq B \int_{-\infty}^{\infty} \frac{d\lambda}{(1+\lambda^2)^b} = \text{const} \times B.$$

Therefore B is of order of magnitude of 1, and formula (7.122) can be written as

$$\phi_{\min} \leq \text{const}\sigma^{4/3} A^{1/3}. \tag{7.123}$$

This estimate holds if $h_{\min} \ll 1$, that is if

$$\sigma \ll A^{1/2}. \tag{7.124}$$

All these estimates are asymptotic as $\sigma \to 0$.

The optimal h satisfies the equation

$$\int_{t-T}^{t} \left[R_s(y-z)+\sigma^2 R(y-z)\right] h(x,z)dz = f(y,x), \quad t-T \le y \le t, \tag{7.125}$$

where

$$f(y,x) := \overline{s^*(y)s'(x)} = \frac{\partial}{\partial x} R_s(y-x) = -R_s'(y-x). \tag{7.126}$$

The error of the optimal estimate can be computed by formula (2.108):

$$\epsilon(x) = \overline{|s'(x)|^2} - (f, h_0) = -R_s''(0) - (f, h_0), \tag{7.127}$$

where h_0 is the solution to equation (7.125) of minimal order of singularity.

For simplicity let us assume that $t-T=-\infty$ and $t=\infty$. The error for this physically nonrealizable filter is not more than for the physically realizable filter. The reasons for considering the nonrealizable filter are: 1) this filter is easy to calculate, and 2) its error gives a lower bound for the error of realizable filter.

Let $x=0$ in (7.125). Since the random functions we consider are assumed to be stationary there is no loss in the assumption that $x=0$. Take the Fourier transform of (7.125) with $t-T=-\infty$, $t=+\infty$, and use the theorem about the Fourier transform of convolution to get

$$[\tilde{R}_s(\lambda) + \sigma^2 \tilde{R}(\lambda)]\tilde{h}_0(\lambda) = -i\lambda \tilde{R}_s. \tag{7.128}$$

Thus

$$\tilde{h}_0 = -i\lambda \tilde{R}_s [\tilde{R}_s + \sigma^2 \tilde{R}]^{-1}. \tag{7.129}$$

Use (7.129) and apply Parseval's equality to (7.127) to get

$$\begin{aligned}\epsilon(0) &= \frac{1}{2\pi}\int_{-\infty}^{\infty} \lambda^2 \tilde{R}_s(\lambda)d\lambda - \int_{-\infty}^{\infty} \lambda^2 \frac{\tilde{R}_s^2}{\tilde{R}_s+\sigma^2\tilde{R}} d\lambda \\ &= \frac{\sigma^2}{2\pi}\int_{-\infty}^{\infty} \lambda^2 \tilde{R}_s(\lambda)\tilde{R}(\lambda)[\tilde{R}_s(\lambda)+\sigma^2\tilde{R}(\lambda)]^{-1} d\lambda. \end{aligned} \tag{7.130}$$

It follows from (7.130), (7.116) and (7.117), if we assume that for large λ the sign $\le$ in (7.116) and (7.117) becomes asymptotic equality, that

$$\begin{aligned}\epsilon(0) &\le \text{const} \times \sigma^2 AB \int_{-\infty}^{\infty} \frac{\lambda^2 d\lambda}{(1+\lambda^2)^b A + \sigma^2 B (1+\lambda^2)^a} \\ &\le \text{const} \times \sigma^2 \quad \text{as} \quad \sigma \to 0. \end{aligned} \tag{7.131}$$

The estimate (7.113) gives $O(\sigma^{4/3})$ as $\sigma \to 0$ but if one takes larger m the estimate will be $0\left(\sigma^{\alpha(m)}\right)$, where $\alpha(m) \to 2$ as m grows. One can estimate $\alpha(m)$ using formula (7.102). For small h one has $\phi \leq \gamma_m h^{2m-2} + \text{const}\sigma^2 h^{-1}$, $h_{\min} \sim \sigma^{2/(2m-1)}$ and $\phi_{\min} \sim \sigma^{2\,(2m-2)/(2m-1)}$, so that $\alpha(m) = 2\frac{2m-2}{2m-1}$. Therefore $\alpha(m) \to 2$ as $m \to \infty$.

7.3.4 *Finding critical points*

Before we discuss the case $r > 1$ of random functions of several variables, let us outline briefly an application of the results of section 7.3.2 to the problem of finding the extremum of a random function. Assume that the observed function $\mathcal{U}(t) = s(t)+n(t)$, where $s(t)$ is a smooth function defined on the interval $[0, 1]$ which has exactly one maximum on this interval. Such functions $s(t)$ are called univalent. Suppose that this maximum is attained at a point τ. Assume that

$$|s''| \leq M, \quad |n(t)| \leq \delta. \tag{7.132}$$

The problem is to find τ given the signal $\mathcal{U}(t)$, $0 \leq t \leq 1$.

The solution to this problem is:

1) divide the interval $[0, 1]$ by the points $t_k = kh$, where $h = h(\delta)$ is given by (7.67), $k = 0, 1, 2, \dots$
2) calculate $\hat{\mathcal{U}}_k := (2h)^{-1}\left[\mathcal{U}(t_k + h) - \mathcal{U}(t_k - h)\right] = \Delta_h^{(1)}\mathcal{U}(t_k)$
3) compute $\hat{\mathcal{U}}_k\hat{\mathcal{U}}_{k+1}$, $k = 0, 1, 2, \dots$
4) if

$$|\hat{\mathcal{U}}_k| > \epsilon(\delta) \quad \forall k \tag{7.133}$$

and

$$\hat{\mathcal{U}}_j\hat{\mathcal{U}}_{j+1} < 0 \quad \text{for some } j \tag{7.134}$$

then

$$t_j < \tau < t_j + h. \tag{7.135}$$

Indeed, from (7.67) and (7.71) one concludes that

$$\hat{\mathcal{U}}(t_k) - \epsilon(\delta) \leq s'(t_k) \leq \hat{\mathcal{U}}(t_k) + \epsilon(\delta). \tag{7.136}$$

From (7.133), (7.134) and (7.136) it follows that $s'(t)$ changes sign on the interval (t_j, t_{j+1}). This implies (7.135) since $s(t)$ is univalent.

If (7.133) is not valid for some $k = k_0$, then the maximum may be on the interval $(t_k - h, t_k + h)$. In this case it may happen that for no j condition (7.134) holds. The above method may not work if the derivative of $s(t)$ is very small (smaller than $\epsilon(\delta)$) in a large neighborhood of τ.

Remark 7.1 *Since we used formula (7.71) we assumed that $\mathcal{U}(t)$ is defined on all of R^1. If it is defined only on a bounded interval $[a, b]$ then the expression $\Delta_h^{(1)}\mathcal{U}(t)$ is not defined for $t < a+h$. In this case one can define*

$$\begin{aligned}\tilde{\Delta}_h^{(1)}\mathcal{U}(t) &= h^{-1}\left[\mathcal{U}(t+h) - \mathcal{U}(t)\right], \quad a \le t < a+h \\ &= \Delta_h^{(1)}\mathcal{U}(t), \quad a+h \le t \le b-h \\ &= h^{-1}\left[\mathcal{U}(t) - \mathcal{U}(t-h)\right], \quad b-h \le t \le b. \end{aligned} \tag{7.137}$$

In this case

$$\left|\tilde{\Delta}_h^{(1)}\mathcal{U}(t) - s'(t)\right| \le \frac{2\delta}{h} + \frac{Mh}{2} \tag{7.138}$$

so that the minimizer $\tilde{h}_{min}$ of the right hand side of (7.138) is

$$\tilde{h}_{\min} = 2(\delta/M)^{1/2} \tag{7.139}$$

and the minimum of the right hand side of (7.138) is

$$\tilde{\epsilon}(\delta) = 2(M\delta)^{1/2}. \tag{7.140}$$

Note that $\tilde{\epsilon}(\delta) = 2^{1/2}\epsilon(\delta)$, where $\epsilon(\delta)$ is given by (7.67).

7.3.5 *Derivatives of random fields*

Let us consider the multidimensional case. There are no new ideas in this case but we give a brief outline of the results for convenience of the reader.

Suppose that $\mathcal{U}(x) = s(x) + n(x)$, $x \in R^r$. Let ∇s denote the gradient of $s(x)$ and $\| s \| = \max_{x\in R^r} |s(x)|$. Assume that

$$\| n(x) \| \le \delta, \tag{7.141}$$

and

$$\max_{x\in R^r} \left|(d^2 s(x)\theta, \theta)\right| \le M \quad \forall \theta \in S_1 := \{\theta : \theta \in R^r, \quad \theta\cdot\theta = 1\}, \tag{7.142}$$

where

$$(d^2 s(x)\theta, \theta) = \sum_{i,j=1}^{r} \frac{\partial^2 s(x)}{\partial x_i \partial x_j}\theta_i\theta_j.$$

Define

$$\Delta_h \mathcal{U}(x) := \frac{\mathcal{U}(x+h\theta) - \mathcal{U}(x-h\theta)}{2h}, \quad h > 0. \tag{7.143}$$

Theorem 7.2 *If $h(\delta)$ and $\epsilon(\delta)$ are given by (7.67) then*

$$|\Delta_{h(\delta)}\mathcal{U}(x) - \nabla s(x)\cdot\theta| \le \epsilon(\delta), \quad \forall\theta \in S_1. \tag{7.144}$$

Moreover

$$\inf_{T\in A}\sup_{s,n} \parallel T\mathcal{U}(x) - \nabla s(x)\cdot\theta \parallel = \epsilon(\delta) \tag{7.145}$$

and the infimum is attained at $T = \Delta_{h(\delta)}$. Here the supremum is taken over all $s(x) \in C^2(R^r)$, which satisfy (7.142), and all $n(x)$, which satisfy (7.141), and the infimum is taken over the set A of all operators $T : C(R^r) \to C(R^r)$ linear or nonlinear.

The proof of this theorem is similar to the proof of Theorem 7.1. The role of the function $s_1(t)$ in formula (7.76) is played by the function

$$s_1(x) = \frac{M}{2}\left(|x|^2 - 2h(\delta)x\cdot\theta\right) \quad \text{in} \quad B_\delta, \tag{7.146}$$

where B_δ is the ball, centered at the point $h(\delta)\theta$, with radius $h(\delta)$, $|x|^2 = \sum_{j=1}^r |x_j|^2$. It is clear that $s_1(x)$ vanishes at the boundary ∂B_δ of the ball, that

$$\left|(d^2 s_1(x)\theta,\theta)\right| \le M \quad \forall\theta\in S_1 \tag{7.147}$$

and that

$$|s_1(x)| \le \delta. \tag{7.148}$$

Let $s_2(x) = -s_1(x)$ and argue as in the proof of Theorem 7.1 in order to obtain (7.144) and (7.145).

7.4 Stable summation of orthogonal series and integrals with randomly perturbed coefficients

7.4.1 *Introduction*

Consider an orthogonal series

$$f(x) = \sum_{j=1}^{\infty} c_j\phi_j(x), \quad x \in D \subset R^r, \tag{7.149}$$

where

$$(\phi_j, \phi_m) := \int_D \phi_j(x)\phi_m^*(x)dx = \delta_{jm} \tag{7.150}$$

and

$$c_j := (f, \phi_j). \tag{7.151}$$

Suppose that the data

$$\{b_j := c_j + \epsilon_j\}, \quad 1 \le j < \infty \tag{7.152}$$

are given, that is the Fourier coefficients of f are known with some errors ϵ_j. Assume that

$$\overline{\epsilon}_j = 0, \quad \overline{\epsilon_j^* \epsilon_m} = \sigma^2 \delta_{jm}, \quad \sigma = \text{const} > 0. \tag{7.153}$$

The problem is: given the data (7.152), (7.153), estimate $f(x)$.

From the point of view of systems theory, one can interpret this problem as follows. Suppose that the system's response to the signal $\phi_j(x)$ is $K_j\phi_j(x)$, where K_j is a generalized transmission coefficient of the system. For example, if ω is a continuous analogue of j and $\phi_j(x) = \exp(i\omega x)$ then $K(i\omega)$ is the usual transmission coefficient of the linear system. If there is a noise at the output of the system then one actually receives $\sum_{j=1}^{\infty}(K_j c_j + \epsilon_j)\phi_j(x)$ at the output, where ϵ_j is the noise comlponent corresponding to the j-th generalized harmonic $\phi_j(x)$.

Let us consider two methods for solving the problem. These methods are easy to use in practice.

The first method is to define

$$f_N := \sum_{j=1}^{N} b_j \phi_j(x) \tag{7.154}$$

and to choose $N = N(\sigma)$ so that

$$\overline{\| f_{N(\sigma)}(x) - f(x) \|^2} = \min, \quad \| f \|^2 := \int_D |f|^2 dx. \tag{7.155}$$

The second method is to define

$$g(x) := \sum_{j=1}^{\infty} \rho_j(\nu) b_j \phi_j(x), \tag{7.156}$$

where $\nu > 0$ is a parameter, and to choose the multipliers $\rho_j(\nu)$ so that

$$\overline{\| g - f \|^2} = \min. \tag{7.157}$$

The same problem can be formulated for orthogonal integrals that is for continuous analogues of orthogonal series:

$$f(x) = \int_{-\infty}^{\infty} c(\lambda)\phi(x,\lambda)d\lambda \tag{7.158}$$

and

$$\int_{R^r} \phi(x,\lambda)\phi^*(x,\lambda')dx = \delta(\lambda - \lambda'). \tag{7.159}$$

We assume that $b(\lambda)$ are given, $b(\lambda) = c(\lambda) + \epsilon(\lambda)$, $\overline{\epsilon^*(\lambda)\epsilon(\lambda')} = \sigma^2\delta(\lambda - \lambda')$ and the problem is to estimate $f(x)$. This can be done in the same way as for the problem for series.

7.4.2 *Stable summation of series*

Let us consider the first method. Assume that

$$|\phi_j(x)| \le c, \quad |c_j| \le Aj^{-a}, \quad a > \frac{1}{2}, \tag{7.160}$$

where c and A are positive constants which do not depend on x and j. Then

$$\begin{aligned}\overline{\| f_N - f \|^2} &= \sum_{j,j'=1}^{N} \overline{\epsilon_j \epsilon_{j'}^*}(\phi_j, \phi_{j'}) + \sum_{j=N+1}^{\infty} |c_j|^2 \\ &\le \sigma^2 N + A^2 \sum_{j=N+1}^{\infty} j^{-2a} \\ &\le \sigma^2 N + A^2 \frac{N^{-2a+1}}{2a-1} := \gamma(N), \quad a > \frac{1}{2}.\end{aligned} \tag{7.161}$$

Here we used (7.150), (7.152), (7.153) and (7.160). Let us find N_m for which $\gamma(N) = \min$, σ and A assumed fixed. One has

$$N(\sigma) := N_m = \left(\frac{2a}{2a-1}\right)^{1/(2a)} \left(\frac{A}{\sigma}\right)^{1/a} \tag{7.162}$$

and

$$\gamma_m := \gamma(N_m) = \text{const} A^{1/a} \sigma^{(2a-1)/a}, \tag{7.163}$$

where const depends on a but not on A and σ. We have proved

Proposition 7.1 *If $N(\sigma)$ is given by (7.162) then*

$$\overline{\| f_{N(\sigma)}(x) - f(x) \|^2} \leq constA^{1/a}\sigma^{(2a-1)/a}. \tag{7.164}$$

Therefore formula (7.154) with $N = N(\sigma)$ gives an estimate of $f(x)$ such that the error of this estimate goes to zero according to (7.164) as $\sigma \to 0$.

7.4.3 *Method of multipliers*

Let us consider the second method. Take

$$\rho_j(\nu) := \exp(-\nu j). \tag{7.165}$$

These are multipliers of convergence used in Abel's summation of series. Then

$$\begin{aligned}\mathcal{J} &:= \left\| \sum_{j=1}^{\infty} \exp(-\nu j) b_j \phi_j(x) - \sum_{j=1}^{\infty} c_j \phi_j(x) \right\|^2 \\ &= \sum_{j=1}^{\infty} |1 - \exp(-j\nu)|^2 |c_j|^2 + \sigma^2 \sum_{j=1}^{\infty} \exp(-2j\nu) \\ &\leq A^2 \sum_{j=1}^{\infty} \frac{|1 - \exp(-j\nu)|^2}{j^{2a}} + \sigma^2 \frac{\exp(-2\nu)}{1 - \exp(-2\nu)}.\end{aligned} \tag{7.166}$$

For fixed A and σ one can find ν_m which minimizes the right side of (7.166). If γ_m is the minimum of the right side of (7.166), then the error estimate of the method is

$$\mathcal{J} \leq \gamma_m. \tag{7.167}$$

One can see that $\gamma_m \to 0$ as $\sigma \to 0$.

7.5 Resolution ability of linear systems

7.5.1 *Introduction*

Let us briefly discuss the notions of resolution ability of a linear system. In optics Rayleigh gave an intuitive definition of resolution ability: if one has two bright points $[\delta(x-a) + \delta(x+a)]/2$ as the input signal and if the optical system is described by the transmission function $h(x, y)$, so that the output signal is $[h(x, a) + h(x, -a)]/2$ then the two points can be resolved according to the Rayleigh criterion if $[h(0, a) + h(0, -a)]/2 \lesssim 0.8h(0)$.

Note that we took the signal $[\delta(x - a) + \delta(x + a)]/2$ rather than $\delta(x - a) + \delta(x + a)$ in order to compare this signal with a bright point at the origin $\delta(x)$. The sum of the coefficients in front of delta-functions should be therefore equal to 1, the coefficient in front of $\delta(x)$. The factor 0.8 in the Rayleigh criterion is an empirical one. The transmission function $h(x, y)$ is defined as follows: if $s_{\text{in}}(x)$ is the input signal then the output signal of the linear system is given by

$$s_{\text{out}}(x) = \int_D h(x, y) s_{\text{in}}(y) dy. \tag{7.168}$$

The domain D in optics is usually the input pupil of the system, or its input vision domain. The optical system is called isoplanatic if $h(x, y) = h(x - y)$. In describing the Rayleigh criterion we assume that $h(x, y)$ has absolute maximum at $x = y$, that the distance $2a$ between two points is small, so that both points lie in the region near origin in which $h(x, y)$ is positive.

Suppose now that

$$s_1(x) = \frac{\delta(x-a)+\delta(x+a)}{2}, \tag{7.169}$$

$$s_0(x) = \delta(x), \tag{7.170}$$

and one observes the signal

$$\mathcal{U}_j(x) = \int_D h(x, y) s_j(y) dy + n(x), \tag{7.171}$$

where $n(x)$ is the output Gaussian noise, $\overline{n} = 0$, $\overline{|n|^2} = \sigma^2 < \infty$, and $j = 0$ (hypothesis H_0) or $j = 1$ (hypothesis H_1).
The problem:

given the observed signal $\mathcal{U}_j(x)$ decide whether H_0 or H_1 occured. (7.172)

If with the probability 1 one can make a correct decision no matter how small $a > 0$ is, then one says that the resolution ability of the system in the sense of Rayleigh is infinite. The traditional intuition (which says that, for a fixed size of the input pupil, an optical system can resolve the distances of order of magnitude of the wavelength) is based on the calculation of diffraction of a plane wave by a circular hole in a plane. In [Ramm (1970c)] and [Ramm (1970d)] it was proved that, in the absence of noise, the transmission function of a linear system can be made as close to the $\delta(x-y)$ as one wishes, by means of apodization. This means that there is a sequence $h_m(x,y)$ of the transmission functions which is a delta sequence in the sense that

$$\int_D h_m(x,y)s(y)dy \to s(x) \quad \text{as} \quad m \to \infty$$

for any continuous $s(y)$.

7.5.2 *Resolution ability of linear systems*

In this section we apply the theory developed in Section 7.2 in order to show that there exists a linear system, or what is the same for our purposes, the transmission function $h(x,y)$ such that the resolution ability of this system in the sense of Rayleigh is infinite. More precisely, we will formulate a decision rule for discriminating between the hypoteses $\mathbf{H}_0$ and $\mathbf{H}_1$ such that the error of this rule can be made as small as possible. The error of the rule is defined to be

$$\alpha_m := P(\gamma_1 \mid \mathbf{H}_0) \tag{7.173}$$

that is, the probability to decide that hypothesis $\mathbf{H}_1$ occured when in fact $\mathbf{H}_0$ occured. The meaning of the parameter m, the subscript of α, will be made clear shortly. This parameter is associated with the sequence of linear systems whose resolution ability increases without limit.

First let us choose $h_m(x,y)$ so that the sequences

$$h_m(x,z) = \int_D h_m(x,y)\delta(y-z)dy := \delta_m(x,z) \tag{7.174}$$

are delta-sequences. Then, by formula (7.171), the observed signals became

$$\mathcal{U}_1(x) = \frac{\delta_m(x,a) + \delta_m(x,-a)}{2} + n(x) := s_1(x) + n(x) \tag{7.175}$$

or

$$\mathcal{U}_0(x) = \delta_m(x,0) + n(x) := s_0(x) + n(x). \tag{7.176}$$

Let us apply to the problem (7.172) the decision rule based on formula (7.41).

First one should solve the equation (7.29):

$$RV := \int_D R(x,y)V(y)dy = \frac{\delta_m(x,a) + \delta_m(x,-a)}{2} - \delta_m(x,0)$$
$$= s_1(x) - s_0(x) := f, \tag{7.177}$$

where $R(x,y) := \overline{n^*(x)n(y)}$ is the covariance function of noise. We assume that $R(x,y) \in \mathcal{R}$, and, for simplicity, that $P(\lambda) = 1$. In this case $R(x,y)$ solves the equation

$$Q(\mathcal{L})R(x,y) = \delta(x-y). \tag{7.178}$$

We also assume that $\delta_m(x,y)$ is negligibly small for $|x-y| > \frac{|a|}{2}$, and that points 0, a, and $-a$ are inside of D and

$$\rho(0,\Gamma) > |a|, \quad \rho(a,\Gamma) > |a|, \quad \rho(-a,\Gamma) > |a|, \tag{7.179}$$

where $\rho(x,\Gamma)$ is the distance between point x and $\Gamma = \partial D$. In this case one can neglect the singular boundary term of the solution to equation (7.177) and write this solution as

$$V(x) = Q(\mathcal{L})f. \tag{7.180}$$

Let us write inequality (7.39):

$$Re \int_D \mathcal{U}(x)Q(\mathcal{L})f^*(x)dx \geq \frac{1}{2}\int_D \delta_m(x,0)Q(\mathcal{L})f^*(x)dx$$
$$+ \frac{1}{2}\int_D \frac{\delta_m^*(x,a) + \delta_m^*(x,-a)}{2} Q(\mathcal{L})f(x)dx, \tag{7.181}$$

where we assume that the coefficients of $Q(\mathcal{L})$ are real. Otherwise one would write $[Q(\mathcal{L})f]^*$ in place of $Q(\mathcal{L})f^*$.

Let us write the expression (7.173) for α_m:

$$\alpha_m := P(\gamma_1 \mid H_0) = P\left\{2Re \int_D \mathcal{U}_0(x)V^*dx \geq \int_D (s_0V^* + s_1^*V)dx\right\}, \tag{7.182}$$

where V, s_1 and s_0 are given by (7.180), (7.175) and (7.176) respectively.

It follows from (7.182) that

$$\alpha_m = P\left\{2Re\int_D n(x)V^*(x)dx \geq \int_D (s_1^* - s_0^*)Q(\mathcal{L})(s_1 - s_0)dx\right\}$$
$$= P\left\{2Re\int_D n(x)V^*(x)dx \geq \int_D (s_1^*Q(\mathcal{L})s_1 + s_0^*Q(\mathcal{L})s_0)dx\right\}. \tag{7.183}$$

Here we took into account that

$$\int_D s_j^* Q(\mathcal{L})s_i dx = 0 \quad \text{for} \quad i \neq j \tag{7.184}$$

because of the assumption that for m sufficiently large the functions $S_1(x)$ and $S_0(x)$ have practically nonintersecting supports: each of them with derivatives of order sq is negligibly small in the region where the other is not small. One can write (7.183) as

$$\alpha_m = P\left\{Re\int_D n(x)V^*(x)dx \geq \frac{3}{2}A_m\right\}, \tag{7.185}$$

where we denote by A_m the positive quantity of the type

$$A_m = \int_D \delta_m(x,0)Q(\mathcal{L})\delta_m^*(x,0)dx, \tag{7.186}$$

and one can write in place of $\delta_m(x,0)$ the functions $\delta_m(x,a)$ or $\delta_m(x,-a)$. The basic property of A_m is

$$A_m \to +\infty \quad \text{as} \quad m \to \infty. \tag{7.187}$$

Indeed, the elliptic operator $Q(\mathcal{L})$ is positive definite on $H^{sq/2}(R^r)$, and one can assume that $\delta_m(x,0) \in \dot{H}^{sq/2}(B_a)$, $B_a := \{x : x \in R^r, |x| \leq a\}$. Therefore

$$\int_D \delta_m(x,0)Q(\mathcal{L})\delta_m^*(x,0)dx \geq c\int_D |\delta_m(x,0)|^2 dx \to +\infty. \tag{7.188}$$

Here c is a positive constant which does not depend on $\delta_m(x,0)$ (it depends on $Q(\mathcal{L})$ only) and the integral in (7.188) tends to infinity because $\delta_m(x,0)$ is a delta-sequence by construction (see formula (7.174) and the line below it).

Let us apply the Chebyshev inequality to get:

$$P\left(\left|\int_D h(x)V^*(x)dx\right| \geq A_m\right) \leq \frac{\mathcal{D}\left[\int_D n(x)V^*(x)dx\right]}{A_m^2}. \tag{7.189}$$

Here we took into account that $\overline{n} = 0$, and $\mathcal{D}[n]$ stands for the variance of random quantity n. One has

$$\begin{aligned}\mathcal{D}\left[\int_D n(x)V^*(x)dx\right] &= \int_D\int_D R(x,y)V(x)V^*(y)dxdy \\ &= \int_D\int_D R(x,y)Q(\mathcal{L})f(x)Q(\mathcal{L})f^*(y)dxdy \\ &= \int_D\left(\int_D Q(\mathcal{L})R(x,y)f(x)dx\right)Q(\mathcal{L})f^*(y)dy \\ &= \int_D\left(\int_D \delta(x-y)f(x)dx\right)Q(\mathcal{L})f^*(y)dy \\ &= \int_D f(y)Q(\mathcal{L})f^*(y)dy = \frac{3}{2}A_m. \end{aligned} \tag{7.190}$$

Here we used definition (7.177) of f, formula (7.186) and equalities of the type (7.184) for the functions $\delta_m(x,0)$, $\delta_m(x,a)$ and $\delta_m(x,-a)$. From (7.189) and (7.190) it follows that

$$P\left\{\left|\int_D n(x)V^*(x)dx\right| \geq A_m\right\} \leq \frac{3}{2A_m} \to 0 \quad \text{as} \quad \beta_m \to +\infty. \tag{7.191}$$

If X is an arbitrary random variable then

$$P\left\{ReX \geq A_m\right\} \leq P\left\{|X| \geq A_m\right\}. \tag{7.192}$$

From (7.185), (7.187), (7.191) and (7.192) it follows that

$$\alpha_m \leq \frac{2}{3A_m} \to 0 \quad \text{as} \quad m \to \infty. \tag{7.193}$$

Let us compute the probability to take the decision that the hypothesis H_0 occurred while in fact H_1 occurred. We have

$$\begin{aligned}\beta_m &:= P(\gamma_0 \mid H_1) = P\left\{2Re\int_D \mathcal{U}_1V^*dx \leq \int_D (s_0V^* + s_1^*V)dx\right\} \\ &= P\left\{2Re\int_D nV^*dx \leq \int_D (s_0V^* - s_1V^*)dx\right\} \\ &= P\left\{Re\int_D nV^*dx \leq -\frac{1}{2}\int_D fQ(\mathcal{L})f^*dx\right\} \\ &= P\left\{Re\int_D nV^*dx \leq -\frac{3}{4}A_m\right\}, \end{aligned} \tag{7.194}$$

where we used formula (7.190).

If $ReX < -A$ then $|X| > A$. Therefore

$$P(|X| \geq A) \geq P\{ReX \leq -A\}. \tag{7.195}$$

From (7.191), (7.194) and (7.195) one concludes that

$$\beta_m \leq P\left\{\left|\int_D hV^* dx\right| \geq \frac{3}{4}A_m\right\} \leq \frac{3}{2}\frac{16}{9A_m} = \frac{8}{3A_m} \to 0 \quad \text{as} \quad m \to \infty. \tag{7.196}$$

We have proved the following

Theorem 7.3 *The problem (7.172) can be solved by the decision rule from Section 7.2 and $\alpha_m \to 0$, $\beta_m \to 0$, where α_m is defined in (7.182) and β_m is defined in (7.194).*

7.5.3 *Optimization of resolution ability*

In this section we give a theory of optimization of resolution ability of linear optical instruments.

Let us consider the resolution ability for the problem of discriminating between two arbitrary signals $s_1(x)$ and $s_0(x)$ which are deterministic functions of $x \in R^r$. The observed signal in a bounded domain $D \subset \mathbf{R}^r$ is

$$\mathcal{U}(x) = s_j(x) + n(x), \quad j = 0 \quad \text{or} \quad j = 1, \tag{7.197}$$

$n(x)$ is Gaussian noise,

$$\overline{n} = 0, \quad \mathcal{D}[n] = \sigma^2, \quad \overline{n^*(x)n(y)} = R(x-y). \tag{7.198}$$

Assume that the linear optical instrument is isoplanatic. This means that its transmission function, $h(x,y)$, is $h(x-y)$. In optics $r = 2$ and D is the input pupil, or entrance pupil, of the instrument.

Consider the case of incoherent signals. For incoherent signals the transmission power function is $|h(x-y)|^2$. This means that if $I_{\text{in}}(x) := \overline{|s(x)|^2}$ is the intensity of the signal $s(x)$ in the object plane then in the image plane one has the distribution of the intensity $I_{\text{out}}(x)$ given by

$$I_{\text{out}}(x) = \int |h(x-y)|^2 I_{\text{in}}(y)dy, \quad \int = \int_{R^2}. \tag{7.199}$$

The bar denotes averaging in phases. This will be explained soon. Let us briefly derive (7.199). One has

$$I_{\text{out}}(x) = \overline{\left|\int h(x-y)s_{\text{in}}(y)dy\right|^2} = \iint h(x-y)h^*(x-y')\overline{s_{\text{in}}^*(y')dydy'}. \tag{7.200}$$

Let us now explain the meaning of the average. For incoherent signals by definition we have

$$s_{\text{in}}(y) = \sum_j A_j(y)e^{i\phi_j(y)}, \tag{7.201}$$

where $A_j(y)e^{i\phi_j}$ are sources which form the signal $s_{\text{in}}(y)$ at the point y, and ϕ_j are their phases which assumed random, uniformly distributed in the interval $[-\pi, \pi]$ and statistically independent, so that $\overline{\phi_j(y)\phi_j(y')} = \delta_{jj'}\delta(y-y')$ $\overline{\phi}_j = 0$. Under these assumptions one has

$$\begin{aligned}\overline{s_{\text{in}}(y)s_{\text{in}}^*(y')} &= \sum_{j,j} A_j(y)A_{j'}^*(y'\overline{\exp\{i[\phi_j(y)-\phi_{j'}(y')]\}} \\ &= \sum_{j,j'} A_j(y)A_{j'}^*(y')\delta_{jj'}\delta(y-y') = \sum_j |A_j(y)|^2\delta(y-y') \\ &= I_{\text{in}}(y)\delta(y-y'). \end{aligned} \tag{7.202}$$

Here we took into account that

$$I_{\text{in}}(y) = \overline{|s_{\text{in}}(y)|^2} = \sum_{j,j'} A_j(y)A_{j'}^*(y)\overline{\exp\{i[\phi_j(y)-\phi_{j'}(y)]\}} = \sum_j |A_j(y)|^2. \tag{7.203}$$

From (7.202) and (7.200) one obtains (7.199). Denote

$$H(x) := |h(x)|^2 \geq 0, \tag{7.204}$$

where $h(x)$ is the transmission function for an isoplanatic linear instrument. Then

$$I_{\text{out}}(x) = \int H(x-y)I_{\text{in}}(y)dy, \quad \int := \int_{\mathbf{R}^2} \tag{7.205}$$

for incoherent signals. Let us assume that

$$\int H^2(x)dx := \epsilon < \infty. \tag{7.206}$$

One often assumes in applications that

$$\tilde{H}(\lambda) \quad \text{is negligible outside } \Delta, \tag{7.207}$$

where $\Delta \subset \mathbf{R}^2$ is a finite region. The assumption (7.178) means that the instrument filters out the spatial frequencies which do not belong to Δ. Let us finally assume that the noise is a real-valued function which is a perturbation of the observed intensity in the image plane. One can think of $n(x)$, for example, as of the noise of the receiver of the intensity.

The problem is: given the observed signal

$$\mathcal{U}(x) = I_j(x) + n(x), \quad j = 1 \quad or \quad j = 0,$$
$$I_j(x) = \int H(x-y)s_j(y)dy, \quad s_j(x) := I_{inj}(x) \tag{7.208}$$

decide whether it is of the form (7.179) with $j = 1$ (hypothesis H_1) or with $j = 0$ (hypothesis H_0).

Applying the decision rule from Section 7.2 and taking into account that the signals are real valued, one solves the equation (7.29)

$$\int_D R(x,y)V(y)dy = I_1(x) - I_0(x) := I(x) \tag{7.209}$$

and then checks inequality (7.39) with real-valued signals:

$$\int_D \mathcal{U}(x)V(x)dx \geq \frac{1}{2}\int_D [I_0(x) + I_1(x)]\, V(x)dx. \tag{7.210}$$

If (7.210) holds then the decision is that hypothesis H_1 occurred. Otherwise one concludes that hypothesis H_0 occurred.

The error of the first kind of this rule is the probability to decide that H_1 occurred when in fact H_0 occurred:

$$\begin{aligned}\alpha &:= P(\gamma_1 \mid H_0) \\ &= P\left\{\int_D [I_0(x) + n(x)]V(x)dx \geq \frac{1}{2}\int_D [I_0(x) + I_1(x)]V(x)dx\right\} \\ &= P\left\{\int_D n(x)V(x)dx \geq \frac{1}{2}\int_D I(x)V(x)dx\right\},\end{aligned} \tag{7.211}$$

where $I(x)$ is given by (7.209). Note that

$$\overline{\int_D n(x)V dx} = 0, \tag{7.212}$$

$$d^2 := \mathcal{D}\left[\int_D n(x)V(x)dx\right] = \int_D \int_D R(x,y)V(y)V(x)dydx. \qquad (7.213)$$

We assume here that $R(x, y)$ and $V(x)$ are real-valued. Since $\int_D n(x)V(x)dx$ is Gaussian (because $n(x)$ is) one concludes from (7.211), (7.212) and (7.213) that

$$\alpha = \frac{1}{d\sqrt{2\pi}} \int_{d^2/2}^{\infty} \exp\left(-\frac{t^2}{2d^2}\right) dt = \text{erf}(d/2), \qquad (7.214)$$

where $\text{erf}(x)$ is defined in (7.61) and we used the equation

$$d^2 = \int_D I(x)V(x)dx \qquad (7.215)$$

which can be easily checked:

$$d^2 := \int_D \left(\int_D R(x,y)V(y)dy\right) V(x)dx = \int_D I(x)V(x)dx$$

because of (7.209).

Therefore

$$\alpha = \min \quad \textit{for those } H(x) \textit{ for which} \quad d = \max. \qquad (7.216)$$

For the error of the second kind which is the probability to decide that H_0 occurred while in fact H_1 occurred one has

$$\begin{aligned}\beta &:= P(\gamma_0 \mid H_1) = P\left\{\int_D (I_1 + n)Vdx < \frac{1}{2}\int_D (I_0 + I_1)Vdx\right\} \\ &= P\left\{\int_D nVdx < -\frac{1}{2}\int_D I(x)V(x)dx\right\} \\ &= \frac{1}{d\sqrt{2\pi}} \int_{-\infty}^{-d^2/2} \exp\left(-\frac{t^2}{2d^2}\right) dt \\ &= \frac{1}{\sqrt{2\pi}} \int_{-\infty}^{-d/2} \exp\left(-\frac{t^2}{2}\right) dt = \text{erf}(d/2) = \alpha. \end{aligned} \qquad (7.217)$$

Therefore both α and $\beta = \alpha$ will be minimized if $H(x)$ solves the following optimization problem

$$d^2 := \int_D \int_D R(x,y)V(y)V(x)dydx = \max \qquad (7.218)$$

subject to the conditions

$$\int_D R(x,y)V(y)dy = I_1(x) - I_0(x) := I(x) \tag{7.219}$$

and

$$I(x) = \int H(x-y)s(x)dx, \quad s(x) := s_1(x) - s_0(x). \tag{7.220}$$

Let us assume for simplicity that D is the whole plane:

$$D = \mathbf{R}^2. \tag{7.221}$$

Then, taking the Fourier transform of (7.219) and (7.220) yields

$$\tilde{R}(\lambda)\tilde{V}(\lambda) = \tilde{I}(\lambda) = \tilde{H}(\lambda)\tilde{s}(\lambda). \tag{7.222}$$

Here the last assumption (7.198) was used. Thus

$$\tilde{V}(\lambda) = \tilde{H}(\lambda)\tilde{s}(\lambda)\tilde{R}^{-1}(\lambda). \tag{7.223}$$

Write (7.218) as

$$d^2 = \int_{\mathbf{R}^2} I(x)V(x)dx = \frac{1}{(2\pi)^2}\int_{\mathbf{R}^2} \tilde{R}^{-1}|\tilde{H}(\lambda)|^2|\tilde{s}|^2 d\lambda = \max \tag{7.224}$$

and (7.206) as

$$\epsilon = (2\pi)^{-2}\int_{\mathbf{R}^2} |\tilde{H}|^2 d\lambda. \tag{7.225}$$

The instrument with the power transmission function $H(x)$ which maximizes functional (7.224) under the restriction (7.225) will have the maximum resolution power for the problem of discriminating two given signals $s_1(x)$ and $s_0(x)$.

The solution to (7.224) is easy to find: $|\tilde{H}|^2$ should be parallel to the vector $\tilde{R}^{-1}|\tilde{s}|^2$, so

$$|\tilde{H}|^2 = \tilde{R}^{-1}|\tilde{s}|^2 \cdot \text{const}, \tag{7.226}$$

where the constant is uniquely determined by condition (7.225):

$$|\tilde{H}(\lambda)|^2 = \tilde{R}^{-1}(\lambda)|\tilde{s}(\lambda)|^2 \cdot \frac{\epsilon^{1/2}}{\frac{1}{2\pi}\int_{\mathbf{R}^2} \tilde{R}^{-1}|\tilde{s}|^2 d\lambda}. \tag{7.227}$$

Formula (7.227) determines uniquely $|\tilde{H}| := A(\lambda)$, so that

$$\tilde{H}(\lambda) = A(\lambda)\exp[i\phi(\lambda)], \tag{7.228}$$

where $\phi(\lambda)$ is the unknown phase of $\tilde{H}(\lambda)$:

$$\int_{\mathbf{R}^2} H(x) \exp(i\lambda \cdot x)dx = A(\lambda) \exp[i\phi(\lambda)], \quad H(x) \geq 0. \tag{7.229}$$

Formula (7.224) shows that the resolution power of the optimal instrument depends on $A(\lambda) = |\tilde{H}(\lambda)|$ only. The phase $\phi(\lambda)$ does not influence the resolution ability but influences the size of the region D out of which $H(x)$ is negligibly small. If one writes the equation

$$\int_{D} H(x) \exp(i\lambda \cdot x)dx = A(\lambda) \exp[i\phi(\lambda)], \quad H(x) \geq 0 \tag{7.230}$$

and consider it as an equation for $H(x)$ and $\phi(\lambda)$, given $A(\lambda)$ and $\arg H(x) = 0$, then one has a phase retrieval problem. In (7.230) $D \subset \mathbf{R}^2$ is assumed to be a finite region with a smooth boundary. This problem has been studied in [Kl], where some uniqueness theorems are established. However, the numerical solution to this problem has not been studied sufficiently.

The condition (7.206) does not seem to have physical meaning. One could assume that

$$\int H(x)dx = E. \tag{7.231}$$

In this case the const in (7.226) cannot be found explicitly, in contrast to the case when condition (7.206) is assumed. If (7.231) is assumed then it follows from (7.176) that

$$I_{\text{out}}(x) \leq \max_{y \in \mathbf{R}^2} I_{\text{in}}(y)E. \tag{7.232}$$

7.5.4 *A general definition of resolution ability*

In this section a general definition of resolution ability is suggested. The classical Rayleigh definition deals with very special signals: two bright points versus one bright point. Suppose that the set M of signals, which one wishes to resolve is rather large. For example, one can assume that M consists of all functions belonging to $L^2(D)$ or to $C_0^1(D)$, the space of functions which have one continuous derivative in a bounded domain $D \subset R^r$ and are compactly supported in D. Suppose that the linear system is described by its transmission function

$$Lf = \int_D h(x,y)f(y)dy. \tag{7.233}$$

Assume that actually one observes the signal

$$\mathcal{U}(x) = Lf + n(x), \quad \overline{n} = 0, \quad \mathcal{D}[n] = \sigma^2, \tag{7.234}$$

where $n(x)$ is noise. Let

$$B_\sigma \mathcal{U} := f_\sigma \tag{7.235}$$

denote a mapping which recovers f given $\mathcal{U}(x)$. Let us assume that the operator L^{-1} exists but is unbounded. Then in the absence of noise one can recover f exactly by the formula $f = L^{-1}\mathcal{U}$, but L^{-1} cannot serve as B_σ in (7.235): first, because in the presence of noise $\mathcal{U}$ may not belong to the domain of L^{-1}, secondly, because if $L^{-1}\mathcal{U}$ is well defined in the presence of noise it may give a very poor estimate of f due to the fact that L^{-1} is unbounded. Let us define the resolution ability of the procedure B_σ on the class M of input signals as a maximal number $r \in [0,1]$ for which

$$\lim_{\sigma \to 0} \sup_{f \in M} \frac{\{\mathcal{D}[B_\sigma \mathcal{U} - f]\}}{\sigma^r} < \infty. \tag{7.236}$$

This definition takes into account the class M of the input signals, the procedure B for estimating f, and the properties of the system (see the definition (7.234) of $\mathcal{U}$). Therefore all the essential data are incorporated in the definition. The definition makes sense for nonlinear injective mappings L as well.

Roughly speaking the idea of this definition is as follows. If the error in the input signal is $O(\sigma)$ and the identification procedure B_σ produces $f_\sigma = B\mathcal{U}$ such that, in some suitable norm, $\| f_\sigma - f \| = O(\sigma^r)$ as $\sigma \to 0$, then the large r is, the better the resolution ability of the procedure B_σ is.

Example 7.5 Assume that the transmission function is

$$h(x-y) = \left(\frac{2}{\pi}\right)^{1/2} \frac{\sin(x-y)}{x-y}. \tag{7.237}$$

Note that

$$\tilde{h}(\lambda) := \frac{1}{\sqrt{2\pi}} \int_{-\infty}^{\infty} h(x) e^{-i\lambda x} dx = \begin{cases} 1 & |\lambda| \le 1 \\ 0 & |\lambda| > 1. \end{cases} \tag{7.238}$$

Let

$$B_\sigma \mathcal{U} = \frac{1}{\sqrt{2\pi}} \int_{-\infty}^{\infty} \frac{\tilde{\mathcal{U}}(\lambda) \exp(i\lambda x) d\lambda}{\tilde{h}(\lambda) + a(\sigma)}, \tag{7.239}$$

where $a(\sigma) > 0$ for $\sigma > 0$ will be chosen later. Let M be the set of functions $f(x) \in L^2(-\infty, \infty)$ such that

$$\| \tilde{f} \|_{L^2[-1,1]} \leq m, \quad \tilde{f}(\lambda) = 0 \quad \text{for} \quad |\lambda| > 1, \tag{7.240}$$

where $m > 0$ is a constant. Assume that $n(x)$ is an arbitrary function such that

$$n(x) \in L^2(-\infty, \infty), \quad \| n \| := \left(\int_{-\infty}^{\infty} |h(x)|^2 dx \right)^{1/2} \leq \sigma. \tag{7.241}$$

By Parseval's equality, one has

$$\begin{aligned} \| B_\sigma \mathcal{U} - f \|^2 &= \left\| \frac{\tilde{h}\tilde{f} + \tilde{n}}{a(\sigma) + \tilde{h}} - \tilde{f} \right\|^2 = \left\| \frac{\tilde{n} - a(\sigma)\tilde{f}}{a(\sigma) + \tilde{h}} \right\|^2 \\ &\leq 2 \left(\frac{\| \tilde{h} \|^2}{a^2(\sigma)} + a^2(\sigma) \left\| \frac{\tilde{f}}{\tilde{h}} \right\|^2 \right) \leq 2 \left(\frac{\sigma^2}{a^2(\sigma)} + a^2(\sigma) m \right). \end{aligned} \tag{7.242}$$

Choose

$$a(\sigma) = \sigma^{1/2}. \tag{7.243}$$

Then (7.242) and (7.243) yield

$$\| B_\sigma \mathcal{U} - f \| \leq 2(1 + m)\sigma. \tag{7.244}$$

Therefore $r = 1$ for the procedure B_σ defined by formula (7.239):

$$\lim_{\sigma \to 0} \sup_{f \in M} \frac{\| B_\sigma \mathcal{U} - f \|}{\sigma} \leq 2(1 + m) < \infty. \tag{7.245}$$

7.6 Ill-posed problems and estimation theory

7.6.1 *Introduction*

In this section we define the notion of ill-posed problem and give examples of ill-posed problems. Many problems of practical interest can be reduced to solving an operator equation

$$Au = f, \tag{7.246}$$

where $u \in U$, $f \in F$, $A : U \to F$, U and F are Banach spaces, A is an injective mapping with discontinuous inverse and domain $D(A) \subset U$. Such problems are called ill-posed because small perturbations of f may lead to large perturbations of the solution u due to the fact that A^{-1} is not continuous. In many cases $R(A)$, the range of A, is not the whole F. In this case small perturbations of f may lead to an equation which does not have a solution due to the fact that the perturbed f does not belong to $R(A)$.

The problem (7.246) is said to be well-posed in the sense of Hadamard if $A : D(A) \to F$ is injective, surjective, that is $R(A) = F$, and $A^{-1} : F \to U$ is continuous.

The formulation of an ill-posed problem (7.246) can often be given as follows. Assume that the data are $\{\delta, A, f_\delta\}$, where $\delta > 0$ is a given number, f_δ is a δ-approximation of f in the sense

$$\| f - f_\delta \| \leq \delta. \tag{7.247}$$

The problem is: given the data $\{\delta, A, f_\delta\}$, find $u_\delta \in U$ such that

$$\| u_\delta - u \| \to 0 \quad as \quad \delta \to 0. \tag{7.248}$$

We use $\| \cdot \|$ for norms in U and F.

A more general formulation of the ill-posed problem is the one when the data are $\{\delta, \eta, A_\eta, f_\delta\}$ where δ and f_δ are as above, $\eta > 0$ is a positive number and A_η is an approximation of A in a suitable sense, for example, if A is a linear bounded operator one can assume $\| A_\eta - A \| \leq \eta$. We will not discuss this more general formulation here. (See [I], for example.)

Ill-posed problems can be formulated as estimation problems. For example, suppose that A is a linear operator, u solves equation (7.246), and one knows a randomly perturbed right hand side of (7.246), namely

$$f + n, \tag{7.249}$$

where n is a random variable,

$$\overline{n} = 0, \quad \overline{n^*(x)n(y)} = \sigma^2 \delta(x - y). \tag{7.250}$$

The problem is to estimate u given the data (7.249), (7.250).

Let us give a few examples of ill-posed problems of interest in applications.

Example 7.6 *Numerical differentiation.*

Let

$$Au := \int_a^x u(t)dt = f(x). \tag{7.251}$$

Assume that $\delta > 0$ and f_δ are given such that $\| f_\delta - f \| \leq \delta$. Here $\| f \| = \max_{a \leq x \leq b} |f(x)|$, $F = U = C([a, b])$. The problem (7.251) is ill-posed. Indeed, the linear operator A is injective: if $Au = 0$ then $u = 0$. Its range consists of the functions $f \in C^1[a, b]$ such that $f(a) = 0$. Therefore equation (7.251) with f_δ in place of f has no solutions in $C[a, b]$ if $f_\delta \notin C^1[a, b]$ or $f_\delta(0) \neq 0$. If one takes $f_\delta = f + \delta \sin[\omega(x-a)]$, then the solution to equation (7.251) with f_δ in place of f exists: $u_\delta = f' + \delta\omega \cos[\omega(x-a)]$. Since $u = f'$ one has $\| u_\delta - u \| = \delta\omega \gg 1$ if ω is sufficiently large. Therefore, a small in the norm of U perturbation of f resulted in a large in the same norm perturbation of u. Therefore the formula $u_\delta = A^{-1} f_\delta = f'_\delta$ in this example does not satisfy condition (7.248). In Section 7.3 a stable solution to the problem (7.251) is given:

$$u_\delta(x) = \frac{f_\delta(x + h(\delta)) - f_\delta(x - h(\delta))}{2h(\delta)}. \tag{7.252}$$

It is proved that

$$\| f' - u_\delta \| = \| u - u_\delta \| \leq \epsilon(\delta) \to 0 \quad \text{as} \quad \delta \to 0, \tag{7.253}$$

where $h(\delta)$ and $\epsilon(\delta)$ are given by formulas (7.67) (see Theorem 7.1). Formula (7.252) should be modified for $x < a + h(\delta)$ and $x > b - h(\delta)$ (see (7.107)).

Remark 7.2 *The notion of ill-posedness depends on the topology of F. For example, if one considers as F the space $C^1[a, b]$ of functions, which satisfy condition $A(0) = 0$, then problem (7.251) is well posed and A is an isomorphism of $C[a, b]$ onto F.*

Example 7.7 *Stable summation of orthogonal series with perturbed coefficients.*

Let

$$u(x) = \sum_{j=1}^{\infty} c_j \phi_j(x), \quad x \in D \subset R^r. \tag{7.254}$$

Suppose $(\phi_j, \phi_m) = \delta_{jm}$ is a basis of $L^2(D)$, the parentheses denote the inner product in $L^2(D)$. Assume that $u \in L^2(D)$. This happens if and only if

$$\sum_{j=1}^{\infty} |c_j|^2 < \infty. \tag{7.255}$$

Suppose the perturbed coefficients are given

$$c_{j\delta} = c_j + \epsilon_j, \quad |\epsilon_j| \leq \delta. \tag{7.256}$$

The problem is: given the data $c_{j\delta}$ and $\delta > 0$, find u_δ such that (7.248) holds, the norm in (7.248) being $L^2(D)$ norm.

Consider the map $A : L^2(D) \to \ell^\infty$ which sends a function $u(x) \in L^2(D)$ into a sequence $c = \{c_j\} = (c_1, \ldots, c_j, \ldots) \in \ell^\infty$ by formula (7.254). Since, in general, the perturbed sequence $c_\delta = \{c_{j\delta}\} \notin \ell^2$, the series $\sum_{j=1}^{\infty} c_{j\delta}\phi_j(x)$ diverges in $L^2(D)$, so that the perturbed sequence $\{c_{j\delta}\}$ may not belong to the range of A. It is easy to give examples when $\{c_{j\delta}\} \in R(A)$ but the function $u_\delta := \sum_{j=1}^{\infty} c_{j\delta}\phi_j(x)$ differs from u in $L^2(D)$ norm as much as one wishes no matter how small $\delta > 0$ is.

Exercise. Construct such an example:

Therefore A^{-1} is unbounded from ℓ^∞ into $L^2(D)$, and the problem is ill-posed.

Note that if one changes the topology on the set of sequences $\{c_j\}$ from ℓ^∞ to ℓ^2 then the problem becomes well-posed and the operator $A : L^2(D) \to \ell^2$ is an isomorphsm.

A stable solution to the problem is given in Section 7.4.

Example 7.8 *Integral equation of the first kind.*

Let A be an integral operator $A : H \to H$, $H = L^2(D)$,

$$Au = \int_D A(x,y)u(y)dy = f(x), \quad x \in D. \tag{7.257}$$

If A is injective, that is, $Au = 0$ implies $u = 0$, and A is compact, then A^{-1} is not continuous in H, $R(A)$ is not closed in H, so that small in H perturbations of f may result in large perturbations of u, or may lead to an equation which has no solutions in H (this happens if the perturbed f does not belong to $R(A)$). Therefore, the problem (7.257) is ill-posed.

Example 7.9 *Computation of values of unbounded operators.*

Suppose $B : F \to U$ is an unbounded linear operator densely defined in F (that is, its domain of definition $D(B)$ is dense in F). Suppose $f \in D(B)$,

$Bf = u$. Assume that instead of f we are given a number $\delta > 0$ and f_δ such that $\| f - f_\delta \| \leq \delta$.

The problem is: given f_δ and B compute u_δ such that $\| u_\delta - u \| = \| u_\delta - Bf \| \to 0$ as $\delta \to 0$.

This problem is ill-posed. If A^{-1} is unbounded and $B = A^{-1}$, the problem (7.257) reduces to the above problem.

Example 7.10 *Analytic continuation.*
Suppose $f(z)$ is analytic in a bounded domain D of the complex plane and continuous in $\overline{D}$. Assume that $D_1 \subset D$ is a strictly inner subdomain of D.

The problem is: given $f(z)$ in D_1 find $f(z)$ in D.

By Cauchy's formula one has

$$\frac{1}{2\pi i} \int_{\partial D} \frac{f(t)dt}{t - z} = f(z), \quad z \in D_1. \tag{7.258}$$

This is an integral equation of the first kind for the unknown function $f(t)$ on ∂D. If $f(t)$ is found then $f(z)$ in D is determined by Cauchy's formula. Therefore the problem of analytic continuation from the subdomain D_1 to the domain D is ill-posed.

Example 7.11 *Identification problems.*
Let $s_{\text{in}}(x)$ be the input signal and $s_{\text{out}}(x)$ be the output signal of a linear system with the transmission function $h(x, y)$, that is

$$\int_D h(x, y) s_{\text{in}}(y) dy = s_{\text{out}}(x), \quad x \in \Delta, \tag{7.259}$$

where D and Δ are bounded domains in R^r, and the function h in continuous in $D \times \Delta$.

The identification problem is: given $s_{out}(x)$ and $h(x, y)$ find $s_{in}(y)$.

Equation (7.259) is an integral equation of the first kind. Therefore the above problem is ill-posed.

Example 7.12 Many inverse problems arising in physics are ill-posed. For example, the inverse scattering problem in three dimensions. The problem consists of finding the potential from the given scattering amplitude (see [Ramm (2005)] and Chapter 6). Inverse problems of geophysics are often ill-posed [Ramm (2005)].

Example 7.13 *Ill-posed problems in linear algebra.*

Consider equation (7.246) with $U = R^n$ and $F = R^m$, R^m is the m-dimensional Euclidean space. Let $N(A) = \{u : Au = 0\}$ be the null-space of A, and $R(A)$ be the range of A. If $N(A) \neq \{0\}$ define the *normal* solution u_0 to (7.246) as the solution orthogonal to $N(A)$:

$$Au_0 = f, \quad u_0 \perp N(A). \tag{7.260}$$

This solution is unique: if $\tilde{u}_0$ is another solution to (7.260) then

$$A(u_0 - \tilde{u}_0) = 0, \quad u_0 - \tilde{u}_0 \perp N(A).$$

This implies that $u_0 = \tilde{u}_0$. One can prove that the normal solution can be defined as the solution to (7.246) with the property that its norm is minimal:

$$\min \| u \| = \| u_0 \|, \tag{7.261}$$

where minimum is taken over the set of all solutions to equation (7.246), and the minimizer is unique: $u = u_0$. Indeed, any element $u \in R^n$ can be uniquely represented as:

$$u = u_0 \oplus u_1, \quad u_0 \perp N(A), \quad u_1 \in N(A) \tag{7.262}$$

$$\| u \|^2 = \| u_0 \|^2 + \| u_1 \|^2 . \tag{7.263}$$

If $Au = f$ then $Au_0 = f$, and (7.263) implies (7.261). Moreover, the minimum is $\| u_0 \|$ and is attained if and only if $u_1 = 0$.

The normal solution to the equation $Au = f$ can be defined as the least squares solution:

$$\| Au - f \| = \min, \quad u \perp N(A) \tag{7.264}$$

in the case when $f \notin R(A)$. This solution exists and is unique. Existence follows from the fact that minimum in (7.264) is attained at the element u such that $\| Au - f \| = \mathrm{dist}(f, R(A))$ (note that $R(A)$ is a closed subspace of R^m). Uniqueness of the normal solution to (7.264) is proved as above.

Lemma 7.5 *The problem of finding normal solution to (7.246) is a well-posed problem in the sense of Hadamard: for any $f \in R^m$ there exists and is unique the normal solution u_0 to equation (7.246), and this solution depends continuously on f: if*

$$Au_0 = f, \quad Au_{0\delta} = f_\delta, \quad \| f - f_\delta \| \leq \delta \tag{7.265}$$

then

$$\| u_0 - u_\delta \| \to 0 \quad as \quad \delta \to 0. \tag{7.266}$$

Proof. Existence and uniqueness are proved above. Let us prove (7.266). Let

$$f = f_1 \oplus f_2, \quad f_\delta = f_{\delta 1} \oplus f_{\delta 2}, \tag{7.267}$$

where $f_1 \in R(A)$, $f_2 \perp R(A)$, $f_{\delta 1}$ and $f_{\delta 2}$ are defined similarly. One has

$$Au_0 = f_1 \quad Au_{0\delta} = f_{\delta 1}. \tag{7.268}$$

The operator $A : N(A)^\perp \to R(A)$ is an isomorphism. Therefore $A^{-1} : R(A) \to N(A)^\perp$ is continuous. Lemma 7.5 is proved. □

Definition 7.2 The mapping $A^+ : f \to u_0$ is called pseudoinverse of A. The normal solution u_0 is called sometimes pseudosolution.

Although we have proved that the problem of finding normal solution to equation (7.246) is well-posed in the sense of Hadamard when $A : R^n \to R^m$, we wish to demonstrate that practically this problem should be considered as ill-posed in many cases of interest.

As an example, take $n = m$, $A : R^n \to R^n$, $N(A) = \{0\}$, so that A is injective and, by Fredholm's alternative, A is an isomorphism of R^n onto R^n. Consider the equations

$$Au = f, \quad Au_\delta = f_\delta. \tag{7.269}$$

Thus

$$A(u - u_\delta) = f - f_\delta, \quad u - u_\delta = A^{-1}(f - f_\delta).$$

Therefore

$$\| u - u_\delta \| \leq \| A^{-1} \| \| f - f_\delta \| . \tag{7.270}$$

Since

$$\| f \| = \| Au \| \leq \| A \| \| u \| \tag{7.271}$$

one obtains

$$\frac{\| u - u_\delta \|}{\| u \|} \leq \| A^{-1} \| \| A \| \frac{\| f - f_\delta \|}{\| f \|}. \tag{7.272}$$

Define $\nu(A)$, the condition number of A, to be

$$\nu(A) := \| A^{-1} \| \| A \| . \tag{7.273}$$

Then (7.272) shows that the relative error of the solution can be large even if the relative error
$\| f_\delta - f \| / \| f \|$ of the data is small provided that $\nu(A)$ is large.

Note that the inequality (7.272) is sharp in the sense that the equality sign can be attained for some u, u_δ, f and f_δ.

The point is that if $\nu(A)$ is very large, then the problem of solving equation (7.246) with $A : R^n \to R^n$, with $N(A) = \{0\}$, is practically ill-posed in the sense that small relative perturbations of f may lead to large relative perturbations of u.

7.6.2 *Stable solution of ill-posed problems*

In this section we sketch some methods for stable solution of ill-posed problems, that is, for finding u_δ which satisfies (7.248).

First let us prove the following known lemma. By a compactum $M \subset U$ we mean a closed set such that any infinite sequence of its elements contains a convergent subsequence.

Lemma 7.6 *Let $M \subset U$ be a compactum. Assume that $A : M \to N$ is closed and injective, $N := AM$. Then $A^{-1} : N \to M$ is continuous.*

Remark 7.3 *We assume throughout that U and F are Banach spaces but often the results and proofs are valid for more general topological spaces. These details are not of prime interest for the theory developed in this work, and we do not give the results in their most general form for this reason.*

Proof. Let $Au_n = f_n$, $f_n \in N$. Assume that

$$\| f_n - f \| \to 0 \quad \text{as} \quad n \to \infty. \tag{7.274}$$

We wish to prove that $f \in N$, that is, there exists a $u \in M$ such that $Au = f$, and

$$\| u_n - u \| \to 0. \tag{7.275}$$

Since $u_n \in M$ and M is a compactum, there exists a convergent subsequence, which is denoted again u_n, with limit u, $\| u_n - u \| \to 0$. Since M is a compactum, it is closed. Therefore $u \in M$. Since $u_n \to u$, $Au_n \to f$, and A is closed, one concludes that $Au = f$. Lemma 7.6 is proved. □

This lemma shows that if one assumes a priori that the set of solutions to equation (7.248) belongs to a compactum M then the operator A^{-1} (which exists on $R(A)$ since we assume A to be injective) is continuous on the set $N := AM$.

Therefore an ill-posed problem which is considered under the condition $f \in N$ becomes conditionally well posed. This leads to the following definition.

Definition 7.3 A quasisolution of equation (7.246) on a compactum M is the solution to the problem

$$\| Au - f \| = \min, \quad u \in M. \tag{7.276}$$

Here $A : U \to F$ is a linear bounded operator.

The functional $\epsilon(u) := \| Au - f \|$ is continuous and therefore it attains its minimum on the compactum M. Thus, a quasisolution exists. In order to prove its uniqueness and continuous dependence on f, one needs additional assumptions. For example, one can prove

Theorem 7.4 *If A is linear, bounded, and injective, M is a convex compactum and F is strictly convex then, for any $f \in F$, the quasisolution exists, is unique, and depends on f continuously.*

The proof of Theorem 7.4 requires some preparation. Recall that F is called strictly convex if and only if $\| u + v \| = \| u \| + \| v \|$ implies that $v = \lambda u$ for some constant λ.

Exercise. Prove that λ has to be positive.

The spaces $L^p(D)$, ℓ^p, Hilbert spaces are strictly convex, while $L^1(D)$, $C(D)$ and ℓ^1 are not strictly convex.

Definition 7.4 If $g \in U$ is a vector and $M \subset U$ is a set, then an element $h \in M$ is called the metric projection of g onto M if and only if $\| g - h \| = \inf_{u \in M} \| g - u \|$. The mapping $P : g \to h$ is called the metric projection mapping, $Pg = h$, or $P_M g = h$.

In general, Pg is a set of elements. Therefore the following lemma is of interest.

Lemma 7.7 *If U is strictly convex and M is convex then the metric projection mapping onto M is single valued.*

Proof. Suppose $h_1 \neq h_2$, $h_j \in Pg$, $j = 1, 2$. Then

$$m := \| h_1 - g \| = \| h_2 - g \| \leq \| u - g \| \quad \forall u \in M.$$

Since M is convex, $(h_1 + h_2)/2 \in M$. Thus

$$m \leq \left\| g - \frac{h_1 + h_2}{2} \right\| \leq \frac{1}{2} \| g - h_1 \| + \frac{1}{2} \| g - h_2 \| = m. \tag{7.277}$$

Therefore, since U is strictly convex, one concludes that $g - h_1 = \lambda(g - h_2)$, λ is a real constant. Since $\| g - h_1 \| = \| g - h_2 \|$, it follows that $\lambda = \pm 1$. If $\lambda = 1$ then $h_1 = h_2$, contrary to the assumption. If $\lambda = -1$, then

$$g = (h_1 + h_2)/2. \tag{7.278}$$

Since M is convex, equation (7.278) implies that $g \in M$. This is a contradiction since $g \in M$ implies $Pg = g$. Lemma 7.7 is proved. □

Lemma 7.8 *If U is strictly convex and M is convex then $P_M : U \to M$ is continuous.*

Proof. Suppose $\| g_n - g \| \to 0$ but $\| h_n - h \| \geq \epsilon > 0$, where $h_n = Pg_n$, $h = Pg$. Since M is a compactum, one can assume that $h_n \to h_\infty$, $h_\infty \in M$. Thus $\| h_\infty - h \| \geq \epsilon > 0$. One has

$$\| g - h \| \leq \| g - h_\infty \| \tag{7.279}$$

and

$$\begin{aligned} \| g - h_\infty \| &\leq \| g - g_n \| + \| g_n - h_n \| + \| h_n - h_\infty \| \\ &\to \| g - h \|, \quad n \to \infty. \end{aligned} \tag{7.280}$$

Indeed $\| g - g_n \| \to 0$, $\| h_n - h_\infty \| \to 0$, and

$$\| g_n - h_n \| = \text{dist}(g_n, M) \to \text{dist}(g, M) = \| g - h \| . \tag{7.281}$$

From (7.279) and (7.280) one obtains $\| g - h_\infty \| = \| g - h \|$. This implies $h_\infty = h$ as in the proof of Lemma 7.7. This contradiction proves Lemma 7.8. □

Exercise. Prove that $\text{dist}(g, M) := \inf_{u \in M} \| g - u \|$ is a continuous function of g.

We are ready to prove Theorem 7.4.

Proof of Theorem 7.4 Existence of the solution to (7.276) is already proved. Since M is convex and A is linear the set $AM := N$ is convex. Since N is convex and F is strictly convex, Lemma 7.7 says that $P_N f$ exists and is unique, while Lemma 7.8 says that $P_N f$ depends on f continuously. Let $Au = P_N f$. Since A is injective $u = A^{-1} P_N f$ is uniquely defined and, by Lemma 7.6, depends continuously on f. Theorem 7.4 is proved. □

It follows from Theorem 7.4 that if $M \subset U$ is a convex compactum which contains the solution u to equation (7.246), if A is an injective linear bounded operator, and F is strictly convex, then the function

$$u_\delta = A^{-1} P_{AM} f_\delta \tag{7.282}$$

satisfies (7.248). The function u_δ can be found as the unique solution to optimization problem (7.276) with f_δ in place of f. One could assume A closed, rather than bounded, in Theorem 7.4. Uniqueness of the quasi solution is not very important for practical purposes. If there is a set $\{u_\delta\}$ of the solutions to (7.276) with f_δ in place of f, if A is injective and $Au_0 = f_0$, and if $\| f_\delta - f_0 \| \leq \delta$, then $\| u_\delta - u_0 \| \to 0$ as $\delta \to 0$ for any of the elements of the set $\{u_\delta\}$. Indeed $\| Au_\delta - f_\delta \| \leq \| Au_0 - f_\delta \| = \| f_0 - f_\delta \| \leq \delta$. Therefore $\| Au_\delta - Au_0 \| \leq \| Au_\delta - f_\delta \| + \| f_\delta - f_0 \| \leq 2\delta$. Since M is compact and u_δ, $u_0 \in M$ the inequality $\| Au_\delta - Au_0 \| \leq 2\delta$ implies $\| u_\delta - u_0 \| \to 0$ as $\delta \to 0$ (see Lemma 7.6).

We have finished the description of the *first method, the method of quasisolutions*, for finding a stable solution to problem (7.246).

How does one choose M? The choice of M is made in accordance to the a priori knowledge about the solution to (7.246). For instance, in Example 7.6 one can take as M the set of functions u, which satisfy the condition

$$|u(a)| \leq M_1, \quad |u'(x)| \leq M_2, \tag{7.283}$$

where M_j are constants, $j = 1, 2$. Then

$$Au = \int_a^x u dt = f$$

and

$$|f''| \le M_2, \quad |f'(a)| \le M_1, \quad f(a) = 0. \tag{7.284}$$

Inequality (7.283) defines a convex compactum M in $L^2[a,b]$ and (7.284) defines $N = AM$ in $L^2[a,b]$. Theorem 7.4 is applicable (since $L^2[a,b]$ is strictly convex) and guarantees that $\| u - u_\delta \|_{L^2[a,b]} \to 0$. A stable approximation of $u = f'(x)$ in $C[a,b]$ norm is given in Theorem 7.1.

Let us now turn to the *second method* for constructing u_δ which satisfies equation (7.248). This will be a *variational method* which is also known as a *regularization method.* While in the first method one needs to solve the variational problem (7.276) with the restriction $u \in M$, in the second method one has to solve a variational problem without restrictions.

Consider the functional

$$F(u) := \| Au - f_\delta \|^2 + \gamma\phi^2(u), \quad \| f_\delta - f \| \le \delta, \tag{7.285}$$

where $0 < \gamma = \text{const}$ is a parameter, A is a linear bounded operator and $\phi(u)$ is a positive strictly convex densely defined functional which defines a norm:

$$\phi(u) \ge 0, \quad \phi(u) = 0 \Rightarrow u = 0, \quad \phi(\lambda u) = |\lambda|\phi(u) \tag{7.286}$$

$$\frac{\phi(u_1 + u_2)}{2} < \frac{\phi(u_1) + \phi(u_2)}{2} \quad \text{if} \quad u_1 \ne \lambda u_2, \quad \lambda = \text{const}. \tag{7.287}$$

We also assume that the set of $u \in U$, which satisfy the inequality

$$\phi(u) \le c, \tag{7.288}$$

is compact in U. In other words the closure of the set $\text{Dom}\phi(u)$ in the norm $\phi(u)$ is a Banach space $U_\phi \subset U$ which is dense in U, and the imbedding operator $i : U_\phi \to U$ is compact.

One often takes

$$\phi(u) = \| \mathcal{L}u \|, \tag{7.289}$$

where $\mathcal{L} : U \to U$ is a linear densely defined boundedly invertible operator with compact inverse. An operator is called boundedly invertible if its inverse is a bounded operator defined on all of U. Let us assume that U is reflexive so that from any bounded set of U one can select a weakly convergent in U subsequence.

We will need a few concepts of nonlinear analysis in order to study the minimization problem $F(u) = \min$.

Definition 7.5 A functional $F : U \to R^1$ is called convex if $D(F) :=$ Dom$F$ is a linear set and for all $u, v \in D(F)$ one has

$$F(\lambda u + (1-\lambda)v) \le \lambda F(u) + (1-\lambda)F(v), \quad 0 \le \lambda \le 1. \tag{7.290}$$

Definition 7.6 A functional $F(u)$ is called weakly lower semicontinuous from below if

$$u_n \rightharpoonup u \Rightarrow \liminf_{n\to\infty} F(u_n) \ge F(u), \tag{7.291}$$

where $\rightharpoonup$ denotes weak convergence in U.

Lemma 7.9 *A weakly lower semicontinuous from below, functional $F(u)$, in a reflexive Banach space U is bounded from below on any bounded weakly closed set $M \subset DomF$ and attains its minimum on M at a point of M.*

Note that the set is weakly closed if $u_n \in M$ and $u_n \rightharpoonup u$ implies $u \in M$.

Proof of Lemma 7.9 Let

$$-\infty \le d := \inf_{u\in M} F(u), \quad F(u_n) \to d, \quad u_n \in M. \tag{7.292}$$

Since M is bounded and U is reflexive, there exists a weakly convergent subsequence of u_n which we denote u_n again, $u_n \rightharpoonup u$. Since M is weakly closed one concludes that $u \in M$. Since F is weakly lower semicontinuous from below, one concludes that

$$d \le F(u) \le \liminf_{n\to\infty} F(u_n) = d. \tag{7.293}$$

Therefore $d > -\infty$, and $F(u) = d$. Lemma 7.9 is proved. □

Lemma 7.10 *A weakly lower semicontinuous from below functional $F(u)$ in a reflexive Banach space attains its minimum on every bounded, closed, and convex set M.*

Proof. Any such set M in a reflexive Banach space is weakly closed. Thus Lemma 7.10 follows from Lemma 7.9. □

Exercise. Prove the following lemmas.

Lemma 7.11 *If $F(u)$ is weakly lower semicontinuous from below functional in a reflexive Banach space such that*

$$F(u) \to +\infty \quad as \quad \| u \| \to \infty \tag{7.294}$$

then $F(u)$ attains its minimum on any closed convex set $M \subset U$.

Lemma 7.12 *A weakly lower semicontinuous from below functional in a reflexive Banach space attains its minimum on every compactum.*

Lemma 7.13 *A convex Gateux differentiable functional $F(u)$ is weakly lower semicontinuous from below in a reflexive Banach space.*

Definition 7.7 A functional $F(u)$ is Gateux differentiable in U if and only if

$$\lim_{t\to+0} t^{-1}[F(x+th)-F(x)] = Ah \tag{7.295}$$

for any $x, h \in U$, where $A : U \to R^1$ is a linear bounded functional on U.

Proof of Lemma 7.13 If $u_n \rightharpoonup u$ then convexity of $F(u)$ implies

$$F(u) \le F(u_n) + F'(u)(u-u_n). \tag{7.296}$$

Pass to the limit infimum in (7.296) to get

$$F(u) \le \liminf_{n\to\infty} F(u_n). \tag{7.297}$$

Lemma 7.13 is proved. □

Exercise. Prove that if $F(u)$ is Gateux differentiable and convex in the sense (7.290) then

$$F(u) - F(v) \le F'(u)(u-v) \quad \forall u, v \in \mathrm{Dom}F. \tag{7.298}$$

In fact, if $F(u)$ is Gateux differentiable then

$$(7.290) \Leftrightarrow (7.298) \Leftrightarrow (F'(u) - F'(v), u - v) \ge 0. \tag{7.299}$$

The last inequality means that $F'(u)$ is monotone. The parentheses in (7.299) denote the value of the linear functional $F'(u) - F'(v) \in U^*$ at the element $u - v \in U$. By U^* the space of linear bounded functionals on U is denoted.

We are now ready to prove the following theorem.

Theorem 7.5 *Assume that A is a linear bounded injective operator defined on the reflexive Banach space U, and $\phi(u)$ is a strictly convex weakly lower semicontinuous from below functional such that the set (7.288) is compact in U and $u_f := A^{-1}f \in D(\phi)$. Then the minimization problem*

$$F(u) = \min, \tag{7.300}$$

where $F(u)$ is defined in (7.285), has the unique solution $u_{\delta,\gamma}$ for any $\gamma > 0$, and if one chooses $\gamma = \gamma(\delta)$ so that

$$\gamma(\delta) \to 0, \quad \delta^2\gamma^{-1}(\delta) \le m < \infty \quad as \quad \delta \to 0, \tag{7.301}$$

where $m = const > 0$, then $u_\delta := u_{\delta,\gamma(\delta)}$ satisfies (7.248).

Proof. The functional $F(u) \ge 0$. Let $0 \le d := \inf_{u\in U} F(u)$, and let u_n be the minimizing sequence

$$F(u_n) \to d. \tag{7.302}$$

Then

$$d \le F(u_n) \le d + \epsilon \le \delta^2 + \gamma\phi^2(u_f), \quad u_f := A^{-1}f, \quad \forall n > n(\epsilon), \tag{7.303}$$

where $\epsilon > 0$ is a sufficiently small number, see (7.317) below. From (7.301) and (7.303) one concludes that

$$\gamma\phi^2(u_n) \le d + \epsilon \le \gamma\left[\gamma^{-1}\delta^2 + \phi^2(u_f)\right] \le c\gamma \tag{7.304}$$

so that $\phi^2(u_n) \le c$, $c := m + \phi^2(u_f)$. Therefore one can choose a convergent subsequence from the sequence u_n. This subsequence is denoted also u_n:

$$u_n \underset{U}{\to} u_0, u_0 = u_{0\delta} := u_\delta. \tag{7.305}$$

Since A is continuous one has

$$\| Au_n - f \| \to \| Au_0 - f \| . \tag{7.306}$$

The lower weak semicontinuity of $\phi(u)$ and (7.305) imply

$$\liminf_{n\to\infty} \phi(u_n) \ge \phi(u_0). \tag{7.307}$$

Thus

$$d \le F(u_0) \le \liminf_{n\to\infty} F(u_n) = d. \tag{7.308}$$

Therefore the solution to (7.300) exists and the limit (7.305) is a solution.

Suppose v is another solution:

$$d = F(v) = F(u_0). \tag{7.309}$$

Then, since $F(u)$ is convex, one has

$$d \le F(\lambda u_0 + (1-\lambda)v) \le \lambda F(u_0) + (1-\lambda)F(v) = d, \quad \forall\lambda \in [0,1]. \tag{7.310}$$

Therefore

$$F\left(\frac{u_0+v}{2}\right)=\frac{F(u_0)+F(v)}{2}. \tag{7.311}$$

This implies that

$$\frac{\| Au_0-f_\delta \|}{2}+\frac{\| Av-f_\delta \|}{2}=\left\| A\frac{u_0+v}{2}-f\right\| \tag{7.312}$$

and

$$\phi\left(\frac{u_0+v}{2}\right)=\frac{\phi(u_0)+\phi(v)}{2}. \tag{7.313}$$

From (7.287) and (7.312) one concludes that $v = cu_0$, $c = \text{const}$. Since $\phi(cu) = |c|\phi(u)$, equation (7.313) implies that $c \geq 0$. From (7.310) it follows that

$$F\left(\lambda cu_0+(1-\lambda)u_0\right)=d \quad \forall\lambda\in[0,1]. \tag{7.314}$$

Let $\mu := \lambda(c-1)$. If $c \neq 1$, then for all real sufficiently small μ one obtains from (7.314) that

$$\| Au_0-f_\delta \|^2+\gamma\phi^2(u_0)=\| Au_0-f_\delta+\mu Au_0 \|^2+\gamma(1+\mu)^2\phi^2(u_0).$$

Thus

$$0=\mu^2 \| Au_0 \|^2+2\mu Re(Au_0-f_\delta, Au_0)+2\gamma\mu\phi^2(u_0)+\gamma\mu^2\phi^2(u_0). \tag{7.315}$$

Since (7.315) is a quadratic equation it cannot be satisfied for all small μ, since its coefficients are not all zeros. Therefore $c = 1$, and uniqueness of the solution to (7.300) is established.

Let us prove the last statement of Theorem 7.5. Assume that (7.301) holds. Let $Au = f$ $u = A^{-1}f := u_f$. One has

$$F(u)\geq F(u_0). \tag{7.316}$$

Therefore

$$\| Au_0-f_\delta \|^2+\gamma\phi^2(u_0)\leq\delta^2+\gamma\phi^2(u), \quad u=A^{-1}f. \tag{7.317}$$

Thus, by (7.301),

$$\phi^2(u_0)\leq\phi^2(u)+m:=c \tag{7.318}$$

and, using (7.301) again, one obtains

$$\| Au_0-f_\delta \|^2\leq\gamma\left(\gamma^{-1}\delta^2+\phi^2(u)\right)\leq c\gamma\to 0 \quad \text{as} \quad \delta\to 0. \tag{7.319}$$

Similarly, for sufficiently small δ, one has

$$F(u) \geq F(u_\delta), \quad \phi^2(u_\delta) \leq c, \tag{7.320}$$

$$\| Au_\delta - f_\delta \|^2 \leq c\gamma(\delta) \to 0 \quad \text{as} \quad \delta \to 0. \tag{7.321}$$

From (7.321) it follows that

$$\begin{aligned} \| Au_\delta - Au \| &\leq \| Au_\delta - f_\delta \| + \| f_\delta - Au \| \\ &\leq \| Au_\delta - f_\delta \| + \delta \leq c^{1/2}\gamma^{1/2}(\delta) + \delta \to 0 \\ &\qquad \text{as} \quad \delta \to 0. \end{aligned} \tag{7.322}$$

Let $M := \{v : v \in U, \phi^2(v) \leq c\}$. Then M is a compactum, $u_\delta \in M$, $u \in M$ by (7.320) and (7.318). Therefore, (7.322) and Lemma 7.6 imply (7.248). Theorem 7.5 is proved. □

Remark 7.4 *If one finds for $\gamma = \gamma(\delta)$ not the minimizer $u_\delta = u_{\delta,\gamma(\delta)}$ itself but an approximation to it, say v_δ, such that $F(v_\delta) \leq F(u)$, then, as above,*

$$\phi^2(v_\delta) \leq c, \quad \| Av_\delta - f_\delta \|^2 \leq c\gamma(\delta). \tag{7.323}$$

Thus

$$\| Av_\delta - Au \| \leq c^{1/2}\gamma^{1/2}(\delta) + \delta \to 0. \tag{7.324}$$

As above, from (7.323) and (7.324) it follows that $\| v_\delta - u \| \to 0$ as $\delta \to 0$. Therefore one can use an approximate solution to the minimization problem (7.300) as long as the inequality $F(v_\delta) \leq F(u)$ holds.

Remark 7.5 *One can assume in Theorem 7.5 that A is not bounded but closed linear operator, $D(A) \supset D(\phi)$ and the set (7.288) is compact in the space G_A which is $D(A)$ equipped with the graph norm $\| u \|_A := \| u \| + \| Au \|$. One can also assume that $\phi(u)$ is a convex lower weakly semi-continuous functional, the set $\phi(u) \leq c$ is not necessarily compact. The change in the proof of Theorem 7.5 under this assumption is as follows. From (7.304) it follows that $u_n \rightharpoonup u_0$ in U. It follows [Ru, p. 65, Theorem 3.13] that there exists a convex combination $\hat{u}_n$ of u_n such that $\hat{u}_n \to u_0$. The sequence $\hat{u}_n$ is minimizing if u_n is. Therefore equations (7.306)-(7.308) hold with $\hat{u}_n$ in place of u_n. The proof of the uniqueness of the solution to (7.300) is the same as above. One can prove that $u_\delta \rightharpoonup u$ as $\delta \to 0$ and that there is a convex combination $\hat{u}_\delta$ of u_δ such that $\hat{u}_\delta \to u$ as $\delta \to 0$. However, there is no algorithm to compute $\hat{u}_\delta$ given u_δ.*

If $R_{\alpha,\delta} : F \to U$ is the mapping which sends f_δ into $u_{\alpha,\delta}$, the solution to (7.300), then

$$R_\delta f_\delta := R_{\alpha(\delta),\delta} f_\delta = u_\delta \tag{7.325}$$

satisfies (7.248), that is

$$\| R_\delta f_\delta - A^{-1} f \| \to 0 \quad \text{as} \quad \delta \to 0. \tag{7.326}$$

A construction of such a family $R_{\alpha,\delta}$ of operators, that there exists $\alpha(\delta)$ such that (7.326) holds, is used for solving an ill-posed problem (7.246). The family $R_{\alpha,\delta}$ is called a regularizing family for problem (7.246).

The error estimate of the approximate solution $u_{\alpha,\delta} := R_\alpha f_\delta$ can be given as follows:

$$\| u_{\alpha,\delta} - u \| \leq \| R_\alpha (f_\delta - f) \| + \| R_\alpha A u - u \| \leq \omega(\alpha)\delta + \eta(\alpha) := \epsilon(\epsilon, \delta).$$

Here we assumed that R_α is a linear operator and that $\| R_\alpha \| \leq \omega(\alpha)$. One assumes that R_α and u are such that $\eta(\alpha) \to 0$ and $\omega(\alpha) \to +\infty$ as $\alpha \to 0$. Then there exists $\alpha = \alpha(\delta)$ such that $\alpha(\delta) \to 0$ and $\epsilon\,(\alpha(\delta), \delta) := \epsilon(\delta) \to 0$ as $\delta \to 0$.

Therefore R_α is a regularizing family for problem (7.246) provided that $\omega(\alpha) \to +\infty$ and $\eta(\alpha) \to 0$ as $\alpha \to 0$. The stable approximation to the solution u of equation (7.246) is $u_\delta := R_{\alpha(\delta)} f_\delta$, where $\alpha(\delta)$ is chosen so that $\epsilon(\alpha, \delta) \geq \epsilon\,(\alpha(\delta), \delta) := \epsilon(\delta)$ for any $\delta > 0$. One has the error estimate

$$\| u - u_\delta \| \leq \epsilon(\delta).$$

We gave two general methods for constructing such families. The theory can be generalized in several directions:

1) One can consider nonlinear A; unbounded A, for example, closed densely defined A; A given with some error, say A_ϵ is given such that $\| A - A_\epsilon \| < \epsilon$.
2) One can consider special types of A, for example convolution and other special kernels; in this case one can often give a more precise error estimate for approximate solution.
3) One can study the problem of optimal choice of γ and of the stabilizing functional $\phi(u)$ in (7.285).
4) One can study finite dimensional approximations for solving ill-posed problem (7.246).
5) One can study methods of solving problem (7.246) which are optimal in a suitable sense.

These questions are studied in many books and papers, and we refer the reader to [Ivanov *et. al.* (1978); Lavrentiev and Romanov (1986); Morozov (1984); Ramm (1968); Ramm (1973b); Ramm (1975); Ramm (1980); Ramm (1981); Ramm (1984); Ramm (1985b); Ramm (1987b); Ramm (1987c); Tanana (1981); Tikhonov (1977)].

In [Ramm (2003a)] and [Ramm (2005)] a new definition of the regularizing family is given. The new definition operates only with the data $\{\delta, f_\delta, \mathcal{K}\}$, and does not use the unknown f. The compact $\mathcal{K}$ in this definition is the compact to which the unknown solution belongs. The knowledge of this compact is an a priori information about the solutions of ill-posed problems. One calls $R_{\alpha,\delta}$ a regularizing family, if $\lim_{\delta\to 0}\sup_{\{u:u\in\mathcal{K},||Au-f_\delta||\leq\delta\}}||R_\delta f_\delta - u|| = 0$, where u solves the equation $Au = f$, and $R_\delta = R_{\alpha(\delta),\delta}$ for some $0 < \alpha(\delta) \to 0$ as $\delta \to 0$.

7.6.3 *Equations with random noise*

In this section we look at the problem (7.246) with linear bounded injective operator from the point of view of estimation theory. Let us consider the equation

$$Aw = f + n, \tag{7.327}$$

where A is an injective linear operator on a Hilbert space H, and n is noise. Let us assume for simplicity that noise takes values in H. In practice this is not always the case. For example, if $H = L^2(R^r)$ then a sample function $n(x)$ may belong to $L^2(R^r)$ locally, but not globally if $n(x)$ does not decay sufficiently fast as $|x| \to \infty$. Therefore the above assumption simplifies the theory. Assume that

$$\overline{n} = 0, \quad \overline{n^*(x)n(y)} = \sigma^2 R(x,y), \tag{7.328}$$

where $\sigma^2 > 0$ is a parameter which characterizes the power of noise. Let us assume that the solution to (7.327) exists and $f \in \text{Ran}A$. One may try to suggest the following definition.

Definition 7.8 The solution to (7.327) is statistically stable if

$$\mathcal{D}[w-u] \to 0 \quad \text{as} \quad \sigma \to 0, \tag{7.329}$$

where u solves the equation

$$Au = f. \tag{7.330}$$

We will also use this definition with (7.329) substituted by

$$\| w - u \|^2 \to 0 \quad \text{as} \quad \sigma \to 0, \tag{7.331}$$

where $\| \cdot \|$ denotes the norm in H.

This definition is very restrictive. First, the assumption that equation (7.327) is solvable means that a severe constraint is imposed on the noise. Secondly, the requirement (7.329) is rather restrictive.

Let us illustrate this by examples and then consider some less restrictive definition of the stable solution to (7.327). Note that, under the above assumption,

$$w = A^{-1} f + A^{-1} n = u + A^{-1} n. \tag{7.332}$$

Thus

$$\mathcal{D}[w - u] = \mathcal{D}[A^{-1} n]. \tag{7.333}$$

Let us assume that $H = L^2(D)$, $D \subset R^r$ is a finite region, and A is a selfadjoint compact operator on H with kernel $A(x, y)$,

$$A\phi_j = \lambda_j \phi_j, \quad \lambda_1 \geq \lambda_2 \geq \cdots > 0, \tag{7.334}$$

where

$$\int_D \phi_j(x) \phi_i^*(x) dx := (\phi_j, \phi_i) = \delta_{ji}. \tag{7.335}$$

Then

$$A^{-1} n = \sum_{j=1}^{\infty} \lambda_j^{-1} (n, \phi_j) \phi_j. \tag{7.336}$$

Therefore

$$\begin{aligned} \mathcal{D}[A^{-1} n] &= \sum_{i,j=1}^{\infty} \lambda_i^{-1} \lambda_j^{-1} \overline{\int_D n(t) \phi_j^*(t) dt \int_D n^*(z) \psi_i(z) dz \phi_j(x) \phi_i^*(x)} \\ &= \sigma^2 \sum_{i,j=1}^{\infty} \lambda_i^{-1} \lambda_j^{-1} \int_D \int_D R(z, t) \phi_i(z) \phi_j^*(t) dz dt \phi_j(x) \phi_i^*(x) \\ &= \sigma^2 \int_D \int_D A^{-1}(x, z) R(z, t) A^{-1}(t, x) dz dt, \end{aligned} \tag{7.337}$$

where $A^{-1}(x,y)$ is the kernel of the operator A^{-1} in the sense of distributions and is given by the formula

$$A^{-1}(x,y) = \sum_{j=1}^{\infty} \lambda_j^{-1} \phi_j^*(x)\phi_j(y). \tag{7.338}$$

For the right hand side of (7.337) to converge to zero it is necessary and sufficient that the kernel $B(x,y)$ of the operator $A^{-1}RA^{-1}$ be finite for all x and $y = x$, $B(x,x) < \infty$.

If one requests in place of (7.329) that

$$\overline{\| w - u \|^2} \to 0 \quad \text{as} \quad \sigma \to 0, \tag{7.339}$$

where the bar denotes statistical average and

$$\| w \| := (w,w)^{1/2}, \tag{7.340}$$

then the following condition (7.342) will imply (7.339).

One has

$$\overline{(A^{-1}n, A^{-1}n)} = \int_D dx \mathcal{D}[A^{-1}n] = \sigma^2 Tr(A^{-1}RA^{-1}) \to 0 \quad \text{as} \quad \sigma \to 0 \tag{7.341}$$

provided that A is selfadjoint, positive, and

$$Tr(A^{-1}RA^{-1}) < \infty. \tag{7.342}$$

Condition (7.342) is a severe restriction on the correlation function $R(x,y)$ of the noise.

Example 7.14 Consider the case $H = L^2(R^1)$,

$$Au := \int_{-\infty}^{\infty} A(x-y)u(y)dy. \tag{7.343}$$

Then

$$A^{-1}f := \frac{1}{2\pi}\int_{-\infty}^{\infty} \exp(i\lambda x)\tilde{A}^{-1}(\lambda)\tilde{f}(\lambda)d\lambda, \tag{7.344}$$

where

$$f(x) = \frac{1}{2\pi}\int_{-\infty}^{\infty} \tilde{f}(\lambda)\exp(i\lambda x)dx \tag{7.345}$$

one obtains (a derivation is given below) the formula

$$\overline{A^{-1}n(A^{-1}n)^*} = \sigma^2 \int_{-\infty}^{\infty} \frac{\tilde{R}(\lambda)}{|\tilde{A}(\lambda)|^2} dx. \tag{7.346}$$

Here we have assumed that

$$n(x) = \int_{-\infty}^{\infty} \exp(i\lambda x) d\zeta(\lambda), \tag{7.347}$$

where $\zeta(\lambda)$ is a random process with orthogonal increments such that

$$\overline{\zeta(\lambda)} = 0, \quad \overline{d\zeta^*(\lambda)d\zeta(\lambda)} = \sigma^2 \tilde{R}(\lambda) d\lambda, \tag{7.348}$$

$$\overline{d\zeta^*(\lambda)d\zeta(\mu)} = 0 \quad \text{for} \quad \lambda \neq \mu. \tag{7.349}$$

If B is a linear integral operator with convolution kernel:

$$Bn = \int_{-\infty}^{\infty} B(x-y)n(y)dy,$$

and the spectral representation for $n(x)$ is (7.347) then the spectral representation for Bn is

$$\begin{aligned} Bn &= \int_{-\infty}^{\infty} \left\{ \int_{-\infty}^{\infty} B(x-y) \exp(i\lambda y) dy \right\} d\zeta(\lambda) \\ &= \int_{-\infty}^{\infty} \exp(i\lambda x) \tilde{B}(\lambda) d\zeta(\lambda). \end{aligned} \tag{7.350}$$

If B_1 and B_2 are two linear integral operators with convolution kernels, then

$$\begin{aligned} \overline{B_1 n(B_2 n)^*} &= \iint_{-\infty}^{\infty} \exp(i\lambda x - i\mu x) \tilde{B}_1(\lambda) \tilde{B}_2^*(\mu) \overline{d\zeta(\lambda) d\zeta^*(\mu)} \\ &= \sigma^2 \int_{-\infty}^{\infty} \tilde{B}_1(\lambda) \tilde{B}_2^*(\lambda) \tilde{R}(\lambda) d\lambda. \end{aligned} \tag{7.351}$$

Equation (7.346) is a particular case of (7.351). Note that stationary random functions (7.347) have mean value zero and

$$\begin{aligned} \mathcal{D}[n(x)] := \overline{|n(x)|^2} &= \int_{-\infty}^{\infty} \int_{-\infty}^{\infty} e^{ix(\lambda-\mu)} \overline{d\zeta^*(\lambda) d\zeta(\mu)} \\ &= \sigma^2 \int_{-\infty}^{\infty} R(\lambda) d\lambda. \end{aligned} \tag{7.352}$$

It follows from formula (7.346) that the variance of the random process $A^{-1}n$ at any point x is finite if and only if the spectral density $\tilde{R}(\lambda)$ of the noise is such that $\tilde{R}(\lambda)|\tilde{A}(\lambda)|^{-2} \in L^1(-\infty,\infty)$. If $\tilde{A}(\lambda)$ tends to zero as $|\lambda| \to \infty$, the above condition imposes a severe restriction on the noise. For example, if the noise is white, that is $\tilde{R}(\lambda) = 1$, then the condition $|\tilde{A}(\lambda)|^{-2} \in L^1(-\infty,\infty)$ is not satisfied for $\tilde{A}(\lambda) \to 0$ as $|\lambda| \to \infty$.

Example 7.15 Consider the Hilbert space H of 2π-periodic functions

$$f = \sum_{m\neq 0} f_m \exp(imx) \tag{7.353}$$

with the inner product

$$(f,g) := \sum_{m\neq 0} f_m g_m^*. \tag{7.354}$$

Assume that $f(x) + n(x)$ is given, where $n \in H$

$$\overline{n(x)} = 0, \quad \overline{n^*(x)n(y)} = \sigma^2 R(x-y), \tag{7.355}$$

$$R(x+2\pi) = R(x). \tag{7.356}$$

Note that $R(x)$ can be written as

$$R(x) = \sigma^2 \sum_{m\neq 0} r_m \exp(-imx). \tag{7.357}$$

Suppose one wants to estimate $f'(x)$ given $f(x) + n(x)$. If one uses the function $\hat{u} := f'(x) + n'(x)$ as an estimate of f', then the variance of the error of this estimate can be calculated as follows. Let

$$n(x) = \sum_{m\neq 0} n_m \exp(imx), \tag{7.358}$$

where n_m, the Fourier coefficients of $n(x)$, are random variables, such that

$$\overline{n}_m = 0, \quad \overline{n_m^* n_j} = \sigma^2 r_m \delta_{mj}. \tag{7.359}$$

The numbers r_m can be determined easily. Indeed

$$\begin{aligned} R(x-y) = \overline{n^*(x)n(y)} &= \textstyle\sum_{m,j} \overline{n_m^* n_j} \exp(ijy - imx) \\ &= \sigma^2 \textstyle\sum r_m \exp\{-im(x-y)\}. \end{aligned} \tag{7.360}$$

From (7.356) and (7.360) one can see that the numbers r_m in (7.359) can be calculated by the formula

$$r_m = \frac{1}{2\pi}\int_{-\pi}^{\pi} R(x)\exp(imx)dx. \tag{7.361}$$

Since $R(x)$ is a covariance function, the numbers r_m are nonnegative as it should be. From (7.358) and (7.359) it follows that

$$\mathcal{D}[n'] = \sigma^2 \sum_m m^2 r_m. \tag{7.362}$$

Therefore the numbers r_m have to satisfy the condition that the right side of (7.362) is a convergent series, in order that the estimate $\hat{u}$ be statistically stable in the sense of Definition 7.7.

Example 7.16 Let $\mathcal{L}$ be a selfadjoint positive operator on a Hilbert space H. Assume that the spectrum of $\mathcal{L}$ is discrete:

$$0 \le \lambda_1 \le \lambda_2 \le \cdots, \quad \lambda_m \to \infty \quad \text{as} \quad m \to \infty. \tag{7.363}$$

Consider the problem

$$u_t = \mathcal{L}u, \quad t > 0 \tag{7.364}$$

$$u(0) = f. \tag{7.365}$$

Let ϕ_m be the eigenvectors of $\mathcal{L}$:

$$\mathcal{L}\phi_m = \lambda_m \phi_m \tag{7.366}$$

and assume that the system $\{\phi_m\}$, $1 \le m < \infty$, forms an orthonormal basis of H:

$$(\phi_m, \phi_j) = \delta_{mj}. \tag{7.367}$$

The formal solution to problem (7.364)-(7.365) is

$$u = \sum_{m=1}^{\infty} \exp(\lambda_m t) f_m \phi_m \tag{7.368}$$

where

$$f_m := (f, \phi_m). \tag{7.369}$$

Formula (7.368) gives a formal solution to (7.364)–(7.365) in the sense that formal differentiation in t yields

$$u_t = \sum_{m=1}^{\infty} \lambda_m \exp(\lambda_m t) f_m \phi_m, \tag{7.370}$$

and formal application of the operator $\mathcal{L}$ and formula (7.366) yield

$$\mathcal{L}u = \sum_{m=1}^{\infty} \lambda_m \exp(\lambda_m t) f_m \phi_m, \tag{7.371}$$

so that (7.370) and (7.371) yield (7.364). Put $t = 0$ in (7.368) and get (7.365). Formula (7.368) gives the strong solution to (7.364)-(7.365) in H if and only if the series (7.370) converges in H, that is

$$\sum_{m=1}^{\infty} \lambda_m^2 \exp(2\lambda_m t)|f_m|^2 < \infty, \quad t > 0. \tag{7.372}$$

This implies that the problem (7.364)–(7.365) is very ill-posed. This problem is an abstract heat equation with reversed time. If one takes $-\mathcal{L}$ in place of $\mathcal{L}$ in (7.364) then the problem is analogous to the usual heat equation and is well posed. Suppose that $H = L^2(D)$, and that the noisy data are given in (7.365), the function $f(x) + n(x)$ in place of $f(x)$ where $n(x)$ is noise. Let

$$n(x) = \sum_{m=1}^{\infty} n_m \phi_m(x), \tag{7.373}$$

where

$$n_m := (n(x), \phi(x)) \tag{7.374}$$

where the parentheses denote the inner product in $L^2(D)$. It is clear that if $\overline{n(x)} = 0$, then

$$\overline{n_m} = 0, \quad \forall m. \tag{7.375}$$

Let

$$\overline{n^*(x)n(y)} = \sigma^2 R(x,y). \tag{7.376}$$

Then

$$\sum_{m,j=1}^{\infty} \overline{n_m^* n_j} \phi_m^*(x)\phi_j(y) = \sigma^2 R(x,y). \tag{7.377}$$

The kernel $R(x,y)$ is selfadjoint and nonnegative definite, being a covariance function. Let us assume that the matrix

$$r_{mj} := \overline{n_m^* n_j} \tag{7.378}$$

is such that the series (7.377) converges in $L^2(D) \times L^2(D)$. If one uses the formula

$$\hat{u} := \sum_{m=1}^{\infty} \exp(\lambda_m t)(f_m + n_m)\phi_m \tag{7.379}$$

for the solution of the problem (7.364)-(7.365) with the noisy initial data, then for the variance of the error of this estimate one obtains

$$\mathcal{D}[u - \hat{u}] = \sum_{m,j=1}^{\infty} \exp[(\lambda_m + \lambda_j)t] r_{mj} \phi_m^*(x)\phi_j(x). \tag{7.380}$$

It is clear from (7.380) that the solution $\hat{u}$ is not statisticaly stable since the series (7.380) may diverge although the series (7.377) converges.

Exercise. Let (7.363)-(7.368) hold, $\| u(0) \| < \epsilon$ and $\| u(T) \| \leq c$. Prove that $\| u(t) \| \leq \epsilon^{1-\frac{t}{T}} c^{\frac{t}{T}}$ for $0 \leq t \leq T$.

Hint: Consider $\phi(t) := \| u(t) \|^2 = \sum_{m=1}^{\infty} \exp(-2\lambda_m t)|f_m|^2$. Check that $\phi'' > 0$ and $(\ln \phi)'' \geq 0$. Thus $\ln \phi$ is convex. Therefore

$$\ln \phi\,[(1-\alpha)0 + T] \leq (1-\alpha)\ln \phi(0) + \alpha \ln \phi(T), \quad 0 \leq \alpha \leq 1.$$

Let $\alpha = \frac{t}{T}$. Then the desired inequality follows.

The above examples lead to the following *question: how does one find a statistically stable estimate of the solution to equation (7.327)*?

Let us outline a possible answer to this question. We consider linear estimates, but the approach allows one to generalize the theory and to consider nonlinear estimates as well.

The approach is similar to the one outlined in Section 2.1. Let us look for a linear statistically stable estimate $\hat{u}$ of the solution to equation (7.327) of the form

$$\hat{u} = L(f + n). \tag{7.381}$$

Assume that the injective linear operator A in (7.327) is an integral operator on $H = L^2(D)$, $D \subset R^r$, with the kernel $A(x, y)$, and

$$Lf = \int_D L(x,y)f(y)dy. \tag{7.382}$$

Since $Au = f$, one has:

$$\overline{|\hat{u} - u|^2} = \overline{|(LA - I)u + Ln|^2} = |(LA - I)u|^2 + \overline{|Ln|^2} \tag{7.383}$$

where I is the identity operator and the term linear with respect to n vanishes because $\overline{n} = 0$. Let us calculate the last term in (7.383):

$$\begin{aligned} \overline{|Ln|^2} &= \overline{\int_D L^*(x,y)n^*(y)dy \int_D L(x,z)n(z)dz} = \\ &= \sigma^2 \int_D \int_D L^*(x,y)R(y,z)L'(z,x)dydz. \end{aligned} \tag{7.384}$$

Here we used second formula (7.328) and the standard notation

$$L'(z,x) := L(x,z). \tag{7.385}$$

Remember that star denotes complex conjugate (and not the adjoint operator).

Integrating both sides of (7.383) in x over D yields

$$\epsilon := \overline{\| \hat{u} - u \|^2} = \| (LA - I)u \|^2 + \sigma^2 TrQ \tag{7.386}$$

where Q is an integral operator with the kernel

$$Q(x,\xi) := \int_D \int_D L^*(x,y)R(y,z)L'(z,\xi)dydz \tag{7.387}$$

and TrQ stands for the trace of the operator Q. This operator is clearly nonnegative definite in $H = L^2(D)$:

$$(Q\phi, \phi) = (L^* RL'\phi, \phi) = (RL'\phi, L'\phi) \geq 0. \tag{7.388}$$

Here we used the fact that R is nonnegative definite; that $(L^*)^\dagger = L'$, where $A^\dagger$ denotes the adjoint operator in $L^2(D)$; and we have assumed that the function $Q(x, \xi)$ is continuous in $x, \xi \in \overline{D}$. The last assumption and the fact that the kernel $Q(x, \xi)$ is nonnegative definite imply that

$$TrQ = \int_D Q(x,x)dx. \tag{7.389}$$

One wants to choose L such that

$$\epsilon = \min \tag{7.390}$$

where ϵ is given in (7.386). If one put $L = A^{-1}$ then the first term on the right side of (7.386) vanishes and the second is finite if

$$Tr\left\{(A^{-1})^* R (A^{-1})'\right\} < \infty. \tag{7.391}$$

We assume that (7.391) holds. We claim that if (7.391) holds and $\sigma \to 0$ then one can choose L so that

$$\epsilon_{\min} := \epsilon(\sigma) \to 0 \quad \text{as} \quad \sigma \to 0. \tag{7.392}$$

Any such choice of L yields a statistically stable estimate of the solution u.

Let us prove the claim that the choice of L which implies (7.392) is possible. For simplicity we assume that A is positive selfadjoint operator on H.

Put

$$L = (A + \delta I)^{-1} \tag{7.393}$$

where $\delta > 0$ is a small number. Then the spectral theory yields:

$$\begin{aligned} \|(LA - I)u\|^2 &= \int_0^{\|A\|} \left(\frac{\lambda}{\lambda+\delta} - 1\right)^2 d(E_\lambda u, u) \\ &= \int_0^{\|A\|} \frac{\delta^2}{(\lambda+\delta)^2} d(E_\lambda u, u) \end{aligned} \tag{7.394}$$

where E_λ is the resolution of the identity of the operator A. Since

$$\int_0^{\|A\|} \frac{\delta^2}{(\lambda+\delta)^2} d(E_\lambda u, u) \le \int_0^{\|A\|} d(E_\lambda u, u) = \| u \|^2 < \infty \tag{7.395}$$

and, as $\delta \to 0$, the integrand in (7.394) tends to zero, one can use the Lebesgue dominant convergence theorem and conclude that

$$\|(LA - I)u\|^2 := \eta(\delta, u) \to 0 \quad \text{as} \quad \delta \to 0 \tag{7.396}$$

where L is given by (7.393). The claim is proved.

Lemma 7.14 *If L is defined by (7.393) and (7.391) holds then*

$$\limsup_{\delta \to 0} Tr\left\{L^* R L'\right\} < \infty. \tag{7.397}$$

Proof. One has

$$\left[(A+\delta I)^{-1}\right]^* R\left[(A+\delta I)^{-1}\right]'$$
$$= \left[(A+\delta I)^{-1}\right]^* A^*(A^{-1})^* R(A^{-1})' A' \left[(A+\delta I)^{-1}\right]'. \quad (7.398)$$

The operator $(A^{-1})^*R(A^{-1})'$ is in the trace class by (7.391). Moreover, if $A > 0$ and $\delta > 0$, then

$$\left\| \left[(A+\delta I)^{-1}\right]^* A^* \right\| \le 1, \quad \left\| A' \left[(A+\delta I)^{-1}\right]' \right\| \le 1. \quad (7.399)$$

Both inequalities (7.399) can be proved similarly or reduced one to the other, because $A^* = A'$. Note that $A > 0$ implies

$$A^* > 0. \quad (7.400)$$

Indeed, for any $\phi \in H$, one has

$$(A^*\phi, \phi)^* = (A\phi^*, \phi^*) > 0 \quad (7.401)$$

since $A > 0$ by the assumption and $\phi^* \in H$. Therefore

$$(A^*\phi, \phi)^* = (A^*\phi, \phi) > 0 \quad \forall \phi \in H. \quad (7.402)$$

The desired estimate (7.399) follows from the spectral theorem:

$$\left\| \left[(A+\delta I)^{-1}\right]^* A^* \right\| = \max_{0<\lambda\le\|A^*\|} \frac{\lambda}{\lambda+\delta} \le 1. \quad (7.403)$$

Lemma 7.14 is proved. □

It is now easy to prove the following theorem.

Theorem 7.6 *Let $A > 0$ be a bounded operator on $H = L^2(D)$. Assume that condition (7.328) holds and $TrR < \infty$. Then the estimate*

$$\hat{u} = L(f+n), \quad (7.404)$$

with L given by (7.393), is statistically stable (in the sense (7.331)) estimate of the solution to equation $Au = f$, provided that parameter $\delta = \delta(\sigma)$ in formula (7.393) is chosen so that

$$\epsilon := \sigma^2 Tr(L^*RL') + \eta(\delta, u) = \min. \quad (7.405)$$

Proof. Note that we do not assume in this theorem that condition (7.391) holds. Therefore
$Tr(L^*RL') := \psi(\delta) > 0$ will, in general, satisfy the condition

$$\psi(\delta) \to +\infty \quad \text{as} \quad \delta \to 0. \tag{7.406}$$

Thus

$$\epsilon = \sigma^2\psi(\delta) + \eta(\delta, u). \tag{7.407}$$

From (7.406), (7.407) and (7.396) it follows that the function ϵ considered as a function of δ for a fixed $\sigma > 0$ attains its minimum at $\delta = \delta(\sigma)$ and

$$\delta(\sigma) \to 0 \quad \text{as} \quad \sigma \to 0. \tag{7.408}$$

Therefore

$$\epsilon(\sigma) = \epsilon_{\min} = \epsilon\left(\delta(\sigma)\right) \to 0 \quad \text{as} \quad \sigma \to 0. \tag{7.409}$$

Theorem 7.6 is proved. □

If some estimates for $\psi(\delta)$ and $\eta(\delta, u)$ are found then an estimate of $\epsilon(\sigma)$ can be obtained. This requires some a priori assumptions about the solution.

Example 7.17 A simple estimate for $\psi(\delta)$ is the following one.

$$Tr(L^*RL') \leq TrR \parallel L^* \parallel^2 \leq \frac{TrR}{\delta^2}. \tag{7.410}$$

Here we used the estimates

$$\parallel L^* \parallel = \parallel L' \parallel \leq \delta^{-1}, \tag{7.411}$$

where L is given by (7.393), and the estimate

$$Tr(L^*RL') \leq TrR \parallel L \parallel^2 . \tag{7.412}$$

We will prove inequality (7.412) later. Let us estimate $\eta(\delta, u)$. To do this, assume that

$$\parallel A^{-a}f \parallel \leq c, \quad a = 1 + b, \quad b > 0, \tag{7.413}$$

where $c > 0$ is a constant, and

$$A^{-a}f := \int_0^{\|A\|} \lambda^{-a} dE_\lambda f, \tag{7.414}$$

$$\| A^{-a} f \|^2 = \int_0^{\|A\|} \lambda^{-2a} d(E_\lambda f, f) \leq c^2. \tag{7.415}$$

Since $Au = f$, $u = A^{-1}f$ it follows from (7.396) and (7.415) that

$$\begin{aligned}\eta(\delta, u) := \eta(\delta) &= \int_0^{\|A\|} \frac{\delta^2}{(\lambda+\delta)^2} \lambda^{-2} d(E_\lambda f, f) \\ &= \int_0^{\|A\|} \frac{\delta^2 \lambda^{2b}}{(\lambda+\delta)^2} \lambda^{-2-2b} d(E_\lambda f, f) \leq \delta^{2b} \int_0^{\|A\|} \lambda^{-2a} d(E_\lambda f, f) \\ &\leq c^2 \delta^{2b}.\end{aligned} \tag{7.416}$$

Therefore, under the a priori assumption (7.413) about f, one has

$$\epsilon \leq \frac{\sigma^2 TrR}{\delta^2} + c^2 \delta^{2b}. \tag{7.417}$$

The right hand side in (7.417) attains its minimum in δ ($\sigma > 0$ being fixed) at

$$\delta_{\min} = \delta(\sigma) = \left(\frac{TrR}{c^2 b} \right)^{\frac{1}{2a}} \sigma^{1/a} \tag{7.418}$$

and

$$\epsilon_{\min} = \epsilon(\sigma) \leq \text{const} \sigma^{2\ b/a} \tag{7.419}$$

where const can be written explicitly.

Let us finally prove inequaity (7.412). This inequality follows from

Lemma 7.15 *If B is a linear bounded operator on H and $R \geq 0$ is a trace class operator, then BR and RB are trace class operators and*

$$|Tr(BR)| \leq \| B \| TrR, \quad Tr(RB) \leq \| B \| TrR. \tag{7.420}$$

Proof. Let us recall that a linear operator $T : H \to H$ is in the trace class if and only if

$$\| T \|_1 := \sum_{j=1}^{\infty} s_j(T) < \infty, \tag{7.421}$$

where $s_j(T)$ are the s-numbers of T. These numbers are defined by the equality

$$s_j(R) = \lambda_j \left\{ (T^*T)^{1/2} \right\} \tag{7.422}$$

where $\lambda_1 \geq \lambda_2 \geq \cdots \geq 0$ are the eigenvalues of the nonnegative definite selfadjoint operator $(T^\dagger T)^{1/2}$, and $T^\dagger$ is the adjoint of T in H. The minimax principle for the s-values is

$$s_{j+1}(T) = \min_{\mathcal{L}_j} \max_{\substack{\phi \perp \mathcal{L}_j \\ \phi \neq 0}} \frac{\| T\phi \|}{\| \phi \|} \tag{7.423}$$

where $\mathcal{L}_n$ runs through all j-dimensional subspaces of H, and $\phi \perp \mathcal{L}_j$ means that ϕ is orthogonal to all elements of $\mathcal{L}_j$. If B is a linear bounded operator on H then

$$\begin{aligned} s_{j+1}(BT) &= \min_{\mathcal{L}_j} \max_{\substack{\phi \perp \mathcal{L}_j \\ \phi \neq 0}} \frac{\| BT\phi \|}{\| \phi \|} \leq \| B \| \min_{\mathcal{L}_j} \max_{\substack{\phi \perp \mathcal{L}_j \\ \phi \neq 0}} \frac{\| T\phi \|}{\| \phi \|} \\ &= \| B \| \, s_{j+1}(T). \end{aligned} \tag{7.424}$$

Therefore if (7.421) holds then

$$\| BT \|_1 = \sum_{j=1}^{\infty} s_j(BT) \leq \| B \| \sum_{j=1}^{\infty} s_j(T) = \| B \| \| T \|_1 . \tag{7.425}$$

The first part of Lemma 7.15 is proved since if $R \geq 0$ one has $TrR = \| R \|_1$. The second part can be reduced to the first. Indeed, T and T^* are simultaneously in the trace class since

$$s_j(T) = s_j(T^*), \quad \forall j. \tag{7.426}$$

One has

$$(TB)^* = B^* T^*. \tag{7.427}$$

Since $\| B \| = \| B^* \|$, and $\overline{(TrT)} = TrT^*$, one concludes from (7.427) and (7.425) that

$$|tr(TB)| = |tr(B^*T^*)| \leq \| B^* \| \| T^* \|_1 = \| B \| \| T \|_1 . \tag{7.428}$$

Take $T = R \geq 0$ then $\| T \|_1 = TrR$ and the second inequality (7.420) is obtained. Lemma 7.15 is proved. □

Additional information about s-values one can find in Section 6.3.

7.7 A remark on nonlinear (polynomial) estimates

Let

$$\mathcal{U}(x) = s(x) + n(x), \quad x \in D \subset R^r. \tag{7.429}$$

Consider the polynomial estimate (filter):

$$A\mathcal{U} := \sum_{j=1}^{m} H_j \mathcal{U}^{[j]} \tag{7.430}$$

where

$$H_j \mathcal{U}^{[j]} = \int_D \cdots \int_D h_j(x, \xi_1, \dots, \xi_j) \mathcal{U}(\xi_1) \dots \mathcal{U}(\xi_j) d\xi_1, \dots, d\xi_j \tag{7.431}$$

$$\mathcal{U}^{[j]} = \mathcal{U}(\xi_1) \dots \mathcal{U}(\xi_j). \tag{7.432}$$

The problem is to find A such that

$$\epsilon := \mathcal{D}[A\mathcal{U} - s] = \min. \tag{7.433}$$

Here $\mathcal{D}$ is the symbol of variance, the assumptions about $s(x)$ and $n(x)$ are the same as in Chapter 1, the optimal estimate is defined by n functions $(h_1, \dots, h_n)$, and we could consider by the same method the problem of estimating a known operator on $s(x)$, for example $\partial_j s(x)$. Let us substitute (7.430) in (7.433):

$$\begin{aligned}
\epsilon &:= \sum_{i,j=1}^{m} \overline{H_j \mathcal{U}^{[j]} H_i^* \mathcal{U}^{[i]*}} - 2Re \sum_{i=0}^{m} H_i^* \overline{\mathcal{U}^{[i]*} s(x)} + \overline{|s(x)|^2} \\
&= \sum_{i,j=1}^{m} a_{ij} H_j H_i^* - 2Re \sum_{i=0}^{m} H_i b_i + \overline{|s(x)|^2} \\
&= \min
\end{aligned} \tag{7.434}$$

Here

$$b_i := \overline{\mathcal{U}^{[i]*} s(x)}, \quad b_i = b_i(x, \xi_1', \dots, \xi_i') \tag{7.435}$$

$$a_{ij} := \overline{\mathcal{U}^{[j]} \mathcal{U}^{[i]*}} = \overline{\mathcal{U}(\xi_1) \dots \mathcal{U}(\xi_j) \mathcal{U}^*(\xi_1') \dots \mathcal{U}^*(\xi_i')}, \tag{7.436}$$

$$\sum_{j=1}^{m} a_{ij} H_j$$
$$:= \sum_{j=0}^{m} \int_D \cdots \int_D a_{ij}(\xi'_1, \ldots, \xi'_i, \xi_1, \ldots, \xi_j) h_j(x, \xi_1, \ldots, \xi_j) d\xi_1 \ldots d\xi_j. \tag{7.437}$$

Note that (7.435) implies that

$$a^*_{ij} = a_{ji}, \quad a_{ij} = a_{ij}(\xi'_1, \ldots, \xi'_i, \xi_1, \ldots, \xi_j). \tag{7.438}$$

Let

$$h_j(\xi_1, \ldots, \xi_j) + \epsilon_j \eta_j(\xi_1, \ldots, \xi_j) \tag{7.439}$$

be substituted for h_j in (7.437) and we surpress exibiting dependence on x since x will be fixed. Here ϵ_j are numbers. The condition $\epsilon = \min$ at $\epsilon_j = 0$ implies that

$$\sum_{j=1}^{m} a_{ij} H_j = b_i, \quad 1 \leq i \leq m. \tag{7.440}$$

This is a system of integral equations for the functions $h_j(x, \xi_1, \ldots, \xi_j)$:

$$\sum_{j=1}^{m} \underbrace{\int_D \cdots \int_D}_{j\text{times}} a_{ij}(\xi'_1, \ldots, \xi'_i, \xi_1, \ldots, \xi_j) h_j(x, \xi_1, \ldots, \xi_j) d\xi_1 \ldots$$
$$d\xi_j = b_i(x, \xi'_1, \ldots, \xi'_i). \tag{7.441}$$

Consider as an example the case of polynomial estimates of degree 2. Then

$$\int_D a_{11}(\xi'_1, \xi_1) h_1(\xi_1) d\xi_1 + \int_D \int_D a_{12}(\xi'_1, \xi_1, \xi_2) h_2(\xi_1, \xi_2) d\xi_1 d\xi_2 = b_1(\xi'_1) \tag{7.442}$$

$$\int_D a_{21}(\xi'_1, \xi'_2, \xi_1) h_1(\xi_1) d\xi_1 + \int_D \int_D a_{22}(\xi'_1, \xi'_2, \xi_1, \xi_2) h_2(\xi_1, \xi_2) d\xi_1 d\xi_2$$
$$= b_2(\xi'_1, \xi'_2). \tag{7.443}$$

If $a_{22}(\xi'_1, \xi'_2, \xi_1, \xi_2)$ belongs to $\mathcal{R}$, or to some class of the operators which can be inverted, one can find $h_2(\xi_1, \xi_2)$ from equation (7.443) in terms of $h_1(\xi_1)$ and then (7.442) becomes an equation for one function $h_1(\xi_1)$.

In the framework of correlation theory it is customary to consider only linear estimates because the data (covariance functions) consist of the moments of second order.

Chapter 8

Auxiliary Results

8.1 Sobolev spaces and distributions

8.1.1 *A general imbedding theorem*

Let $D \subset R^r$ be a bounded domain with a smooth boundary Γ. The $L^p(D)$, $p \geq 1$ spaces consist of measurable functions on D such that $\| u \|_{L^p(D)} := (\int_D |u|^p dx)^{1/p} < \infty$. For $p = +\infty$ one has $\| u \|_{L^\infty(D)} := \text{ess sup}_{x \in D} |u(x)|$. If in place of the Lebesgue measure a measure μ is used then we use $L^p(D, \mu)$ as the symbol for the corresponding space. These are Banach spaces. If $C_0^\infty(D)$ is the set of infinitely differentiable functions with compact support in D, then $C_0^\infty(D)$ is dense in $L^p(D)$, $1 \leq p < \infty$. If one defines a mollifier, i.e. a function $0 \leq \rho(x) \in C_0^\infty(R^r)$, $\rho(x) = 0$ for $|x| \geq 1$, $\int \rho(x)dx = 1$, $\int := \int_{R^r}$, e.g.

$$\rho(x) := c\exp\{(|x|^2 - 1)^{-1}\} \quad \text{for} \quad |x| < 1, \rho(x) = 0 \quad \text{for} \quad |x| \geq 1 \quad (8.1)$$

where c is the normalizing constant chosen so that $\int \rho dx = 1$, then the function

$$u_\epsilon(x) := \epsilon^{-n} \int \rho(|x - y|\epsilon^{-1})u(y)dy, \quad \epsilon > 0, \quad (8.2)$$

belongs to C_{loc}^∞ and $\| u_\epsilon - u \|_{L_{\text{loc}}^p} \to 0$ as $\epsilon \to 0$. By L_{loc}^p one means a set of functions which belong to L^p on any compact subset of D or R^r. Convergence in $L_{\text{loc}}^p(D)$ means convergence in $L^p(\tilde{D})$ where $\tilde{D}$ is an arbitrary compact subset in D. By $W^{\ell,p}(D)$, the Sobolev space, one means the Banach space of functions $u(x)$ defined on D with the finite norm

$$\| u \|_{W^{\ell,p}(D)} := \sum_{|j|=0}^{\ell} \| D^j u \|_{L^p(D)} . \quad (8.3)$$

Here j is a multiindex, $j = (j_1, \dots, j_r)$, $D^j = D^{j_1}_{x_1} \dots D^{j_r}_{x_r}$, $|j| = j_1 + \dots + j_r$. The space $C^\infty(D)$ of infinitely differentiable in D functions is dense in $W^{\ell,p}(D)$. By $\dot{W}^{\ell,p}(D)$ we denote the closure of $C_0^\infty(D)$ in the norm (8.3). By $H^\ell(D)$ we denote $W^{\ell,2}(D)$. This is a Hilbert space with the inner product

$$(u, v)_\ell := \int_D \sum_{j=0}^{\ell} D^j u D^j v^* dx, \quad \| u \|_\ell := (u, u)_\ell^{1/2}. \tag{8.4}$$

Let $C^\infty(\overline{D})$ denote the space of restrictions to D of functions in $C^\infty(R^r)$.

If the boundary of D is not sufficiently smooth, then $C^\infty(\overline{D})$ may be not dense in $W^{\ell,p}(D)$. A sufficient condition on Γ for $C^\infty(\overline{D})$ to be dense in $W^{\ell,p}(D)$ is that D is bounded and star shaped with respect to a point. D is called star shaped with respect to a point 0 if any ray issued from 0 intersects $\Gamma := \partial D$ at one and only one point. Another sufficient condition for $C^\infty(\overline{D})$ to be dense in $W^{\ell,p}(D)$ is that every point of Γ has a neighborhood U in which $D \cap U$ is representable in a suitable Cartesian coordinate system as $x_r < f(x_1, \dots x_{r-1})$, where f is continuous.

Any function $u \in W^{\ell,p}(D)$, $p \geq 1$, $\ell \geq 1$, (possibly modified on a set of Lebesgue R^r-measure zero) is absolutely continuous on almost all straight lines parallel to coordinate axes, and its distributional first derivatives coincide with the usual derivatives almost everywhere. The spaces $W^{\ell,p}(D)$ are complete.

We say that a bounded domain $D \subset R^r$ satisfies a uniform interior cone condition if there is a fixed cone C_D such that each point of Γ is the vertex of a cone $C_D(x) \subset \overline{D}$ congruent to C_D. A strict cone property holds if Γ has a locally finite covering by an open set $\{U_j\}$ and a corresponding collection of cones $\{C_j\}$ such that $\forall x \in U_j \cap \Gamma$ one has $x + C_j \in D$. According to Calderon's extension theorem, there exists a bounded linear operator $E : W^{\ell,p}(D) \to W^{\ell,p}(R^r)$ such that $Tu = u$ on D for every $u \in W^{\ell,p}(D)$ provided that $D \subset C^{0,1}$. The class $C^{0,1}$ of domains consists of bounded domains D such that each point $x \in \Gamma$ has a neighborhood U with the property that the set $D \cap U$ is represented by the inequality $x_r < f(x_1, \dots, x_{r-1})$ in a Cartesian coordinate system, and the function f is Lipschitz-continuous. The domains in $C^{0,1}$ have the cone property. The extension theorem holds for a wider class of domains than $C^{0,1}$, but we do not go into details.

Let us formulate a general embedding theorem ([Mazja (1986)]).

Theorem 8.1 *Let $D \subset R$ be a bounded domain with a cone property, and let μ be a measure on D such that*

$$\sup \rho^{-s} \mu(D \cap B(x,\rho)) < \infty, \quad B(x,\rho) = \{y : |x-y| \le \rho, y \in R^r\}, \quad x \in R^r$$

and $s > 0$. (If $s \le n$ is an integer, then μ can be s-dimensional Lebesgue measure on $D \cap \Gamma_s$, where Γ_s is an s-dimensional smooth manifold).

Then, for any $u \in C^\infty(D) \cap W^{\ell,p}(D)$, one has

$$\sum_{j=0}^{k} \| D^j u \|_{L^q(D,\mu)} \le c \| u \|_{W^{\ell,p}(D)} \tag{8.5}$$

where $c = const > 0$ does not depend on u. Here the parameters q, s, ℓ, p, k satisfy one of the following sets of restrictions:

a) $p > 1$, $0 < r - p(\ell - k) < s \le r$, $q \le sp[r - p(\ell - k)]^{-1}$
b) $p = 1$, $0 < r - \ell + k \le s \le r$, $q \le s(r - \ell + k)^{-1}$
c) $p > 1$, $r = p(\ell - k)$, $s \le r$, $q > 0$ *is arbitrary if*
d) $p > 1$, $r < p(\ell - k)$ *or*
e) $p = 1$, $r \le \ell - k$ *then*

$$\sum_{j=0}^{k} \sup_{x \in D} |D^j u| \le c \| u \|_{W^{\ell,p}(D)} . \tag{8.6}$$

If
f) $p \ge 1$, $(\ell - k - 1)p < r < (\ell - k)p$, *and* $\lambda := \ell - k - rp^{-1}$ *then*

$$\sup_{\substack{x,y \in D \\ x \ne y}} |x-y|^{-\lambda} |D^k u(x) - D^k u(y)| \le C \| u \|_{W^{\ell,p}(D)} . \tag{8.7}$$

If
g) $(\ell - k - 1)p = r$ *then (8.7) holds for all* $0 < \lambda < 1$.
The imbedding operator $i : W^{\ell,p}(D) \to W^{k,q}(D_s \cap D)$ is compact if $s > 0$, $r > (\ell - k)p$, $r - (\ell - k)p < s \le r$, and $q < sp[r - (\ell - k)p]^{-1}$.
If $r = (\ell - k)p$, $q \ge 1$, $s \le r$ then the above imbedding operator i is compact.
If $r < (\ell - k)p$ then $i : W^{\ell,p}(D) \to C^k(D)$ is compact.
The trace operator $i : H^\ell(D) \to H^{\ell-1/2}(\Gamma)$ is bounded if $\ell > \frac{1}{2}$.

8.1.2 Sobolev spaces with negative indices

We start with the brief exposition of the facts in distribution theory which will be used later. Let $S(R^r)$ be the Schwartz's space of $C^\infty(R^r)$ functions which decay with all their derivatives faster than any negative power of $|x|$ as $|x| \to \infty$, so that

$$|\phi|_m := \max_{x\in R^n}(1+|x|)^m \sum_{j=0}^{m} |D^j\phi(x)| < \infty, \quad 0 \le m < \infty. \tag{8.8}$$

The set of the norms $|\cdot|_m$ defines the topology of $S(R^r)$. A sequence ϕ_n converges to ϕ in S if $|\phi_n - \phi|_m \to 0$ as $n \to \infty$ for $0 \le m < \infty$. It is easy to see that $C_0^\infty(R^r)$ is dense in S in the above sense. A space S' of tempered distributions is the space of linear continuous functionals on S. Continuity of a linear functional f on S means that $(f, \phi_n) \to (f, \phi)$ if $\phi_n \to \phi$ in S. By (f, ϕ) the value of f at the element ϕ is denoted. A wider class of distributions is used often, the class $\mathcal{D}'$ of linear continuous functionals over the space $\mathcal{D} = C_0^\infty(R^r)$ of test functions. Continuity of $f \in \mathcal{D}'$ means that for any compact set $K \subset R^r$ there exist constants c and m such that $|(f,\phi)| \le c\sum_{|j|\le m}\sup_{x\in K}|D^j\phi|$, $\phi \in C_0^\infty(K)$. By the derivative of a distribution $f \in S'$ one means a distribution $f' \in S'$, such that

$$(f', \phi) = -(f, \phi') \quad \forall \phi \in S. \tag{8.9}$$

Also

$$(D^m f, \phi) = (-1)^{|m|}(f, D^m\phi), \quad \forall\phi \in S.$$

Let D be an open set in R^r, and $S(D)$ be the completion of $C_0^\infty(D)$ in the topology of $S(R^r)$. A continuous linear functional on $S(D)$ is a distribution in D. The space of these distributions is denoted $S'(D)$. If $(F, \phi) = (f, \phi)$ for all $\phi \in S(D)$, some $F \in S'(R^r)$ and some $f \in S'(D)$, then f is called the restriction of F to D and we write $f = pF$. Since $S(D)$ is, by definition, a closed linear subspace of $S(R^r)$, the Hahn-Banach theorem says that a linear continuous functional $f \in S'(D)$ can be extended from $S'(D)$ to $S'(R^r)$. Let us denote this extension by $Ef = F$. The space $S'(R^r)$ is a Frechet space, i.e. a complete locally convex metrizable space, so that Hahn-Banach theorem holds. If $F \in S'(R^r)$ then one says that $F = 0$ in an open set D if $(F, \phi) = 0$ for all $\phi \in S(D)$, or which is equivalent, for all $\phi \in C_0^\infty(D)$. If D is the maximal open set on which $F = 0$, one says that $\overline{\Omega} := R^r \setminus D$ is the support of F. By $\overline{\Omega}$ the closure of Ω is denoted. If F

is locally integrable, i.e. a usual function, then supp F is the complement of the maximal open set on which $F = 0$ almost everywhere in the sense of Lebesgue measure in R^r. Note that supp $D^m F \subseteq$ supp F. The Fourier transform of $f \in S'(R^r)$ is defined by

$$(\tilde{f}, \tilde{\phi}^*) = (2\pi)^r (f, \phi^*) \tag{8.10}$$

where the star stands for complex conjugation, the tilde stands for the Fourier transform, and

$$\tilde{\phi} := \int \phi(x) \exp(i\lambda \cdot x) dx, \int := \int_{R^r}, \phi \in S(R^r).$$

This definition of $\tilde{f}$ is based on the Parseval equality for functions: $(\tilde{\phi}_1, \tilde{\phi}_2^*) = 2\pi(\phi_1, \phi_2^*)$, and on the fact that the Fourier transform is a linear continuous bijection of $S(R^r)$ onto itself. Note that

$$\widetilde{D^m f} = (-i\lambda)^m \tilde{\phi} \tag{8.11}$$

$$\mathcal{F}(f * \phi) = \tilde{f}\tilde{\phi}, \quad \mathcal{F}f := \tilde{f} \tag{8.12}$$

where $f * \phi$ is the convolution of $f \in S'(R^r)$ and $\phi \in S(R^r)$, defined by $f * \phi = (f, \phi(x - y))$. Here $x \in R^r$ is fixed and $(f, \phi(x - y))$ denotes the value of f at the element $\phi(x - y) \in S(R^r)$. The function $f * \phi$ is infinitely differentiable.

Any $f \in S'(R^r)$ can be represented in the form

$$f = \sum_{|j| \leq m} D^j f_j \tag{8.13}$$

where f_j are locally continuous functions which satisfy the inequality

$$|f_j(x)| \leq c(1 + |x|)^N, \quad |j| \leq m, \tag{8.14}$$

m and N are some integers, and $c = \text{const} > 0$. The space $H^\ell(R^r)$ one can define either as the closure of $C_0^\infty(R^r)$ in the norm $\| \cdot \|_\ell := \| \cdot \|_{H^\ell}$, or by noting that, for $\phi \in C_0^\infty(R^r)$, one can define an equivalent norm by the formula:

$$\| \phi \|_\ell^2 = (2\pi)^{-r} \int (1 + |\lambda|^2)^\ell |\tilde{\phi}|^2 d\lambda, \tag{8.15}$$

and

$$(u, v)_\ell := \int (1 + |\lambda|^2)^\ell u\bar{v} d\lambda, \quad (u, v) := (u, v)_0. \tag{8.16}$$

In (8.15) one can assume $-\infty < \ell < \infty$. If $f \in H^\ell(R^r)$ and $\phi \in S(R^r)$ then

$$|(f,\phi)| \leq (2\pi)^{-r} \int |\tilde{f}||\tilde{\phi}|d\lambda \leq \| f \|_\ell \| \phi \|_{-\ell}, \tag{8.17}$$

where

$$\| \phi \|_{-\ell} := (2\pi)^{-r/2} \left(\int (1+|\lambda|^2)^{-\ell} |\tilde{\phi}|^2 d\lambda \right)^{1/2}. \tag{8.18}$$

It is clear from (8.17) that one can define $H^{-\ell}(R^r)$ as the closure of $S(R^r)$ in the norm

$$\| \phi \|_{-\ell} = \sup_{f \in H^\ell(R^r) f \neq 0} \{ |(f,\phi)| / \| f \|_\ell \}. \tag{8.19}$$

Thus $H^{-\ell}(R^r)$ is the dual space to $H^\ell(R^r)$ with respect to the pairing given by the inner product $(\cdot,\cdot)_0$. Consider the spaces $\dot{H}^\ell(D)$ and $\dot{H}^\ell(\Omega)$, $-\infty < \ell < \infty$, of functions belonging to $H^\ell(R^r)$ with support in $\overline{D}$ and $\overline{\Omega}$ respectively, $\Omega := R^3 \setminus D$. These spaces are closed in $H^\ell(R^r)$ norm subspaces of $H^\ell(R^r)$ which can be described as completions of $C_0^\infty(D)$ and $C_0^\infty(\Omega)$ in the norm $H^\ell(R^r)$. If $f \in \dot{H}^\ell(D)$ then $(f,\phi) = 0 \ \forall \phi \in C_0^\infty(\Omega)$.

Consider the space $H^\ell(D)$, where $D \subset EH^\ell$, $\ell \geq 0$. This means that D has the property that there exists an extension operator $E : H^\ell(D) \to H^\ell(R^r)$ with the property $Ef = f$ in D, $\| Ef \|_{H^\ell(R^r)} \leq c \| f \|_{H^\ell(D)}$. A similar definition can be given for $D \subset EW^{\ell,p}$. Bounded domains $D \subset C^{0,1}$ are in the class EH^ℓ (and in $EW^{\ell,p}$). The property $Ef = f$ in D for $\ell < 0$ means that $pEf = f$, where $Ef \in H^\ell(R^r) \subset S'(R^r)$, $\ell < 0$, and p is the restriction operator $p : S'(R^r) \to S'(D)$. Thus, for any ℓ, $-\infty < \ell < \infty$, we consider the linear space of restrictions of elements $f \in H^\ell(R^r)$, $-\infty < \ell < \infty$, to $D \subset C^{0,1}$.

Define a norm on this space by the formula

$$\| f \|_\ell := \| f \|_{H^\ell(D)} = \inf_E \| Ef \|_{H^\ell(R^r)}, \quad -\infty < \ell < \infty, \tag{8.20}$$

where the infimum is taken over all extensions of f belonging to $H^\ell(R^r)$. If Ef is such an extension, then $Ef + f_-$ is also such an extension for any $f_- \in \dot{H}^\ell(\Omega)$.

If $\ell \geq 0$ and $D \subset EH^\ell$ then the norm (8.20) is equivalent to the usual norm (1.4).

Let $(H^\ell(D))'$ denote the dual space to $H^\ell(D)$ with respect to $L^2(D) = H^0(D)$. This notion we introduce in an abstract form. Let H_+ and H_0 be a pair of Hilbert spaces, $H_+ \subset H_0$, H_+ is dense in H_0 in H_0 norm and

$\| f \|_+ \geq \| f \|_0$. Define the dual to H_+ space $H'_+ := H_-$ as follows. Note that if $f \in H_0$ and $\phi \in H_+$ then

$$|(f,\phi)| \leq \| f \|_0 \| \phi \|_0 \leq \| f \|_0 \| \phi \|_+, \tag{8.21}$$

where $(f,\phi) = (f,\phi)_0$ is the inner product in H_0. Define

$$\| f \|_- := \sup_{\substack{\phi \in H_+ \\ \phi \neq 0}} |(f,\phi)| \, \| \phi \|_+^{-1} . \tag{8.22}$$

It follows from (8.21) that $\| f \|_- \leq \| f \|_0$, and that (f,ϕ) is a bounded linear functional on H_+. By Riesz's theorem, $(f,\phi)_0 = (If,\phi)_+$, where $I : H_0 \to H_+$ is a linear bounded operator, $\| If \|_+ \leq \| f \|_0$. Define H_- as the completion of H_0 in the norm (8.22). Clearly $H_+ \subset H_0 \subset H_-$. The triple $\{H_+, H_0, H_-\}$ is called a rigged triple of Hilbert space. The space H_- is called the dual space to H_+ with respect to H_0. The inner product in H_- can be defined by the formula

$$(f,g)_- := (If, Ig)_+. \tag{8.23}$$

Indeed, the operator I was defined as an operator from H_0 into H_+. Therefore, for $f, g \in H_0$ the right side of (8.23) is well defined. Consider the completion of H_0 in the norm $\| f \|_- = \| If \|_+$. Then we obtain H_-. Note that the right side of (8.22) can be written as $\| If \|_+$. Therefore, the operator I is now defined as a linear isometry from H_- into H_+. In fact, this isometry is onto H_+. Indeed, suppose $(If,\phi)_+ = 0 \ \forall f \in H_-$. We wish to prove that $\phi = 0$. If $f \in H_0$, then $0 = (If,\phi)_+ = (f,\phi)_0$. Thus, $\phi = 0$. If one considers H_+ as the space of test functions, then H_- is the corresponding space of distributions.

In this sense, we treat $(H^\ell(D))'$ as the space of distributions corresponding to $H_+ = H^\ell(D)$. Since $\dot{H}^\ell(D) \subset H^\ell(D)$, one has $(H^\ell(D))' \subset (\dot{H}^\ell(D))'$. One of the results we will need is the following:

$$\begin{aligned} (\dot{H}^\ell(D))' &= H^{-\ell}(D), \quad -\infty < \ell < \infty, \\ (H^\ell(D))' &= \dot{H}^{-\ell}(D), \quad -\infty < \ell < \infty. \end{aligned} \tag{8.24}$$

Here we assume that $D \subset C^{0,1}$. Let us recall that $\dot{H}^{-\ell}(D)$ can be defined as the completion of the set $C_0^\infty(D)$ in the norm of $H^{-\ell}(R^r)$, while $H^{-\ell}(D)$ can be defined as the space of restrictions of elements of $H^{-\ell}(R^r)$ to the domain $D \subset C^{0,1}$ with the norm (8.20).

Let us now describe a canonical factorization of the isometric operator $I : H_- \to H_+$ defined above. This factorization is

$$I = p_+ p_-, \tag{8.25}$$

where $p_- : H_- \to H_0$ and $p_+ : H_0 \to H_+$ are linear isometries onto H_0 and H_+ respectively. In order to prove (8.25) let us note that $(If, g)_+ = (f, g)_0$ for all $f, g \in H_+$. Thus I is a selfadjoint positive linear operator on H_+, $(If, f)_+ = (f, f)_0 \geq 0$ ($= 0$ if and only if $f = 0$). Let $\mathcal{J}_+$ be the closure of $I^{1/2} : H_+ \to H_+$ considered as an operator in H_0. Since $\| I^{1/2} f \|_+ = \| f \|_0$ the operator $I^{1/2}$ is closable. If $f_n \in H_+$ and f_n converges in H_0 to $f \in H_0$, then $I^{1/2} f_n \longrightarrow_{H_+} g$. Let us define $\mathcal{J}_+ f = I^{1/2} f = g$. Then $\mathcal{J}_+$ is defined on all of H_0, it is an isometry: $\| \mathcal{J}_+ f \|_+ = \| f \|_0$, and its range is all of H_+. Indeed, suppose $(\mathcal{J}_+ f, \phi)_+ = 0$ for all $f \in H_0$ and some $\phi \in H_+$. Then $(f, \phi)_0 = 0 \ \forall f \in H_0$. Thus $\phi = 0$. Since $\mathcal{J}_+$ is an isometry, its range is a closed subspace in H_+. We have proved above that the orthogonal complement of the range of $\mathcal{J}_+$ is trivial. Therefore the range of $\mathcal{J}_+$ is H_+. So one can take $p_+ = \mathcal{J}_+$. Define $p_- := \mathcal{J}_+^{-1} I$. Then (8.25) holds and $p_- : H_- \to H_0$ is an isometry with domain H_- and range H_0, while $p_+ : H_0 \to H_+$ is an isometry with domain H_0 and range H_+. One has $I = p_+ p_-$, which is the desired factorization (8.25).

If $i : H_+ \to H_0$ is the imbedding operator and I is considered as an operator from H_0 into H_+ then $i = I^*$, where I^* is the operator adjoint to $I : H_0 \to H_+$, $I^* : H_+ \to H_0$. Indeed

$$(If, g)_+ = (f, g)_0 = (f, ig)_0 \quad \forall f, g \in H_+. \tag{8.26}$$

If one assumes that the imbedding operator $i : H_+ \to H_0$ is in the Hilbert-Schmidt class as an operator in H_0, then the p_+ is in the Hilbert-Schmidt class as an operator in H_0. Indeed, $p_+ : H_0 \to H_+$ is an isometry. Therefore it sends a bounded set in H_0, say the unit ball of $H_0 : \| u \|_0 \leq 1$, into a bounded set in H_+, $\| p_+ u \|_+ = \| u \|_0 \leq 1$ if $\| u \|_0 \leq 1$. Since the imbedding $i : H_+ \to H_0$ is Hilbert-Schmidt, the operator p_+ considered as an operator on H_0 is in the Hilbert-Schmidt class:

$$p_+ : H_0 \to H_0 = (i : H_+ \to H_0)(p_+ : H_0 \to H_+). \tag{8.27}$$

The right side of (8.27) is the product of an operator in the Hilbert-Schmidt class and a bounded operator. Therefore the product is in the Hilbert-Schmidt class.

Define $p_+^{-1} := q_+$, $q_+ : H_+ \to H_0$, $\| q_+ f \|_0 = \| f \|_+$, $p_-^{-1} := q_-$, $q_- : H_0 \to H_-$, $\| q_- f \|_- = \| f \|_0$. One has

$$(q_+h, q_+g)_0 = (h, g)_+, \quad h, g \in H_+, \tag{8.28}$$

$$(p_+h, p_+g)_+ = (h, g)_0, \quad h, g \in H_0 \tag{8.29}$$

$$(p_-h, p_-g)_0 = (h, g)_-, \quad h, g \in H_-, \tag{8.30}$$

$$(q_-h, q_-g)_- = (h, g)_0, \quad h, g \in H_0. \tag{8.31}$$

Because $p_+p_- = I$, one has

$$q_-q_+ = I^{-1}. \tag{8.32}$$

Note that

$$(f, q_+u)_0 = (q_-f, u)_0 \quad f \in H_0, \quad u \in H_+, \tag{8.33}$$

so that $q_- = q_+^*$ is the adjoint to q_+ in H_0.

To check (8.33) one writes

$$(f, q_+u)_0 = (q_-f, q_-q_+u)_- = (q_-f, I^{-1}u)_- = (q_-f, u)_0 \tag{8.34}$$

which is equation (8.33).

8.2 Eigenfunction expansions for elliptic selfadjoint operators

8.2.1 *Resoluion of the identity and integral representation of selfadjoint operators*

Every selfadjoint operator A on a Hilbert space H can be represented as

$$A = \int_{-\infty}^{\infty} \lambda dE_\lambda, \tag{8.35}$$

where E_λ is a family of orthogonal projection operators such that

$$E_\lambda^2 = E_\lambda, \quad E_{-\infty} = 0, \quad E_{+\infty} = I, \tag{8.36}$$

$$E_\Delta E_{\Delta'} = E_{\Delta \cap \Delta'} \tag{8.37}$$

where 0 is the zero operator, I is the identity operator, $\Delta = (a, b]$, $-\infty < a < b < \infty$, $E_\Delta := E_b - E_a$. The family E_λ is called the resolution of the identity of A. The domain of definition of A is:

$$\text{Dom } A = \{f : f \in H, \int_{-\infty}^{\infty} \lambda^2 d(E_\lambda f, f) < \infty\}. \tag{8.38}$$

A function $\phi(A)$ is defined as

$$\phi(A) = \int_{-\infty}^{\infty} \phi(\lambda) dE_\lambda, \tag{8.39}$$

where

$$\text{Dom } \phi(A) = \{f : f \in H, \int_{-\infty}^{\infty} |\phi(\lambda)|^2 d(E_\lambda f, f) < \infty\}.$$

The operator integrals (8.35), (8.39) can be understood as improper operator Stieltjes integrals which converge strongly.

8.2.2 *Differentiation of operator measures*

A family $E(\Delta)$ of selfadjoint bounded nonnegative operators from the set of Borel sets $\Delta \in R^1$ into the set of $B(H)$ of bounded linear operators on a Hilbert space H is called an operator measure if $E(\bigcup_{j=1}^{\infty} \Delta_j) = \sum_{j=1}^{\infty} E(\Delta_j)$ where the limit on the right is taken in the sense of weak convergence of operators, $\Delta_i \cap \Delta_j = \emptyset$ for $i = j$, and $E(\emptyset) = 0$. Assume that for Δ bounded one has $Tr\ E(\Delta) < \infty$, where TrA is the trace of A. Then $\rho(\Delta) := Tr\ E(\Delta) \geq 0$ is a usual (scalar) measure on R^1. Let $|\Psi|$ denote the Hilbert-Schmidt norm of a linear operator Ψ, $|\Psi| = \left(\sum_{j=1}^{\infty} \| \Psi e_j \|^2\right)^{1/2}$, where $\{e_j\}$ is an orthonormal basis of H. In this chapter the star will stand for the adjoint operator and the bar for complex conjugate or for the closure.

Lemma 8.1 *For ρ-a.e. (almost everywhere) there exists a HS (Hilbert-Schmidt) operator-valued function $\Psi(\lambda) \geq 0$ with $\| \Psi(\lambda) \| \leq Tr\Psi(\lambda) \leq 1$ such that*

$$E(\Delta) = \int_{\Delta} \Psi(\lambda) d\rho(\lambda). \tag{8.40}$$

The function $\Psi(\lambda)$ is uniquely defined ρ-a.e. and can be obtained as a weak *limit*

$$\Psi(\lambda) = w - \lim E(\Delta_j)\rho^{-1}(\Delta_j) \quad as \quad \Delta_j \to \lambda. \tag{8.41}$$

The integral in (8.40) converges in the norm of operators for any bounded Δ. The limit in (8.41) means that $\lambda \in \Delta_j$ and $|\Delta_j| \to 0$, where $|\Delta_j|$ is the length of Δ_j, and Δ_j is a suitable sequence of intervals.

Let A be a selfadjoint operator on H and $E(\Delta)$ be its resolution of the identity. In general, $trE(\Delta)$ is not finite so that Lemma 8.1 is not applicable. In order to be able to use this lemma, let us take an arbitrary linear densely defined closed operator T on H with the properties: i) Ran$T = H$, T is injective, ii) $T^{-1} \in \sigma_2$ where σ_2 is the class of Hilbert-Schmidt operators.

Definition 8.1 *A linear compact operator K on H belongs to the class σ_p, $K \in \sigma_p$, if and only if $\sum_{n=1}^{\infty} s_n^p(K) < \infty$, where $s_n(K)$ are the s-values of K, which are defined by the formula $s_n(K) = \lambda_n[(K^*K)^{1/2}]$ and $\lambda_n(B)$ are the eigenvalues of a compact selfadjoint nonnegative operator B ordered so that $\lambda_1 \geq \lambda_2 \geq \cdots \geq 0$. If $K \in \sigma_1$ it is called trace class operator; if $K \in \sigma_2$ it is called HS operator.*

More information about trace class operators is given in Section 8.3.3.

Having chosen T as above, define the operator measure $\theta(\Delta) := T^{-1*}E(\Delta)T^{-1}$. If $A \in \sigma_2$ define its HS norm by the formula $|A|^2 := \sum_{j=1}^{\infty} \| Ae_j \|^2$, where $\| Af \|$ is the norm in H of the vector Af, and $\{e_j\}$ is the orthonormal basis of H. By $|A|$ denote the norm of A. Note that $|AB| \leq |A| \| B \|$, $\| A \| \leq |A|$, $|\lambda A| = |\lambda||A|$, $|A+B| \leq |A| + |B|$, and that $|A|$ does not depend on the choice of the orthonormal basis $\{e_j\}$ of H. Since $Tr(B^*AB) \leq |A| \| B \|^2$ one has $Tr\theta(s) \leq \| T^{-1} \|^2 |E(\Delta)| \leq \| T^{-1} \|^2 < \infty$.

Therefore Lemma 8.1 is applicable to $\theta(\Delta)$, so that

$$(E(\Delta)f, g) = (T^{-1*}E(\Delta)T^{-1}Tf, Tg) = \int_\Delta (\Psi(\lambda)Tf, Tg)\, d\rho(\lambda) \tag{8.42}$$

where $d\rho$ is a nonnegative measure, $\rho((-\infty,\infty)) < \infty$, $\Psi(\lambda)$ is a nonnegative operator function, $\Psi(\lambda) \geq 0$, and $|\Psi(\lambda)| \leq Tr\Psi(\lambda) = 1$.

Let, for a fixed λ, $\phi_\alpha(\lambda)$ and $\nu_\alpha(\lambda)$, $\alpha = 1, 2, \ldots N_\lambda \leq \infty$, be respectively the orthonormal system of eigenvectors of $\Psi(\lambda)$ and the corresponding eigenvalues. Then

$$(\Psi(\lambda)Tf, Tg) = \sum_{\alpha=1}^{N(\lambda)} \nu_\alpha(Tf, \phi_\alpha)(Tg, \phi_\alpha) = \sum_{\alpha=1}^{N(\lambda)} (Tf, \psi_\alpha)\overline{(Tg, \psi_\alpha)}, \tag{8.43}$$

where the bar denotes complex conjugate,

$$\psi_\alpha := \nu_\alpha^{1/2}(\lambda)\phi_\alpha, \quad \sum_{\alpha=1}^{N_\lambda} \| \psi_\alpha \|^2 = \sum_{\alpha=1}^{N_\lambda} \nu_\alpha = Tr\Psi(\lambda) = 1. \tag{8.44}$$

One can write

$$(E(\Delta)f, g) = \int_\Delta \sum_{\alpha=1}^{N_\lambda} (Tf, \psi_\alpha(\lambda)) \overline{(Tg, \psi_\alpha(\lambda))} d\rho(\lambda). \tag{8.45}$$

If $F(\lambda)$ is a Borel measureable function on Λ, then

$$(F(A)f, g) = \int_{-\infty}^{\infty} F(\lambda) \, (\Psi(\lambda)Tf, Tg) \, d\rho(\lambda), \tag{8.46}$$

where $f \in D(F(A)) \cap D(T)$, $g \in D(T)$. If one takes $T = q_+$ and if the imbedding $i : H_+ \to H_0$ is in the Hilbert-Schmidt class, then the operator $P(\lambda) := T^*\Psi(\lambda)T$, which appears in (8.42):

$$(E(\Delta)f, g) = \int_\Delta (P(\lambda)f, g) \, d\rho(\lambda) \tag{8.47}$$

and in (8.46):

$$(F(A)f, g) = \int_{-\infty}^{\infty} F(\lambda) \, (P(\lambda)f, g) \, d\rho(\lambda), \tag{8.48}$$

can be considered as an operator from H_+ into H_- for any fixed $\lambda \in R^1$. This operator is in the Hilbert-Schmidt class because it is a product of two bounded operators and a Hilbert-Schmidt operator $\Psi(\lambda)$: $T^* = q_- : H_0 \to H_-$ is bounded, $T = q_+ : H_+ \to H_0$ is bounded, $\Psi(\lambda) : H_0 \to H_0$ is in the Hilbert-Schmidt class. The range of the operator $P(\lambda)$ is a generalized eigenspace of the operator A corresponding to the point λ. This eigenspace belongs to H_-. Formula (8.47) can be written as

$$E(\Delta) = \int_\Delta P(\lambda) d\rho(\lambda) \tag{8.49}$$

where the integral (8.49) converges weakly, as follows from formula (8.47), but also it converges in the Hilbert-Schmidt norm of the operators $\mathcal{L}(H_+, H_-)$, where by $\mathcal{L}(H_+, H_-)$ we denote the set of linear bounded operators from H_+ into H_-. Indeed

$$|P(\lambda)| \leq \| T^* \| \, |\Psi(\lambda)| \, \| T \| \leq |\Psi(\lambda)| \leq 1 \tag{8.50}$$

where we took into account that $\| T \| = \| q_+ \| = 1$, $\| T^* \| = \| q_- \| = 1$.

The operator $P(\lambda)$ is an orthogonal projector in the following sense. If $\phi \in H_+$ and

$$(P(\lambda)u, \phi)_0 = 0 \quad \forall u \in H_+ \tag{8.51}$$

then

$$P(\lambda)\phi = 0. \tag{8.52}$$

Indeed

$$\begin{aligned} 0 &= (P(\lambda)u, \phi)_0 = (q_-\Psi(\lambda)q_+u, \phi)_0 = (\Psi(\lambda)q_+u, q_+\phi)_0 \\ &= (q_+u, \psi(\lambda)q_+\phi)_0 = (u, q_-\Psi(\lambda)q_+\phi)_0 \\ &= (u, P(\lambda)\phi)_0 = 0 \quad \forall u \in H_+. \end{aligned} \tag{8.53}$$

Thus, equation (8.52) follows from (8.53). Therefore, if ϕ is orthogonal to the range of $P(\lambda)$ then the projection of ϕ onto the range of $P(\lambda)$ vanishes.

Let us rewrite formula (8.43) in terms of the generalized eigenvectors. Define

$$T^*\psi_\alpha = q_-\psi_\alpha := \eta_\alpha \in H_-. \tag{8.54}$$

Then (8.43) can be rewritten as

$$(P(\lambda)f, g) = \sum_{\alpha=1}^{N(\lambda)} f_\alpha(\lambda)\overline{g_\alpha}(\lambda), \tag{8.55}$$

where

$$f_\alpha(\lambda) := (f, \eta_\alpha)_0, \quad f \in H_+. \tag{8.56}$$

Formula (8.45) becomes

$$(E(\Delta)f, g) = \int_\Delta \sum_{\alpha=1}^{N(\lambda)} f_\alpha(\lambda)\overline{g_\alpha}(\lambda)d\rho(\lambda). \tag{8.57}$$

Since $P(\lambda)$ is in the Hilbert-Schmidt class, it is an integral operator with a kernel $\Phi(x, y, \lambda)$, the (generalized) spectral kernel (see the Remark at the end of Section 3.6). The operator E_λ is an integral operator with the kernel

$$E(x, y, \lambda) = \int_{-\infty}^{\lambda} \Phi(x, y, \lambda)d\rho(\lambda). \tag{8.58}$$

The operator $F(A)$ is an integral operator with the kernel

$$F(A)(x,y,\lambda)=\int_{-\infty}^{\infty}F(\lambda)\Phi(x,y,\lambda)d\rho(\lambda). \tag{8.59}$$

8.2.3 *Carleman operators*

An integral operator

$$Af=\int A(x,y)f(y)dy, \quad \int=\int_{R^r} \tag{8.60}$$

is called a Carleman operator if

$$\nu(A):=\sup_{x\in R^r}\int |A(x,y)|^2dy<\infty. \tag{8.61}$$

A selfadjoint operator $\mathcal{L}$ is called a Carleman operator if there exists a continuous function $\phi(\lambda)$,

$$\phi(\lambda)\in C(\Lambda), \quad 0<|\phi(\lambda)|\le c \quad \forall\lambda\in\Lambda, \tag{8.62}$$

where c = const >0, Λ is the spectrum of $\mathcal{L}$, such that the operator $A=\phi(\mathcal{L})$ is an integral operator with the kernel $A(x,y)$ which satisfies (8.61).

Let $H_0=L^2(R^r)$ and take $H_+=L^2\left(R^r,p(x)\right)$, where $p(x)\ge 1$ and

$$\int p^{-1}(x)dx<\infty. \tag{8.63}$$

For example, one can take $p(x)=(1+|x|^2)^{(r+\epsilon)/2}$, where $\epsilon>0$ is any positive number. Then the operator A defined by the formula (8.60) is in the Hilbert-Schmidt class $\sigma_2(H_0,H_+)$ if the condition (8.63) holds.

Indeed, if $\{f_j\}$, $1\le j<\infty$, in an orthonormal basis of H_0, then

$$\begin{aligned}\sum_{j=1}^{\infty}\parallel Af_j\parallel_+^2&=\sum_{j=1}^{\infty}\int p^{-1}(x)\left|\int A(x,y)f_j(y)dy\right|^2dx\\&=\int dxp^{-1}(x)\int|A(x,y)|^2dy\\&\le\nu(A)\int dxp^{-1}(x)<\infty,\end{aligned} \tag{8.64}$$

where Parseval's equality was used to get the second equality and the conditions (8.61) and (8.63) were used to get the final inequality (8.64).

If $A = \phi(\mathcal{L})$ and $A \in \sigma_2(H_0, H_+)$ one can use the triple

$$L^2(R^r, p(x)) \subset L^2(R^r) \subset L^2\left(R^r, p^{-1}(x)\right) \tag{8.65}$$

for eigenfunction expansions of the operator $\mathcal{L}$. Here

$$H_+ = L^2(R^r, p(x)), \quad H_0 = L^2(R^r), \quad H_- = L^2\left(R^r, p^{-1}(x)\right). \tag{8.66}$$

Indeed, the basic condition, which has to be satisfied for the theory developed in Section 8.2.2 to be valid, is

$$Tr\left\{(T^{-1})^* E(\Delta) T^{-1}\right\} < \infty, \tag{8.67}$$

provided that Δ is bounded. Since $\phi(\lambda)$ is continuous and (8.62) holds, one has

$$\xi_\Delta(\lambda) \le |\phi(\lambda)|^2 c(\Delta), \quad \lambda \in \Lambda, \tag{8.68}$$

where $\xi_\Delta(\lambda)$ is the characteristic function of Δ:

$$\xi_\Delta(\lambda) = \begin{cases} 1, & \lambda \in \Delta \\ 0, & \lambda \notin \Delta \end{cases} \tag{8.69}$$

and $c(\Delta) = \text{const} > 0$. Inequality (8.68) implies

$$E(\Delta) \le c(\Delta) \int_\Lambda |\phi(\lambda)|^2 dE_\lambda = c(\Delta)\phi^*(\mathcal{L})\phi(\mathcal{L}). \tag{8.70}$$

Therefore

$$\begin{aligned} Tr\left\{(T^{-1})^* E(\Delta) T^{-1}\right\} &\le c(\Delta) Tr\left\{\left[\phi(\mathcal{L})T^{-1}\right]^\dagger \phi(\mathcal{L})T^{-1}\right\} \\ &= c(\Delta)|\phi(\mathcal{L})T^{-1}|. \end{aligned} \tag{8.71}$$

Thus (8.67) holds if

$$|\phi(\mathcal{L})T^{-1}| < \infty. \tag{8.72}$$

If $T^{-1} = p_+$, then condition (8.72) becomes

$$|\phi(\mathcal{L})p_+| < \infty. \tag{8.73}$$

In the case of the triple (8.65)-(8.66) the operator p_+ is a multiplication operator given by the formula

$$p_+ f = p^{-1/2}(x) f(x). \tag{8.74}$$

Condition (8.73) holds if $p(x)$ satisfies condition (8.63) and $\phi(\mathcal{L}) = A$ is a Carleman operator so that condition (8.61) holds.

Indeed, inequality (8.73) holds if

$$\iint |A(x,y)|^2 p^{-1}(y) dy dx < \infty. \tag{8.75}$$

We assume hat the function $\phi(\lambda)$ is such that (8.61) implies

$$\sup_{y \in R^r} \int |A(x,y)|^2 dx < \infty. \tag{8.76}$$

If this is the case then inequality (8.75) holds provided that (8.63) and (8.76) hold. Therefore in this case one can use the triple (8.65) for eigenfunction expansions of the operator $\mathcal{L}$ and the generalized eigenfunctions of $\mathcal{L}$ are elements of $L^2\left(R^r, p^{-1}(x)\right)$, so that they belong to $L^2_{\text{loc}}(R^r)$.

Inequality (8.61) implies (8.76), for example, if $\phi(\lambda)$ is a real-valued function. In this case $A = A^*$, so that

$$A(x,y) = \overline{A(y,x)} \tag{8.77}$$

and if (8.77) holds then clearly (8.61) implies (8.76).

In many applications one takes

$$\phi(\lambda) = (\lambda - z)^{-m}, \tag{8.78}$$

where z is a complex number and m is a sufficiently large positive integer. If ϕ is so chosen then

$$A(x,y;z) = \int_{-\infty}^{\infty} (\lambda - z)^{-m} \Phi(x,y,\lambda) d\rho(\lambda) \tag{8.79}$$

and

$$\begin{aligned} A(y,x,z) &= \int_{-\infty}^{\infty} (\lambda - z)^{-m} \Phi(y,x,\lambda) d\rho(\lambda) \\ &= \int_{-\infty}^{\infty} (\lambda - z)^{-m} \overline{\Phi(x,y,\lambda)} d\rho(\lambda) \\ &= \overline{\left[\int_{-\infty}^{\infty} (\lambda - z^*)^{-m} \Phi(x,y,\lambda) d\rho(\lambda)\right]} \\ &= A(x,y,\overline{z}), \end{aligned} \tag{8.80}$$

where we have used the equation

$$\Phi(x,y,\lambda) = \overline{\Phi(y,x,\lambda)}, \tag{8.81}$$

which follows from the assumed selfadjointness of $\mathcal{L}$. Therefore, if both kernels $A(x, y; z)$ and $A(x, y; \overline{z})$ satisfy inequality (8.61), then (8.76) holds.

8.2.4 *Elements of the spectral theory of elliptic operators in $L^2(R^r)$*

Let

$$\mathcal{L}u = \sum_{|j|\leq s} a_j(x) D^j u, \quad D^j = D_1^{j_1} \cdots D_r^{j_r}, \quad D_p = -i\frac{\partial}{\partial x_p}, \tag{8.82}$$

where $x \in R^r$, j is a multiindex, $a_j(x) \in C^{|j|}(R^r)$,

$$a_s(x, \xi) := \sum_{|j|=s} a_j(x)\xi^j \neq 0 \quad \text{for} \quad (x, \xi) \in R^r \times (R^r \setminus 0), \tag{8.83}$$

and assume that $\mathcal{L}$ is formally selfadjoint

$$\mathcal{L} = \mathcal{L}^* \tag{8.84}$$

that is

$$(\mathcal{L}\phi, \psi) = (\phi, \mathcal{L}\psi) \quad \forall \phi, \psi \in C_0^\infty(R^r). \tag{8.85}$$

The function $a_s(x, \xi)$ is called the symbol of the elliptic operator (8.82), and condition (8.83) is the ellipticity condition. If (8.83) holds and $r \geq 3$ then s is necessarily an even number.

Often one assumes that $\mathcal{L}$ is strongly elliptic. This means that

$$\sum_{|j|=s} Re a_j(x)\xi^j \neq 0 \quad \text{for} \quad (x, \xi) \in R^r \times (R^r \setminus 0). \tag{8.86}$$

The assumptions (8.84) and (8.86) imply that the operator $\mathcal{L}$ is bounded from below on $C_0^\infty(R^r)$.

$$(\mathcal{L}u, u)_0 \geq c_1 \parallel u \parallel^2_{H^s(R^r)} - c_2 \parallel u \parallel^2_{L^2(R^r)} \quad \forall u \in C_0^\infty(R^r), \tag{8.87}$$

where c_1 and c_2 are positive constants.

Define the minimal operator in $L^2(R^r)$ generated by the formally selfadjoint differential expression (8.82) as the closure of the symmetric operator $u \to \mathcal{L}u$ with the domain of definition $C_0^\infty(R^r)$. Any densely defined of a Hilbert space H symmetric operator $\mathcal{L}$ is closable. Recall that $\overline{\mathcal{L}}$, the closure of $\mathcal{L}$, is defined as follows. Let $u_n \in \text{Dom}\mathcal{L} := D(\mathcal{L})$, $u_n \to u$ in H and $\mathcal{L}u_n \to f$ in H. Then one declares that $u \in D(\overline{\mathcal{L}})$ and $\overline{\mathcal{L}}u = f$. This definition implies that $\overline{\mathcal{L}}$ is defined on the closure of $D(\mathcal{L})$ in the graph norm

$\| u \|_{\mathcal{L}} := \| u \| + \| \mathcal{L} u \|$ and the graph of $\overline{\mathcal{L}}$ is the closure in the graph norm of the graph of $\mathcal{L}$, that is of the set of ordered pairs $\{u, \mathcal{L}u\}$, $u \in D(\mathcal{L})$. One says that $\mathcal{L}$ is closable if and only if the set $\{u, \overline{\mathcal{L}}u\}$, $u \in D(\overline{\mathcal{L}})$, is a graph. In other words, $\mathcal{L}$ is closable if and only if there is no pair $\{0, f\}$, $f \neq 0$, in the closure of the graph of $\mathcal{L}$. This means that if

$$u_n \in D(\mathcal{L}), \quad u_n \to 0 \quad \text{and} \quad \mathcal{L}u_n \to f \tag{8.88}$$

then

$$f = 0. \tag{8.89}$$

If $\mathcal{L}$ is symmetric and densely defined and $\phi \in D(\mathcal{L})$, then

$$(f, \phi) = \lim_{n\to\infty} (\mathcal{L}u_n, \phi) = \lim_{n\to\infty} (u_n, \mathcal{L}\phi) = 0 \quad \forall \phi \in D(\mathcal{L}). \tag{8.90}$$

Since $D(\mathcal{L})$ is dense, one concludes that $f = 0$. Therefore $\mathcal{L}$ is closable. We denote $\overline{\mathcal{L}}$ by $\mathcal{L}_m$.

Under some assumptions on the coefficients $a_j(x)$ it turns out that $\mathcal{L}_m$ is selfadjoint. If no assumptions of the growth of $a_j(x)$ as $|x| \to \infty$ are made then $\mathcal{L}_m$ may be not selfadjoint (see [Ma, p 156] for an example).

Let us give some sufficient conditions for $\mathcal{L}_m$ to be selfadjoint.

Note that if $\mathcal{L}$ is densely defined symmetric and bounded from below it has always a (unique) selfadjoint extension with the same lower bound, the Friedrichs extension $\mathcal{L}_F$. This extension is characterized by the fact that its domain of definition belongs to the energy space of the operator $\mathcal{L}$, that is to the Hilbert space $H_{\mathcal{L}}$ defined as the closure of $D(\mathcal{L})$ in the metric

$$[u, u] = (\mathcal{L}u, u) + c(u, u), \tag{8.91}$$

where $c > 0$ is a sufficiently large constant such that $\mathcal{L} + cI$ is positive definite on $D(\mathcal{L})$.

Since the closure $\overline{\mathcal{L}}$ of $\mathcal{L}$ is the minimal closed extension of $\mathcal{L}$ and since $\mathcal{L}_F$ is a closed extension of $\mathcal{L}$, one concludes that if $\mathcal{L}$ is bounded from below in $H = L^2(R^r)$ and $\mathcal{L}_m$ is selfadjoint then

$$\mathcal{L}_m = \mathcal{L}_F. \tag{8.92}$$

In order to give conditions for $\mathcal{L}_m$ to be selfadjoint, consider first the case when

$$a_j(x) = \overline{a}_j(x) = \text{const}, \quad |j| \leq s. \tag{8.93}$$

In this case $\mathcal{L}$ is symmetric on $C_0^\infty(R^r)$.

Lemma 8.2 *If (8.93) holds then $\mathcal{L}_m$ is selfadjoint. Its spectrum consists of the set*

$$\left\{\lambda : \lambda = \sum_{|\alpha|\leq s} a_j \xi^j, \quad \xi \in R^r\right\}. \tag{8.94}$$

Proof. Let

$$\mathcal{F}u = (2\pi)^{-r/2} \int \exp(-i\xi \cdot x) u(x) dx := \hat{u}(\xi) \tag{8.95}$$

$$u(x) = \mathcal{F}^{-1}\hat{u} = (2\pi)^{-r/2} \int \exp(i\xi \cdot x)\hat{u}(\xi) d\xi \tag{8.96}$$

$$\mathcal{F}(D_p u) = \xi_p \hat{u}(\xi), \quad 1 \leq p \leq r, \quad D_p = -i\frac{\partial}{\partial x_p}. \tag{8.97}$$

Therefore

$$\mathcal{F}\left(\sum_{|j|\leq s} a_j D^j u\right) = \sum_{|j|\leq s} a_j \xi^j \hat{u}$$

and

$$\mathcal{L}u = \mathcal{F}^{-1}\mathcal{L}(\xi)\mathcal{F}u, \tag{8.98}$$

where

$$\mathcal{L}(\xi) := \sum_{|j|\leq s} a_j \xi^j. \tag{8.99}$$

The operator $\mathcal{F}$ of the Fourier transform is unitary in $L^2(R^r)$. Formula (8.98) shows that the operator $\mathcal{L}$ is unitarily equivalent to the operator of multiplication by the polynomial $\mathcal{L}(\xi)$ defined by formula (8.99). This multiplication operator is defined on the set $\mathcal{F}(C_0^\infty(R^r))$ of the Fourier transforms of the functions belonging to the set $C_0^\infty(R^r)$, that is to the domain of definition of the operator $\mathcal{L}$. Consider the closure M in $L^2(R_\xi^r)$ of the operator of multiplication by the function $L(\xi)$:

$$M\hat{u} := L(\xi)\hat{u}(\xi). \tag{8.100}$$

The domain of definition of M is

$$D(M) = \left\{\hat{u} : L(\xi)\hat{u}(\xi) \in L^2(R^r)\right\}. \tag{8.101}$$

The operator M is clearly selfadjoint since the function $L(\xi)$ is real-valued. Therefore

$$\mathcal{L}_m = \mathcal{F}^{-1} M \mathcal{F} \tag{8.102}$$

is also selfadjoint. Since the spectra of the unitarily equivalent operators are identical and the spectrum of M is the set (8.94), the spectrum of $\mathcal{L}_m$ is the set (8.94). Lemma 8.2 is proved. □

The following well-known simple lemmas are useful for proving, under suitable assumptions, that $\mathcal{L}_m$ is selfadjoint.

Lemma 8.3 *Let A and B be linear operators on a Hilbert space H, A is selfadjoint, $D(A) \subset D(B)$, B is symmetric, and*

$$\| Bu \| \leq \epsilon \| Au \| + c \| u \|, \quad \forall u \in D(A), \tag{8.103}$$

where $0 < \epsilon < 1$ is a fixed number and c is a positive number. Then the operator $A + B$ with $Dom(A + B) = DomA$ is self-adjoint.

Lemma 8.4 *Let A be a symmetric densely defined in a Hilbert space H operator. Then $\overline{A}$, the closure of A, is selfadjoint if and only if one of the following conditions holds*

$$c\ell Ran(A \pm i\lambda) = H \tag{8.104}$$

or

$$N(A^* \pm i\lambda) = \{0\}. \tag{8.105}$$

Here $\lambda > 0$ is an arbitrary fixed number, $c\ell RanA$ is the closure of the range of A,

$$N(B) = \{u : Bu = 0\} \tag{8.106}$$

and A^* is the adjoint of A.

For convenience of the reader let us prove these lemmas. We start with Lemma 8.4.

Proof of Lemma 8.4 a) The argument is the same for any $\lambda > 0$, so let us take $\lambda = 1$. Suppose that $\overline{A}$ is selfadjoint. Then $A^* = \overline{A}^* = \overline{A}$. If $A^* u = iu$ then

$$(A^* u, u) = i(u, u). \tag{8.107}$$

Since A^* is selfadjoint the quadratic form (A^*u, u) is real valued. Thus, equation (8.107) implies $U = 0$, and (8.105) is established. To derive (8.104) one uses the formula

$$c\ell\text{Ran}(A \pm i) \oplus N(A^* \mp i) = H, \tag{8.108}$$

where $\oplus$ is the symbol of the orthogonal sum of the subspaces. From (8.108) and (8.105) one gets (8.104).

b) Assume now that (8.105) holds and prove that $\overline{A}$ is selfadjoint. If (8.105) holds then (8.104) holds as follows from (8.108). Conversely, (8.104) implies (8.105). Note that

$$c\ell\text{Ran}(A \pm i) = \text{Ran}(\overline{A} \pm i) \tag{8.109}$$

because $\text{Ran}(\overline{A} \pm i)$ are closed subspaces. Indeed, if $(\overline{A} \pm i)u_n = f_n$ and $f_n \to f$, then

$$\| (\overline{A} \pm i)u_{nm} \| \to 0, \quad n, m \to \infty, \tag{8.110}$$

where $u_{nm} := u_n - u_m$. Since $\overline{A}$ is symmetric, one obtains from (8.11) that

$$\| (\overline{A} \pm i)u_{nm} \|^2 = \| \overline{A}u_{nm} \|^2 + \| u_{nm} \|^2 \to 0, \quad n, m \to \infty. \tag{8.111}$$

Therefore u_n is a Cauchy sequence. Let $u_n \to u$. Then, since $\overline{A}$ is closed, one obtains

$$(\overline{A} \pm i)u = f. \tag{8.112}$$

Therefore $\text{Ran}(\overline{A} \pm i)$ are closed subspaces. If (8.105) holds then

$$\text{Ran}(\overline{A} \pm i) = H. \tag{8.113}$$

This and the symmetry of $\overline{A}$ imply that $\overline{A}$ is selfadjoint. Indeed, let

$$((\overline{A} + i)u, v) = (u, f) \quad \forall u \in D(\overline{A}). \tag{8.114}$$

Using (8.113), find w such that

$$(\overline{A} - i)w = f. \tag{8.115}$$

This is possible because of (8.113). Use (8.114) and the symmetry of $\overline{A}$ to obtain

$$((\overline{A} + i)u, v) = (u, (\overline{A} - i)w) = ((\overline{A} + i)u, w)\,. \tag{8.116}$$

By (8.113) one has $\text{Ran}(\overline{A}+i) = H$. This and (8.116) imply $v = w$. Therefore $v \in D(\overline{A})$ and $\overline{A}$ is selfadjoint on $D(\overline{A})$. Lemma 8.4 is proved. □

Proof of Lemma 8.3 It is sufficient to prove that, for some $\lambda > 0$,

$$\text{Ran}(A + B \pm i\lambda) = H. \tag{8.117}$$

One has

$$A + B \pm i\lambda = \left[I + B(A \pm i\lambda)^{-1}\right](A \pm i\lambda). \tag{8.118}$$

Equation (8.117) is established as soon as we prove that

$$\| B(A \pm i\lambda)^{-1} \| < 1. \tag{8.119}$$

Indeed, if (8.119) holds then the operator $I+B(A+i\lambda)^{-1}$ is an isomorphism of H onto H and, since A is selfadjoint, $\text{Ran}(A \pm i\lambda) = H$, so

$$\text{Ran}(A + B \pm i\lambda) = \text{Ran}\left\{\left[I + B(A \pm i\lambda)^{-1}\right](A \pm i\lambda)\right\} = H. \tag{8.120}$$

Use the basic assumption (8.103) to prove (8.119). Let

$$(A + i\lambda)^{-1}u = f, \quad B(A + i\lambda)^{-1}u = Bf. \tag{8.121}$$

Then

$$\begin{aligned} \| B(A + i\lambda)^{-1}u \| &= \| Bf \| \le \epsilon \| Af \| + c \| f \| \\ &= \epsilon \| A(A + i\lambda)^{-1}u \| + c \| (A + i\lambda)^{-1}U \| \\ &\le \epsilon \| u \| + c\lambda^{-1} \| u \| = (\epsilon + c\lambda^{-1}) \| u \| . \end{aligned} \tag{8.122}$$

If $\epsilon < 1$ and $\lambda > 0$ is large enough, then

$$\epsilon + c\lambda^{-1} < 1. \tag{8.123}$$

Thus (8.122) holds, and Lemma 8.3 is proved. □

In deriving (8.122) we have used the inequalities

$$\| (A \pm i\lambda)^{-1} \| \le \lambda^{-1}, \quad \lambda > 0, \tag{8.124}$$

and

$$\| A(A \pm i\lambda)^{-1} \| \le 1, \tag{8.125}$$

both of which follow immediately from the spectral representation of a selfadjoint operator A: if

$$\phi(A) = \int \phi(t) dE_t \tag{8.126}$$

then

$$\| \phi(A) \| \leq \max_t |\phi(t)|. \tag{8.127}$$

Both inequalities (8.124) and (8.125) can be derived in a simple way which does not use the result (8.126)-(8.127). This derivation is left for the reader as an exercise. □

We now return to the question of the selfadjointness of $\mathcal{L}_m$. If $\mathcal{L}_m$ is selfadjoint then $\mathcal{L}$ is called *essentially selfadjoint.* We wish to prove that if the principal part of $\mathcal{L}$ is an operator with constant coefficients and the remaining part is an operator with smooth bounded coefficients then $\mathcal{L}$ is essentially selfadjoint. The principal part of $\mathcal{L}$ is the differential expression

$$\mathcal{L}_0 := \sum_{|j|=s} a_j(x) D^j. \tag{8.128}$$

Let us recall that a polynomial $P(\xi)$ is *subordinate to the polynomial* $Q(\xi)$ if

$$\frac{|P(\xi)|}{1+|Q(\xi)|} \to 0 \quad \text{as} \quad |\xi| \to \infty, \quad \xi \in R^r. \tag{8.129}$$

If (8.129) holds then we write

$$P \prec\prec Q. \tag{8.130}$$

We say that $Q(\xi)$ is stronger than $P(\xi)$ and write

$$P \prec Q \tag{8.131}$$

if

$$\frac{\tilde{P}(\xi)}{\tilde{Q}(\xi)} \leq c, \quad \forall \xi \in R^r, \tag{8.132}$$

where $c > 0$ is a constant, and

$$\tilde{P}(\xi) := \left\{ \sum_{|j|\geq 0} |P^{(j)}(\xi)|^2 \right\}^{1/2}. \tag{8.133}$$

In the following lemma a characterization of elliptic polynomials are given. A homogeneous polynomial $Q(\xi)$ of degree s is called elliptic if

$$Q(\xi) \neq 0 \quad \text{for} \quad \xi \in R^r \setminus 0. \tag{8.134}$$

Lemma 8.5 *A homogeneous polynomial $Q(\xi)$ of degree s is elliptic if and only if it is stronger than every polynomial of degree $\leq s$. In particular*

$$c|\xi|^s \leq |Q(\xi)|, \tag{8.135}$$

where $c = const > 0$.

A proof of this result can be found in [Hörmander (1983-85), vol. II, p. 37].

Lemma 8.6 *Assume that*

$$\mathcal{L}u = \mathcal{L}_0 u + \sum_{|j|<s} a_j(x) D^j u \tag{8.136}$$

and

$$\mathcal{L}_0 u = \sum_{|j|=s} a_j D^j u, \tag{8.137}$$

where

$$\mathcal{L}_0(\xi) := \sum_{|j|=s} a_j \xi^j \quad \text{is an elliptic polynomial} \tag{8.138}$$

and

$$\sup_{x \in R^r} |a_j(x)| \leq c, |j| < s. \tag{8.139}$$

Then $\mathcal{L}$ is essentially selfadjoint on $C_0^\infty(R^r)$ and $\overline{\mathcal{L}} = \mathcal{L}_m$ is selfadjoint on $H^s(R^r)$.

Proof. By Lemma 8.2 $\mathcal{L}_0$ defined on $C_0^\infty(R^r)$ is essentially selfadjoint. If (8.138) holds then inequality (8.135) holds. Therefore the closure of $\mathcal{L}_0$, the operator $\mathcal{L}_{0m}$, is selfadjoint and

$$\text{Dom}\mathcal{L}_{0m} = H^s(R^r). \tag{8.140}$$

Let us apply Lemma 8.3 with $A = \mathcal{L}_{0m}$ and $B = \mathcal{L} - \mathcal{L}_0$. The basic condition to check is inequality (8.103) on $H^s(R^r)$. It is sufficient to check this inequality on $C_0^\infty(R^r)$ since $C_0^\infty(R^r)$ is dense in $H^s(R^r)$. If (8.103) established for any $u \in C_0^\infty(R^r)$ then one takes $u \in H^s(R^r)$ and a sequence

$u_n \in C_0^\infty(R^r)$ such that $\| u_n - u \|_{H^s(R^r)} \to 0$, $n \to \infty$, and passes to the limit $n \to \infty$ in (8.103). This yields inequality (8.103) for any $u \in H^s(R^r)$.

In order to check inequality (8.103) it is sufficient to prove that

$$\| a_j(x) D^j u \| \leq \epsilon \| \mathcal{L}_0 u \| + c(\epsilon) \| u \|, \quad \forall u \in C_0^\infty(R^r) \tag{8.141}$$

for any $\epsilon > 0$, however small, and any j such that $|j| < s$. Using (8.139) and Parseval's equality one obtains:

$$\| a_j(x) D^j u \| \leq c \| D^j u \| = c \| \xi^j \hat{u} \| . \tag{8.142}$$

On the other hand, Parseval's equality, condition (8.138) and inequality (8.135) yield

$$\| \mathcal{L}_0 u \| = \| \mathcal{L}_0(\xi) \hat{u} \| \leq c \| |\xi|^s \hat{u} \| . \tag{8.143}$$

If $|j| < s$ then

$$|\xi|^j \leq \epsilon |\xi|^s \quad |\xi| > R = R(\epsilon). \tag{8.144}$$

In the region $|\xi| \leq R$ one estimates

$$|\xi^j \hat{u}| \leq c(R) |\hat{u}|, \tag{8.145}$$

where, for example, one can take $c(R) = R^{|j|}$. Therefore

$$\int_{|\xi| \leq R} |\xi^j \hat{u}|^2 d\xi \leq c^2(R) \| \hat{u} \|^2 = c^2(R) \| u \|^2, \tag{8.146}$$

and

$$\int_{|\xi| > R} |\xi^j \hat{u}|^2 d\xi \leq \epsilon^2 \int |\mathcal{L}_0(\xi) \hat{u}|^2 d\xi = \epsilon^2 \| \mathcal{L}_0 u \|^2 \tag{8.147}$$

where Parseval's equality and estimates (8.144) and (8.145) were used. From (8.142), (8.146) and (8.147) one obtains the desired inequality (8.141). Lemma 8.6 is proved. □

Remark 8.1 *Note that the method of the proof allows one to relax condition (8.139). For example, one could use some integral inequalities to estimate the Fourier transform.*

Let us now prove the following result.

Lemma 8.7 *If $\mathcal{L}$ defined by (8.82) has smooth and uniformly bounded in R^r coefficients such that the principal part of $\mathcal{L}$ is an elliptic operator* *of uniformly constant strength, then $\mathcal{L}$ is essentially selfadjoint on $C_0^\infty(R^r)$ and $\mathcal{L}_m$ is selfadjoint on $H^s(R^r)$.*

Proof. Let us recall that the principal part $\mathcal{L}_0$ of $\mathcal{L}$,

$$\mathcal{L}_0 u = \sum_{|j|=s} a_j(x) D^j u \tag{8.148}$$

is an elliptic operator of uniformly constant strength if the *principal symbol*

$$a_s(x,\xi) := \sum_{|j|=s} a_j(x)\xi^j$$

satisfies the ellipticity condition (8.83) and the condition of uniformly constant strength

$$\frac{\tilde{a}_s(x,\xi)}{\tilde{a}_s(y,\xi)} \le c \quad \forall x, y \in R^r \tag{8.149}$$

where c does not depend on x, y and

$$\tilde{a}_s(x,\xi) := \left\{ \sum_{|j|\ge 0} |D_\xi^j a_s(x,\xi)|^2 \right\}^{1/2}. \tag{8.150}$$

By Lemma 8.4 the operator $\mathcal{L}_m$ is selfadjoint on $H^s(R^r)$ if the equations

$$(\mathcal{L}_m \pm i\lambda)u = f, \tag{8.151}$$

are solvable in $H^s(R^r)$ for any $f \in C_0^\infty(R^r)$ and for some $\lambda > 0$. Indeed, in this case $\mathrm{Ran}(\mathcal{L}_m \pm i\lambda) \supset C_0^\infty(R^r)$ and therefore is dense in $H = L^2(R^r)$. Since $\mathcal{L}_m$ is symmetric $\mathrm{Ran}(\mathcal{L}_m \pm i\lambda)$ is closed in $L^2(R^r)$ and, being dense in $L^2(R^r)$, has to coincide with $L^2(R^r)$. This implies, by Lemma 8.4 that $\mathcal{L}_m$ is selfadjoint.

Existence of the solution to (8.151) in $H^s(R^r)$ for any $f \in C_0^\infty(R^r)$ follows from the existence of the fundamental solution $\mathcal{E}(x,y,\lambda)$:

$$(\mathcal{L}_m \pm i\lambda)\mathcal{E}(x,y,\lambda) = \delta(x-y) \quad \text{in} \quad R^r \tag{8.152}$$

and the estimate

$$|\mathcal{E}(x,y,\lambda)| \le \begin{cases} c|x-y|^{s-r}, & \text{if } r \text{ is odd or } r > s, \\ & |x-y| \le 1 \\ c|x-y|^{s-r} + c_1 \log\frac{1}{|x-y|}, & \text{if } r \text{ is even and } r \le s, \\ & |x-y| \le 1 \end{cases} \tag{8.153}$$

where c and c_1 are positive constants, and

$$|\mathcal{E}(x,y,\lambda)| \le c\exp(-a(\lambda)|x-y|) \quad \text{if} \quad |x-y| \ge 1 \tag{8.154}$$

where $c > 0$ is a constant and $a(\lambda) > 0$ is a constant, $a(\lambda) \to +\infty$ as $\lambda \to +\infty$. Also $\mathcal{E}(x,y,\lambda)$ is smooth away from the diagonal $x = y$, and the following estimates hold

$$|D^j\mathcal{E}(x,y,\lambda| \le c\exp\left(-a(\lambda)|x-y|\right) \quad \text{if} \quad |x-y| \ge 1 \tag{8.155}$$

$$|D^j\mathcal{E}(x,y,\lambda)| \le \begin{cases} c_0 + c_1|x-y|^{s-r-|j|} & \text{if } s \ne r+|j|, |x-y| \le 1 \\ c_0 + c_1\left|\log|x-y|\right| & \text{if } s = r+|j|, |x-y| \le 1. \end{cases} \tag{8.156}$$

Indeed, if there exists the fundamental solution with the properties (8.152)-(8.156) then

$$u = \int_{R^r} \mathcal{E}(x,y,\lambda)f(y)dy \tag{8.157}$$

solves (8.151) and $u \in H^s(R^r)$, so that $\updownarrow_m$ is selfadjoint.

Existence of the fundamental solution with the properties (8.152)-(8.156) for elliptic selfadjoint operators with constant coefficients can be established if one uses the Fourier transform [Hörmander (1983-85), vol. I, p. 170], and for the operators of uniformly constant strength it is established in [Hörmander (1983-85), vol. II, p. 196]. Thus, Lemma 8.7 is proved. □

It is not difficult now to establish that the function $(\mathcal{L}^N - i\lambda)^{-1} := \phi(\mathcal{L})$, $\lambda > 0$, has a kernel which is a Carleman operator if $N > \frac{r}{2s}$. Indeed, it follows from the estimate (8.156) that the singularity of the kernel of the operator $\phi(\mathcal{L})$ is $0\left(|x-y|^{Ns-r}\right)$ so that this kernel is locally in L^2 if $N > \frac{r}{2s}$. On the other hand, the estimate (8.154) implies that the kernel of $\phi(\mathcal{L})$ is in $L^2(R^r)$ globally. Since the constants in the inequalities (8.154) and (8.156) do not depend on x, one concludes that $\phi(\mathcal{L})$ is a Carleman operator. Let us formulate this result as Lemma 8.8.

Lemma 8.8 *Suppose that $N > \frac{r}{2s}$ and the assumptions of Lemma 8.7 hold. Then the operator $(\mathcal{L}^N - i\lambda)^{-1}$, $\lambda > 0$, is a Carleman operator.*

8.3 Asymptotics of the spectrum of linear operators

In this section we develop an abstract theory of perturbations preserving asymptotics of the spectrum of linear operators. As a by-product a proof of Theorem 2.3 is obtained.

8.3.1 *Compact operators*

8.3.1.1 *Basic definitions*

Let H be a separable Hilbert space and $A : H \to H$ be a linear operator. The set of all bounded operators we denote $L(H)$. The set of all linear compact operators on H we denote σ_∞. Let us recall that an operator A is called compact if it maps bounded sets into relatively compact sets. It is well known that A is compact if and only if one of the following conditions hold

1) $f_n \rightharpoonup f$, $g_n \rightharpoonup g$ implies $(Af_n, g_n) \to (Af, g)$,
2) $f_n \rightharpoonup f$ implies $Af_n \to Af$,
3) from any bounded sequence of elements of H one can select a subsequence f_n such that

$$(Af_{nm}, f_{nm}) \quad \text{converges as} \quad n, m \to \infty$$

where

$$f_{nm} := f_n - f_m.$$

By $\rightharpoonup$ we denote weak convergence and $\to$ stands for convergence in the norm of H (strong convergence).

If A is compact and B is a bounded linear operator then AB and BA are compact. A linear combination of compact operators is compact. If $A_n \in \sigma_\infty$ and $\| A_n - A \| \to 0$ as $n \to \infty$, then $A \in \sigma_\infty$. The operator A is compact if and only if A^* is compact. *In this section we denote the adjoint operator by* A^*. If H is a separable Hilbert space and A is compact then there exists a sequence A_n of finite rank operators such that $\| A_n - A \| \to 0$. An operator B is called a finite rank operator if rank $A := \dim \operatorname{Ran} B < \infty$. If A is compact then $A^*A \geq 0$ is compact and selfadjoint. The spectrum of a selfadjoint compact operator is discrete, the eigenvalues $\lambda_n(A^*A)$ are nonnegative and have at most one limit point $\lambda = 0$. We define the singular values of a compact operator A (s-values of A) by the equation

$$s_j(A) = \lambda_j^{1/2}(A^*A). \tag{8.158}$$

One has

$$s_1(A) \geq s_2(A) \geq \cdots \geq 0. \tag{8.159}$$

Note that

$$s_1(A) = \| A \| \tag{8.160}$$

and if $A = A^*$ then

$$s_j(A) = |\lambda_j(A)|. \tag{8.161}$$

The following properties of the s-values are known

$$s_j(A) = s_j(A^*), \tag{8.162}$$
$$s_j(BA) \leq \| B \| s_j(A), \tag{8.163}$$
$$s_j(AB) \leq \| B \| s_j(A) \tag{8.164}$$

for any bounded linear operator B. Obviously

$$s_j(cA) = |c| s_j(A), \quad c = \text{const}. \tag{8.165}$$

Any bounded linear operator A can be represented as

$$A = U|A| \tag{8.166}$$

where

$$|A| := (A^*A)^{1/2} \tag{8.167}$$

and U is a partial isometry which maps $\text{Ran}(A^*)$ onto $\text{Ran}A$. Representation (8.166) is called polar representation of A.

The operator $|A|$ is selfadjoint. If A is compact then $|A|$ is compact. Let ϕ_j be its eigenvectors and $s_j = s_j(A)$ be its eigenvalues:

$$|A|\phi_j = s_j \phi_j, \quad (\phi_j, \phi_m) = \delta_{jm}. \tag{8.168}$$

Then

$$|A| = \sum_{j=1}^{\infty} s_j(A)(\cdot, \phi_j)\phi_j \tag{8.169}$$

where the series (8.169) converges to $|A|$ in the norm of operators:

$$\left\| |A| - \sum_{j=1}^{n} s_j(A)(\cdot, \phi_j)\phi_j \right\| \to 0 \quad \text{as} \quad n \to \infty. \tag{8.170}$$

Let

$$\psi_j := U\phi_j. \tag{8.171}$$

Then

$$A = \sum_{j=1}^{\infty} s_j(\cdot, \phi_j)\psi_j. \tag{8.172}$$

Formula (8.172) is the canonical representation of a compact operator A. Note that

$$(\psi_j, \psi_m) = \delta_{jm}$$

since U is a partial isometry.

It follows from (8.172) that A is a limit in the norm of operators of the finite rank operator $\sum_{j=1}^{n} s_j(\cdot, \phi_j)\psi_j$. Moreover

$$A^* = \sum_{j=1}^{\infty} s_j(A)(\cdot, \psi_j)\phi_j. \tag{8.173}$$

In the formulas (8.172) and (8.175) the summation is actually taken over all j for which $s_j(A) \neq 0$. If rank $A < \infty$, then $s_j(A) = 0$ for $j >$ rank A. If A is compact and normal, that is $A^*A = AA^*$, then its eigenvectors form an orthonormal basis of H

$$A = \sum_{j=1}^{\infty} \lambda_j(A)(\cdot, \phi_j)\phi_j, \quad A\phi_j = \lambda_j(A)\phi_j \tag{8.174}$$

and

$$s_j(A) = |\lambda_j(A)|. \tag{8.175}$$

8.3.1.2 *Minimax principles and estimates of eigenvalues and singular values*

Lemma 8.9 *Let A be a selfadjoint and compact operator on H. Let*

$$\lambda_1^+ \geq \lambda_2^+ \geq \cdots \tag{8.176}$$

be its positive eigenvalues counted according to their multiplicities and ϕ_j *are corresponding eigenvectors*

$$A\phi_j = \lambda_j^+ \phi_j. \tag{8.177}$$

Then

$$\lambda_{j+1}^+ = \min_{L_j} \max_{\phi \perp L_j} \frac{(A\phi, \phi)}{(\phi, \phi)} \tag{8.178}$$

where $L_n \subset H$ *is an n-dimensional subspace. Maximum in (8.178) is attained on the subspace*

$$\mathcal{L}_j(A) := span\{\phi_1, \dots \phi_j\} \tag{8.179}$$

spanned by the first j eigenvectors of A corresponding to the positive eigenvalues.

Remark 8.2 *Maximum may be attained not only on the subspace (8.179). The sign* $\phi \perp L$ *means that* ϕ *is orthogonal to the subspace L.*

Lemma 8.10 *If* $A \in \sigma_\infty$ *then*

$$s_{j+1}(A) = \min_{L_j} \max_{\phi \perp L_j} \frac{\| A\phi \|}{\| \phi \|}. \tag{8.180}$$

Lemma 8.10 follows immediately from Lemma 8.9 and from the definition of the s-values given by formula (8.158).

Lemma 8.11 *If* $A \in \sigma_\infty$ *then*

$$s_{j+1}(A) = \min_{K \in \mathcal{K}_j} \| A - K \| \tag{8.181}$$

where $\mathcal{K}_j$ *is the set of operators of rank* $\leq j$.

The following inequalities for eigenvalues and singular values of compact operators are known.

Lemma 8.12 *If A and B are selfadjoint compact operators and* $A \geq B$, *that is*

$$(A\phi, \phi) \geq (B\phi, \phi) \quad \forall \phi \in H, \tag{8.182}$$

then

$$\lambda_j^+(A) \geq \lambda_j^+(B). \tag{8.183}$$

Lemma 8.13 *If A and B are selfadjoint and compact operators, then*

$$\lambda^+_{m+n-1}(A+B) \le \lambda^+_m(A) + \lambda^+_n(B), \tag{8.184}$$

and

$$\lambda^-_{m+n-1}(A+B) \ge \lambda^-_n(A) + \lambda^-_m(B). \tag{8.185}$$

Moreover

$$\left|\lambda^+_j(A) - \lambda^+_j(A+B)\right| \le \| B \| \tag{8.186}$$

and

$$\left|\lambda^-_j(A) - \lambda^-_j(A+B)\right| \le \| B \| \tag{8.187}$$

where $\lambda^-_1(A) \le \lambda^-_2(A) \le \cdots < 0$ are the negative eigenvalues of a selfadjoint compact operator A counted according to their multiplicities.

Lemma 8.14 *If A is compact and B is a finite rank operator, rank $B = \nu$, then*

$$s_{j+\nu}(A) \le s_j(A+B) \le s_{j-\nu}(A). \tag{8.188}$$

Lemma 8.15 *If $A, B \in \sigma_\infty$ then*

$$s_{m+n-1}(A+B) \le s_m(A) + s_n(B), \tag{8.189}$$

$$s_{m+n-1}(AB) \le s_m(A)s_n(B),$$

$$|s_n(A) - s_n(B)| \le \| A - B \| . \tag{8.190}$$

Lemma 8.16 *If $A \in \sigma_\infty$ then*

$$\sqcap^n_{j=1}|\lambda_j(A)| \le \sqcap^n_{j=1} s_j(A). \tag{8.191}$$

Lemma 8.17 *If $A, B \in \sigma_\infty$ and $f(x)$, $0 \le x < \infty$, is a real-valued nondecreasing, convex, and continuous function vanishing at $x = 0$, then*

$$\sum_{j=1}^{n} f\left(s_j(A+B)\right) \le \sum_{j=1}^{n} f\left(s_j(A) + s_j(B)\right) \tag{8.192}$$

for all $n = 1, 2, \ldots, \infty$.

In particular, if $f(x) = x$, one obtains

$$\sum_{j=1}^{n} s_j(A+B) \le \sum_{j=1}^{n} s_j(A) + \sum_{j=1}^{n} s_j(B) \tag{8.193}$$

for all $n = 1, 2, \dots, \infty$. If $f(x)$, $f(0) = 0$, $0 \le x < \infty$, is such that the function $\phi(t) := f(\exp(t))$ is convex, $-\infty < t < \infty$, then

$$\sum_{j=1}^{n} f(s_j(AB)) \le \sum_{j=1}^{n} f(s_j(A)s_j(B)) \tag{8.194}$$

for all $n = 1, 2, \dots, \infty$.

In particular, if $f(x) = x$, then

$$\sum_{j=1}^{n} s_j(AB) \le \sum_{j=1}^{n} s_j(A)s_j(B). \tag{8.195}$$

for all $n = 1, 2, \dots, \infty$.

Lemma 8.18 *Let $A, B \in \sigma_\infty$ and*

$$\lim_{n\to\infty} n^a s_n(A) = c, \tag{8.196}$$

where $a > 0$ and $c = const > 0$. Assume that

$$\lim_{n\to\infty} n^a s_n(B) = 0. \tag{8.197}$$

Then

$$\lim_{n\to\infty} n^a s_n(A+B) = c.$$

Proofs of the above results can be found in [GK].

8.3.2 *Perturbations preserving asymptotics of the spectrum of compact operators*

8.3.2.1 *Statement of the problem*

Here we are interested in the following question. Suppose that A and Q are linear compact operators on H, and B is defined by the formula

$$B = A(I+Q). \tag{8.198}$$

Question 1: *Under what assumptions are the singular values of B asymptotically equivalent to the singular values of A in the following sense:*

$$\lim_{n\to\infty} \frac{s_n(B)}{s_n(A)} = 1. \tag{8.199}$$

Assume now that

$$s_n(A) = cn^{-p}\left[1 + O(n^{-p_1})\right], \quad n \to \infty, \tag{8.200}$$

where p and p_1 are positive numbers, and $c > 0$ is a constant.

Question 2: *Under what assumptions is the asymptotics of the singular values of B given by the formula*

$$s_n(B) = cn^{-p}\left[1 + O(n^{-q})\right], \quad n \to \infty? \tag{8.201}$$

When is $q = p_1$?

We will answer these questions and give some applications of the results.

8.3.2.2 *A characterization of the class of linear compact operators*

We start with a theorem which gives a characterization of the class of linear compact operators on H.

In order to formulate this theorem let us introduce the notion of limit dense sequence of subspaces. Let

$$L_n \subset L_{n+1} \subset, \dots, \dim L_n = n \tag{8.202}$$

be a sequence of finite-dimensional subspaces of H such that

$$\rho(f, L_n) \to 0 \quad \text{as} \quad n \to \infty \quad \text{for any} \quad f \in H \tag{8.203}$$

where $\rho(f, L)$ is the distance from f to the subspace L.

Definition 8.2 A sequence of the subspaces L_n is called limit dense in H if the conditions (8.202) and (8.203) hold.

Theorem 8.2 *A linear operator $A : H \to H$ is compact if and only if there exists a limit dense in H sequence L_n of subspaces L_n such that*

$$\sup_{h \perp L_n} \frac{\| Ah \|}{\| h \|} \to 0 \quad as \quad n \to \infty, \quad h \in H. \tag{8.204}$$

If (8.204) holds for a limit dense in H sequence L_n then it holds for every limit dense in H sequence of subspaces.

Proof. *Sufficiency.* Assume that L_n is a limit dense in H sequence of subspaces and condition (8.204) holds. We wish to prove that A is compact. Let P_n denote the orthoprojector in H onto L_n. Condition (8.204) can be written as

$$\gamma_n := \sup_{\substack{\|h\|=1 \\ h \perp L_n}} \| Ah \| \to 0 \quad \text{as} \quad n \to \infty. \tag{8.205}$$

Therefore

$$\begin{aligned} \| A - AP_n \| &= \sup_{\|h\| \leq 1} \| Ah - AP_n h \| = \sup_{\|h\| \leq 1} \| A(I - P_n)h \| \\ &= \sup_{\substack{g=(I-P_n)h \\ \|g\|=(1-\|P_n h\|^2)^{1/2} \leq 1, \ \|h\| \leq 1}} \|Ag\| \leq \sup_{g \perp L_n} \| Ag \| = \gamma_n \to 0. \end{aligned} \tag{8.206}$$

Therefore A is the norm limit of the sequence of the operators AP_n. The operator AP_n is of finite rank $\leq n$. Therefore A is compact. Note that in the sufficiency part of the argument the assumption that the sequence L_n is limit dense in H is not used. In fact, if condition (8.204) holds for any sequence of subspaces $L_n \subset L_{n+1}$ then A is compact as we have proved above. □

Necessity. Assume now that A is compact and $\{L_n\}$ is a limit dense in H sequence of subspaces. We wish to derive (8.204). We have

$$\begin{aligned} \sup_{\substack{h \perp L_n \\ \|h\|=1}} \| Ah \| &= \sup_{\substack{P_n h=0 \\ \|h\|=1}} \| Ah - AP_n h \| \leq \sup_{\|h\|=1} \| A(I - P_n)h \| \\ &= \| A(I - P_n) \| \to 0 \quad \text{as} \quad n \to \infty. \end{aligned} \tag{8.207}$$

The last conclusion follows from the well known result which is formulated as Proposition 8.1.

Proposition 8.1 *If A is compact and the selfadjoint orthoprojection P_n converges strongly to the identity operator I then*

$$\| A(I - P_n) \| \to 0 \quad \textit{as} \quad n \to \infty. \tag{8.208}$$

Note that $P_n \to I$ strongly if and only if the sequence L_n is limit dense in H. Let us prove Proposition 8.1. Let A be compact and $B_n^* = B_n \to 0$

strongly. In our case $B = I - P_n$. Represent $A = K + F_\epsilon$, where K is a finite rank operator and $\| F_\epsilon \| < \epsilon$. Then

$$\| AB_n \| \leq \| F_\epsilon B_n \| + \| KB_n \| \leq c\epsilon + \| KB_n \| . \tag{8.209}$$

Here $c \geq \| B_n \|$ does not depend on n and ϵ. Choose n sufficiently large. Then

$$\| KB_n \| < \epsilon \tag{8.210}$$

since $B_n \to 0$ strongly and K is a finite rank operator. Indeed

$$KB_n h = \sum_{j=1}^{m} s_j (B_n h, \phi_j) \psi_j \tag{8.211}$$

where

$$K := \sum_{j=1}^{m} s_j (\cdot, \phi_j) \psi_j, \quad s_j = \text{const.} \tag{8.212}$$

It is known that $B_n \to 0$ strongly *does not imply* $B_n^* \to 0$ strongly, in general. But since we have assumed that $B_n = B_n^*$, we have

$$\| KB_n h \| \leq \sum_{j=1}^{m} |s_j| \| \psi_j \| \| h \| \| B_n \phi_j \| \leq \epsilon(n) \| h \| \tag{8.213}$$

where $\epsilon(n) \to 0$ as $n \to \infty$ because

$$\| B_n \phi_j \| \to 0 \quad \text{as} \quad n \to \infty, \quad 1 \leq j \leq m. \tag{8.214}$$

Proposition 8.1 and Theorem 8.2 are proved.

Note that in the proof of the necessity part the assumption that the sequence L_n is limit dense in H plays the crucial role: it allows one to claim that $P_n \to I$ strongly. If the sequence L_n is not limit dense then there may exist a fixed vector $h \neq 0$ such that $\| Ah \| > 0$ and $h \perp \bigcup_{n=1}^{\infty} L_n$. In this case condition (8.204) does not hold. This is the case, for example, if h is the first eigenvector of a selfadjoint compact operator A, and $L_n := \text{span}\{\phi_2, \dots, \phi_{n+1}\}$ where $A\phi_j = \lambda_j \phi_j$.

8.3.2.3 *Asymptotic equivalence of s-values of two operators*

We are now ready to answer Question 1. Recall that $N(A) := \{u : Au = 0\}$.

Theorem 8.3 *Assume that* $A,\ Q \in \sigma_\infty$, $N(I+Q) = \{0\}$ *and rank* $A = \infty$. *Then*

$$\lim_{n\to\infty} \frac{s_n\{A(I+Q)\}}{s_n(A)} = 1 \tag{8.215}$$

and

$$\lim_{n\to\infty} \frac{s_n\{(I+Q)A\}}{s_n(A)} = 1. \tag{8.216}$$

Proof. By the minimax principle for singular values one has

$$\begin{aligned} s_{n+1}\{A(I+Q)\} &= \min_{L_n} \max_{\phi\perp L_n} \frac{\| A(I+Q)\phi \|}{\| \phi \|} \\ &= \min_{L_n} \max_{\phi\perp L_n} \left\{ \frac{\| A(I+Q)\phi \|}{\| (I+Q)\phi \|} \frac{\| (I+Q)\phi \|}{\| \phi \|} \right\} \\ &\le \max_{\phi\perp M_n} \frac{\| A(I+Q)\phi \|}{\| (I+Q)\phi \|} \left(1 + \max_{\phi\perp M_n} \frac{\| Q\phi \|}{\| \phi \|} \right) \\ &= s_{n+1}(A)(1+\epsilon_n), \end{aligned} \tag{8.217}$$

where

$$\epsilon_n := \max_{\phi\perp M_n} \frac{\| Q\phi \|}{\| \phi \|} \to 0 \quad \text{as} \quad n\to\infty. \tag{8.218}$$

Here M_n is so chosen that the condition $\phi \perp M_n$ is equivalent to the condition $(I+Q)\phi \perp \mathcal{L}_n(A)$, where $\mathcal{L}_n(A)$ is the linear span of the first n eigenvectors of the operator $(A^*A)^{1/2}$

$$M_n := (I+Q^*)\mathcal{L}_n(A). \tag{8.219}$$

Since $N(I+Q) = \{0\}$ and Q is compact, the operator $I+Q$ is an isomorphism of H onto H and so is $I+Q^*$. Therefore the limit dense in H sequence of the subspaces $\mathcal{L}_n(A)$ is mapped by the operator $I+Q^*$ into a limit dense in H sequence of the subspaces M_n. Indeed, suppose that $f \perp M_n\ \forall n$, that is

$$(f, (I+Q^*)\phi_j) = 0 \quad \forall j \tag{8.220}$$

where $\{\phi_j\}$ is the set of all eigenvectors of the operator $(A^*A)^{1/2}$ including the eigenvectors corresponding to the eigenvalue $\lambda = 0$ if zero is an eigenvalue of $(A^*A)^{1/2}$. Then

$$((I+Q)f, \phi_j) = 0 \quad \forall j. \tag{8.221}$$

Since the set of all the eigenvectors of the operator $(A^*A)^{1/2}$ is complete in H, we conclude that

$$(I+Q)f = 0. \tag{8.222}$$

This implies that $f = 0$ since $I + Q$ is an isomorphism.

The fact that $\epsilon_n \to 0$ follows from the compactness of Q and Theorem 8.2. Let $B = A(I+Q)$. Since $(I+Q)^{-1} = I+Q_1$ where $Q_1 := -Q(I+Q)^{-1}$ is a compact operator, one has $A = B(I+Q_1)$. Therefore one obtains as above the inequality:

$$s_{n+1}(A) \le s_{n+1}(B)(1+\delta_n), \quad \delta_n \to 0 \quad \text{as} \quad n \to \infty. \tag{8.223}$$

From (8.217) and (8.223)equation (8.215) follows. The proof of (8.216) reduces to (8.215) if one uses property (8.162) of s-values. Theorem 8.3 is proved. □

The result given in Theorem 8.3 is optimal in some sense. Namely, if Q is not compact but, for example, an operator with small norm, then the conclusion of Theorem 8.3 does not hold in general (take, for instance, $Q = \epsilon I$ where I is the identity operator). The assumption rank $A = \infty$ is necessary since if rank $A < \infty$ one has only a finite number of nonzero singular values. The assumption $N(I+Q) = \{0\}$ is often easy to verify and it is natural. It could be dropped if the assumption about the rate of decay of $s_n(A)$ is

$$s_n(A) \sim cn^{-p}, \quad p > 0$$

but we do not go into detail.

8.3.2.4 *Estimate of the remainder*

Let us now answer the second question.

Theorem 8.4 *Assume that A and Q are linear compact operators on H, $N(I+Q) = \{0\}$, $B := A(I+Q)$,*

$$s_n(A) = cn^{-p}\left[1 + O(n^{-p_1})\right] \quad \textit{as} \quad n \to \infty, \tag{8.224}$$

where p, p_1 *and* c *are positive numbers, and*

$$\| Qf \| \leq c \| Af \|^a \| f \|^{1-a}, \quad a > 0. \tag{8.225}$$

Then

$$s_n(B) = cn^{-p}\left[1 + O(n^{-q})\right] \tag{8.226}$$

where

$$q := \min\left(p_1, \quad \frac{pa}{1+pa}\right). \tag{8.227}$$

In particular,

$$\textit{if} \quad \frac{pa}{1+pa} > p_1 \quad \textit{then} \quad q = p_1 \tag{8.228}$$

and therefore not only the main term of the asymptotics of $s_n(A)$ *is preserved but the order of the remainder as well.*

Remark 8.3 *The estimate (8.227) of the remainder in (8.226) is sharp in the sense that it is attained for some* Q.

Proof. Let n and m be integers. It follows from (8.180) that

$$\begin{aligned} s_{n+m+1}(B) &= \min_{L_{n+m}} \max_{h \perp L_{n+m}} \frac{\| Bh \|}{\| h \|} \\ &\leq \max_{h \perp M_n} \frac{\| A(I+Q)h \|}{\| (I+Q)h \|} \cdot \left(1 + \max_{h \perp \mathcal{L}_m(A)} \frac{\| Qh \|}{\| h \|}\right). \end{aligned} \tag{8.229}$$

Here, as in the proof of Theorem 8.3, M_n is defined by formula (8.219), and $\mathcal{L}_m(A)$ is the linear span of first m eigenvectors of the operator $(A^*A)^{1/2}$. This means that we have chosen L_{n+m} to be the direct sum of the subspaces $M_n + \mathcal{L}_m(A)$. Since the sequence $\mathcal{L}_m(A)$ is limit dense in H one can use Theorem 8.2 and conclude from (8.229) that

$$s_{n+m+1}(B) \leq s_{n+1}(A)(1 + \epsilon_m), \quad \epsilon_m \to 0 \quad \text{as} \quad m \to \infty \tag{8.230}$$

and

$$\epsilon_m = \max_{h \perp \mathcal{L}_m(A)} \frac{\| Qh \|}{\| h \|} \leq c \max_{h \perp \mathcal{L}_m(A)} \frac{\| Ah \|^a}{\| h \|^a} = cs^a_{m+1}(A). \tag{8.231}$$

Therefore

$$s_{n+m+1}(B) \leq s_{n+1}(A)\left[1 + cs^a_{m+1}(A)\right]. \tag{8.232}$$

Unfortunately our assumptions now do not allow to use the argument similar to the one used at the end of the proof of Theorem 8.3. The reason is that our assumptions now are no longer symmetric with respect to A and B. For example, inequality (8.225) is not assumed with B in place of A. In applications it is often possible to establish the inequality (8.225) with B in place of A, and in this case the argument can be simplified: one can use by symmetry the estimate (8.232) in which B and A exchange places.

With the assumption formulated in Theorem 8.4 we proceed as follows. Write

$$A = B(I + Q_1), \quad Q_1 = -Q(I+Q)^{-1}. \tag{8.233}$$

Choose

$$M_{1n} := (I + Q_1^*)\mathcal{L}_n(B) \tag{8.234}$$

and use the inequalities similar to (8.229)-(8.232) to obtain

$$s_{n+m+1}(A) \le s_n(B)\left[1 + c_1 s_{m+1}^a(A)\right]. \tag{8.235}$$

It follows from (8.232) and (8.235) that

$$\begin{aligned} s_{n+2m+1}(A) &\le s_{n+m+1}(B)\left[1 + c_1 s_{m+1}^a(A)\right] \\ &\le s_{n+1}(A)\left[1 + c s_{m+1}^a(A)\right]\left[1 + c_1 s_{m+1}^a(A)\right] \\ &\le s_{n+1}(A)\left[1 + c_2 s_{m+1}^a(A)\right], \end{aligned} \tag{8.236}$$

where we took into account that

$$0 < s_m(A) \to 0 \quad \text{as} \quad m \to \infty \tag{8.237}$$

so that $s_m^2(A) \le s_m(A)$ for all sufficiently large m.

Therefore, for all sufficiently large m one has

$$\begin{aligned} \frac{s_{n+2m+1}(A)}{s_{n+m+1}(A)}\left[1 + 0\left(s_m^a(A)\right)\right] &\le \frac{s_{n+m+1}(B)}{s_{n+m+1}(A)} \\ &\le \frac{s_{n+1}(A)}{s_{n+m+1}(A)}\left[1 + O\left(s_m^a(A)\right)\right]. \end{aligned} \tag{8.238}$$

Choose

$$m = n^{1-x}, \quad 0 < x < 1. \tag{8.239}$$

It follows from (8.224) and (8.239) that

$$\frac{s_{n+m}(A)}{s_n(A)} = \left(\frac{n+m}{n}\right)^{-p} \left[1 + 0(n+m)^{-p_1} + O(n^{-p_1})\right]$$
$$= 1 + 0\left(\frac{m}{n}\right) + O(n^{-p_1}) = 1 + O(n^{-x}) + O(n^{-p_1}). \quad (8.240)$$

From (8.240) and (8.238) one obtains

$$\frac{s_{n+m+1}(B)}{s_{n+m+1}(A)} = 1 + O(n^{-p_1}) + O(n^{-x}) + O(n^{-(1-x)pa}). \quad (8.241)$$

Let

$$q := \min\{p_1, x, (1-x)pa\}. \quad (8.242)$$

Then

$$\frac{s_{n+m+1}(B)}{s_{n+m+1}(A)} = 1 + O(n^{-q}). \quad (8.243)$$

Since

$$\frac{n+m+1}{n} \sim 1 \quad \text{as} \quad n \to \infty \quad (8.244)$$

it follows from (8.243) that formula (8.226) holds. Choose now x, $0 < x < 1$ such that

$$\min(x, (1-x)pa) = \max. \quad (8.245)$$

An easy calculation shows that (8.245) holds if

$$x = \frac{pa}{1+pa}. \quad (8.246)$$

Therefore

$$q = \min\left(p_1, \frac{pa}{1+pa}\right).$$

This is formula (8.227). Theorem 8.4 is proved. □

We leave for the reader as an exercise to check the validity of the Remark.

Hint. A trivial example, which shows that for some Q the order of the remainder in (8.226) is attained, is the following one. Let $A > 0$ be a selfadjoint compact operator. In this case $s_j(A) = \lambda_j(A)$. Take $Q = \phi(A)$. Then

$$\lambda_n(B) = \lambda_n \{A[1 + \phi(A)]\} = \lambda_n(A)[1 + \phi(\lambda_n)] \tag{8.247}$$

by the spectral mapping theorem. If one chooses $\phi(\lambda)$ such that

$$\phi(\lambda_n) = o(n^{-p_1}). \tag{8.248}$$

Then $q = p_1$ is the order of the remainder in the formula

$$\lambda_n(B) = cn^{-p}\left[1 + O(n^{-p_1})\right]. \tag{8.249}$$

8.3.2.5 *Unbounded operators*

Note that the results of Theorems 8.3, 8.4 can be used in the cases when the operators we are interested in are unbounded. For example, suppose that $\mathcal{L}$ is an elliptic selfadjoint operator in $H = L^2(D)$ of order s and ℓ is a selfadjoint in H differential operator of lower order, $\mathrm{ord}\ell = m$. We wish to check that

$$\lim_{n\to\infty} \frac{\lambda_n(\mathcal{L}+\ell)}{\lambda_n(\mathcal{L})} = 1. \tag{8.250}$$

Since $\lambda_n(\mathcal{L}) \to +\infty$ and $\lambda_n(\mathcal{L}+cI) = \lambda_n(\mathcal{L}) + c$ where c is a constant, one can take $\mathcal{L} + cI$ in place of $\mathcal{L}$ in (8.250) and choose $c > 0$ such that the operator $\mathcal{L}+cI$ is positive definite in H. Then the operator $A := (\mathcal{L}+cI)^{-1}$ is compact in H. Moreover

$$\mathcal{L} + cI + \ell = \left[I + \ell(\mathcal{L}+cI)^{-1}\right](\mathcal{L}+cI)$$

so that

$$B := (\mathcal{L} + cI + \ell)^{-1} = (\mathcal{L}+cI)^{-1}\left[I + \ell(\mathcal{L}+cI)^{-1}\right]^{-1}. \tag{8.251}$$

If $\mathrm{ord}\ell < \mathrm{ord}\mathcal{L}$ then the operator

$$S := \ell(\mathcal{L}+cI)^{-1} \quad \text{is compact in } H. \tag{8.252}$$

One can always choose the constant $c > 0$ such that $N(I+S) = \{0\}$, so that

$$(I+S)^{-1} = I + Q \tag{8.253}$$

where Q is compact in H. Then (8.251) can be written as

$$B = A(I+Q) \tag{8.254}$$

and the assumptions of Theorem 8.3 are satisfied. In fact, since A and B are selfadjoint and $\lambda_n(A)$ and $\lambda_n(B)$ are positive for all sufficiently large n, one has

$$s_n(B) = \lambda_n(B), \quad s_n(A) = \lambda_n(A), \quad \forall n > n_0. \tag{8.255}$$

By Theorem 8.3 one has

$$\lim_{n\to\infty} \frac{\lambda_n(B)}{\lambda_n(A)} = 1. \tag{8.256}$$

Since $\lambda_n(B^{-1}) = \lambda_n^{-1}(B)$, it follows from (97) that

$$\lim_{n\to\infty} \frac{\lambda_n(B^{-1})}{\lambda_n(A^{-1})} = 1. \tag{8.257}$$

This is equivalent to (8.250) because, as was mentioned above,

$$\lim_{n\to\infty} \frac{\lambda_n(\mathcal{L} + \ell + cI)}{\lambda_n(\mathcal{L} + \ell)} = 1 \tag{8.258}$$

for any constant c.

8.3.2.6 *Asymptotics of eigenvalues*

In this section we prove some theorems about perturbations preserving asymptotics of the spectrum. In order to formulate these theorems in which unbounded operators appear, we need some definitions.

Let A be a closed liner densely defined in a Hilbert space H operator, $D(A)$ is its domain of definition, $R(A)$ is its range, $N(A) = \{u : Au = 0\}$ is its null-space, $\sigma(A)$ is its spectrum.

Definition 8.3 We say that the spectrum of A is discrete if it consists of isolated eigenvalues with the only possible limit point at infinity, each of the eigenvalues being of finite algebraic multiplicity and for each eigenvalue λ_j the whole space can be represented as a direct sum of the root subspace $\mathcal{M}_j$ corresponding to λ_j, and a subspace H_j which is invariant with respect to A and in which the operator $A - \lambda_j I$ has bounded inverse. In this case λ_j is called a normal eigenvalue.

The root linear manifold of the operator A corresponding to the eigenvalue λ is the set of vectors which solve the equation

$$\mathcal{M}_\lambda := \{u : (A - \lambda I)^n u = 0 \quad \text{for some } n\}. \tag{8.259}$$

The algebraic multiplicity $\nu(\lambda)$ of the eigenvalue λ is

$$\nu(\lambda) := \dim \mathcal{M}_\lambda. \tag{8.260}$$

If $\mathcal{M}_\lambda$ is closed in H it is called root subspace. The geometric multiplicity $n(\lambda)$ of λ is the dimension of the eigenspace corresponding to λ, $n(\lambda) = \dim N(A-\lambda I)$. If $\epsilon > 0$ is small enough so that there is only one eigenvalue in the disc $|z-\lambda| < \epsilon$ then

$$P_\lambda := -\frac{1}{2\pi i}\int_{|z-\lambda|=\epsilon} R(z)dz, \quad R(z) := (A-zI)^{-1} \tag{8.261}$$

is the projection, that is $P^2 = P$. The subspace $P_\lambda H$ is invariant for A, P_λ commutes with A, $P_\lambda A = AP_\lambda$, the spectrum of the restriction of A onto $P_\lambda H$ consists of only one point λ, which is its eigenvalue of algebraic multiplicity $\nu(\lambda)$.

An example of operators with discrete spectrum is the class of operators for which the operator $(A-\lambda_0 I)^{-1}$ is compact for some $\lambda_0 \in \mathbf{C}$. Such are elliptic operators in a bounded domain.

If A is an operator with discrete spectrum then

$$|\lambda_n(A)| \to \infty \quad \text{as} \quad n \to \infty. \tag{8.262}$$

Thus, for any constant c,

$$\lim_{n\to\infty} \frac{\lambda_n(A+cI)}{\lambda_n(A)} = 1. \tag{8.263}$$

Therefore it is not too restrictive to assume that A^{-1} exists and is compact: if A^{-1} does not exist then choose c such that $(A+cI)^{-1}$ exists and study the asymptotics of $\lambda_n(A+cI) = \lambda_n(A) + c$. Note that if $(A-\lambda_0 I)^{-1}$ is compact for some λ_0, then $(A-\lambda I)^{-1}$ is compact for any λ for which $A-\lambda I$ is invertible. This follows from the resolvent identity

$$(A-\lambda I)^{-1} = (A-\lambda_0)^{-1} + (\lambda-\lambda_0)(A-\lambda I)^{-1}(A-\lambda_0 I)^{-1}. \tag{8.264}$$

If A^{-1} is compact we define the singular values of A by the formula

$$s_n(A) = \left[s_n(A^{-1})\right]^{-1}. \tag{8.265}$$

If

$$A = A^* \geq m > 0 \tag{8.266}$$

then we denote by H_A the Hilbert space which is the completion of $D(A)$ in the norm $\| u \|_A = (Au, u)^{1/2}$. Clearly $H_A \subset H$, $\| u \| \leq m^{-1} \| u \|_A$, and $(u, v)_A := (Au, v)$ is the inner product in H_A. The inner product can be written also as $(u, v)_A = (A^{1/2}u, A^{1/2}v)$. If $B = B^* \geq -m$, then by H_B we mean the Hilbert space which is the completion of $D(B)$ in the norm $\| u \|_B := ((B + m + 1)u, u)^{1/2}$. *All unbounded operators we always assume densely defined in H.*

Theorem 8.5 *Let $A = A^* \geq m > 0$ be a linear closed operator with discrete spectrum, T be a linear operator, $D(A) \subset D(T)$, $B := A + T$, $D(B) = D(A)$. Assume that $A^{-1}T$ is compact in H_A, $B = B^*$ and $H_A \subset D(T)$. Then*

$$\lim_{n\to\infty} \frac{\lambda_n(B)}{\lambda_n(A)} = 1. \tag{8.267}$$

The conclusion (8.267) remains valid if $A \geq -m$ and $[A + (m+1)I]^{-1}T$ is compact in H_A.

Remark 8.4 *If $T > 0$ then $A^{-1}T$ is compact in H_A if and only if the imbedding operator $i : H_A \to H_T$ is compact. By H_T we mean the Hilbert space which is the completion of $D(T)$ in the norm $(Tu, u)^{1/2}$. If T is not positive but $|(Tf, f)| \leq (Qf, f)$ for some $Q > 0$ and all $f \subset D(T)$, $D(T) \subset D(Q)$, and if the imbedding $i : H_A \to H_Q$ is compact then $A^{-1}T$ is compact.*

The reader can prove these statements as an exercise or find a proof in [Glazman (1965), §4].

To prove Theorem 8.5 we need a lemma.

Lemma 8.19 *If the operator $A^{-1}T$ is compact in H_A then $H_A = H_B$ and the spectrum of B is discrete.*

Assuming the validity of this lemma let us prove Theorem 8.5 and then prove the Lemma.

Proof of Theorem 8.5 Let us use the symbol Ⅱ for orthogonality in H_A and $\perp$ for orthogonality in H. If $\mathcal{L}_n(A)$ is the linear span of the first n eigenvectors of A^{-1} then $f \perp \mathcal{L}_n(A)$ is equivalent to $f \,\amalg\, \mathcal{L}_n(A)$. Indeed, if $A^{-1}\phi_j = \lambda_j \phi_j$, $\lambda_j \neq 0$, then

$$0 = (f, \phi_j) = \lambda_j (f, A\phi_j) = \lambda_j (f, \phi_j)_A \Leftrightarrow (f, \phi_j)_A = 0.$$

Note also that

$$\inf\{\alpha(1+\beta)\} \geq (1-\sup\beta)\inf\alpha \quad \text{if} \quad \alpha \geq 0 \quad \text{and} \quad -1 < \beta < 1. \tag{8.268}$$

We will use the following statement

$$\gamma_n :\equiv \sup_{f \perp \mathcal{L}_n(A)} \frac{|(Tf,f)|}{(f,f)_A} = \sup_{f \amalg \mathcal{L}_n(A)} \frac{|(A^{-1}Tf,f)_A|}{(f,f)_A} \to 0 \quad \text{as} \quad u \to \infty \tag{8.269}$$

which follows from Theorem 8.2 and the assumed compactness of $A^{-1}T$ in H_A.

We are now ready to prove Theorem 8.5. By the minimax principle one has

$$\begin{aligned} \lambda_{n+1}(B) &= \sup_{L_n} \inf_{f \perp L_n} \frac{(Bf,f)}{(f,f)} \\ &\geq \inf_{f \perp \mathcal{L}_n(A)} \frac{(Bf,f)}{(f,f)} = \inf_{f \perp \mathcal{L}_n(A)} \left\{ \frac{(Af,f)}{(f,f)} \left[1 + \frac{(Tf,f)}{(Af,f)} \right] \right\} \\ &\geq \lambda_{n+1}(A)(1-\gamma_n), \quad \gamma_n \to 0 \quad \text{as} \quad n \to \infty, \end{aligned} \tag{8.270}$$

where we have used (8.268) and (8.269).

By symmetry, for all sufficiently large n, one has

$$\lambda_{n+1}(A) \geq \lambda_{n+1}(B)(1-\delta_n), \quad \delta_n \to 0 \quad \text{as} \quad n \to \infty. \tag{8.271}$$

We left for the reader as an exercise to check that under the assumptions of Theorem 8.5 the operator $(B+cI)^{-1}T$ is compact if $B+cI$ is invertible.

From (8.270) and (8.271) the desired conclusion (8.267) follows. If $A \geq -m$ and $[A+(m+1)I]^{-1}T$ is compact in H_A then we argue as above and obtain in place of (8.270) the following inequality

$$\begin{aligned} \lambda_{n+1}\{B+(m+1)I\} \geq \lambda_{n+1}\{A+(m+1)I\}(1-\gamma_n), \\ \gamma_n \to 0 \quad \text{as} \quad n \to \infty. \end{aligned} \tag{8.272}$$

Since $\lambda_n(A+cI) = \lambda_n(A)+c$, $\lambda_n(A) \to +\infty$ and $\lambda_n(B) \to +\infty$, inequality (8.272) implies

$$\lambda_{n+1}(B)\,[1+o(1)] \geq \lambda_{n+1}(A)\,[1+o(1)]\,(1-\gamma_n) \quad \text{as} \quad n \to \infty, \tag{8.273}$$

and the rest of the argument is the same as above. Theorem 8.5 is proved.

□

Proof of the Lemma 8.19 Note that $B = A(I+S)$, where $S := A^{-1}T$ is compact in H_A. Let us represent S in the form $S = Q+F$ where $\| Q \|_A < 1$ and F is a finite rank operator. The operator S is selfadjoint in H_A. Indeed

$$(Sf,g)_A = (Tf,g) = (f,Tg) = (f,Sg)_A$$

where we used the symmetry of $T = B - A$ on H. We choose Q and F to be selfadjoint in H_A. The operator $I+Q$ is positive definite in H_A while

$$(Fu,u)_A = \sum_{j=1}^{N} a_j |(u,\phi_j)_A|^2$$

for some orthonormal in H_A set of functions ϕ_j, for some constants a_j, and some number $N = \text{rank}\ F$. Since $D(A)$ is dense in H_A, one can find $v_j \in D(A)$ such that $\| \phi_j - v_j \|_A < \epsilon$, where $\epsilon > 0$ is arbitrarily small. Then

$$\begin{aligned} |(Fu,u)_A| &\le \sum_{j=1}^{N} |a_j| \, |(u, \phi_j - v_j)_A + (u, Av_j)|^2 \\ &\le c_1 \epsilon \| u \|_A^2 + c_2 \| u \|^2 \end{aligned} \tag{8.274}$$

where c_1 and c_2 are some positive constants which do not depend on u. It follows from (8.274) that

$$\begin{aligned} (Bu,u) &= (A(I+Q)u,u) + (AFu,u) \\ &= ((I+Q)u,u)_A + (Fu,u)_A \\ &\ge c_0 \| u \|_A^2 - c_1 \epsilon \| u \|_A^2 - c_2 \| u \|^2 \end{aligned} \tag{8.275}$$

where $c_0 > 0$. It follows from (8.275), if one chooses ϵ so small that $c_0 - \epsilon c_1 > 0$, that B is bounded from below in H. Since, clearly, $(Bu,u) \le c(u,u)_A$ one concludes that the metrics of H_A and H_B are equivalent, so that $H_A = H_B$.

It remains to be proved that the spectrum of B is discrete. Since B is selfadjoint it is sufficient to prove that no point of spectrum $\sigma(B)$ of B belongs to the essential spectrum $\sigma_{ess}(B)$. Recall that $\lambda \in \sigma_{ess}(B)$, where $B = B^*$, if and only if $\dim E(\Delta_\epsilon)H < \infty$ for any $\epsilon > 0$, where $\Delta_\epsilon = (\lambda - \epsilon, \lambda + \epsilon)$ and $E(\Delta)$ is the resolution of the identity corresponding to the selfadjoint operator B. Assume that $\lambda \in \sigma(B)$ and $\dim E(\Delta_{\epsilon_n})H = \infty$ for some sequence $\epsilon_n \to 0$. Then there exists an orthonormal sequence $u_n \in H$, such that $\| Bu_n - \lambda u_n \| \to 0$ as $n \to \infty$. Thus

$$\| u_n + A^{-1}Tu_n - \lambda A^{-1} u_n \| \to 0 \quad \text{as} \quad n \to \infty. \tag{8.276}$$

Since $\| u_n \|= 1$ we have

$$(Au_n, u_n) + (Tu_n, u_n) - \lambda(u_n, u_n) \to 0, \quad n \to \infty. \tag{8.277}$$

If $A^{-1}T$ is compact in H_A, we have proved that, for any $\epsilon > 0$,

$$|(Tu, u)| \leq \epsilon \| u \|_A^2 + c(\epsilon) \| u \|^2, \quad u \in H_A. \tag{8.278}$$

It follows from (8.277) and (8.278) that

$$\| u_n \|_A \leq c \tag{8.279}$$

where $c > 0$ is a constant which does not depend on n. Since $A^{-1}T$ is compact in H_A, inequality (8.279) implies that a subsequence of the sequence u_n exists (we denote this subsequence again by u_n) such that $A^{-1}Tu_n$ converges in H_A and, therefore, in H. Since the set u_n is orthonormal, u_n converges weakly to zero in H

$$u_n \rightharpoonup 0, \quad n \to \infty. \tag{8.280}$$

Therefore

$$\| A^{-1}Tu_n \| \to 0 \quad \text{as} \quad n \to \infty. \tag{8.281}$$

From (8.281) and (8.276) it follows that

$$\| u_n - \lambda A^{-1} u_n \| \to 0 \quad \text{as} \quad n \to \infty \tag{8.282}$$

where u_n is an orthonormal subsequence. This means that if $\lambda \neq 0$ then $\lambda \in \sigma_{ess}(A)$ which is a contradiction since, by assumption, A does not have essential spectrum. If $\lambda = 0$ then (8.282) cannot hold since $\| u_n \|= 1$. Therefore B does not have essential spectrum and its spectrum is discrete. Lemma 8.19 is proved. □

Example 8.1 Let A be the Dirichlet Laplacian $-\Delta$ in a bounded domain $D \subset R^r$ and $B = -\Delta + q(x)$, where $q(x)$ is a real-valued function. In this case $Tu = q(x)u$ is a multiplication operator. The condition $A^{-1}T$ is compact in H_A means that $(-\Delta)^{-1}q$ is compact in $\overset{o}{H}{}^1(D)$. This condition holds if and only if $A^{-1/2}T$ is compact in $H = L^2(D)$, $D \subset R^r$, that is $(-\Delta)^{-1/2}q$ is compact in $L^2(D)$. If, for example, $q \in L^p(D)$ then the operator $(-\Delta)^{-1/2}q(x)$ is compact in $L^2(D)$ provided that $q \in L^\gamma(D)$, $\gamma > r$, and Theorem 8.5 asserts that, in this case,

$$\lim_{n\to\infty} \frac{\lambda_n(-\Delta + q)}{\lambda_n(-\Delta)} = 1. \tag{8.283}$$

In the calculation of the L^p class to which q belongs we have used the known imbedding theorem which says that the imbedding $i : W^{k,p}(D) \to L^{\frac{rp}{r-kp}}(D)$ is compact for $kp < r$, where $W^{k,p}(D)$ is the Sobolev space of functions with derivatives of order $\leq k$ belonging to $L^p(D)$. If $q \in L^\gamma(D)$ and $u \in L^2(D)$ then $qu \in L^p(D)$,

$$\int_D |qu|^p dx \leq \left(\int_D |q|^{p\alpha} dx\right)^{\frac{1}{\alpha}} \left(\int_D |u|^{p\beta}\right)^{\frac{1}{\beta}}$$

where $\alpha > 1$, $\beta = \frac{\alpha}{\alpha-1}$, $p\beta = 2$, $p\alpha = \gamma$. Thus

$$p = \frac{2(\alpha-1)}{\alpha}, \quad p = \frac{\gamma}{\alpha}, \quad \text{so that} \quad \alpha = 1 + \frac{\gamma}{2}$$

and $p = \frac{2\gamma}{\gamma+2}$. On the other hand, if $qu \in L^p(D)$ then $\Delta^{-\frac{1}{2}} qu \in W^{1,p}(D) \subset L^{\frac{rp}{r-p}}(D)$. If $\frac{rp}{r-p} > 2$, that is $p > \frac{2r}{r+2}$, then $\gamma > r$. The condition on $q(x)$ for which (8.283) holds can be relaxed.

In the next theorem we assume compactness of $A^{-1}T$ and TA^{-1} in H rather than in H_A.

Theorem 8.6 *Assume that $A = A^* \geq m > 0$ is an operator with discrete spectrum, $D(A) \subset D(T)$, $B = A + T$, $D(B) = D(A)$, B is normal, $0 \notin \sigma(B)$, and the operator $A^{-1}T$ is compact in H. Then the spectrum of B is discrete and*

$$\lim_{n\to\infty} \frac{\lambda_n(B)}{\lambda_n(A)} = 1. \tag{8.284}$$

Proof. First we prove that the spectrum of B is discrete. Since A is selfadjoint positive definite and its spectrum is discrete it follows that A^{-1} is compact. Let

$$Au + Tu = \lambda u + f, \quad u + A^{-1}Tu = \lambda A^{-1}u + A^{-1}f. \tag{8.285}$$

Since $0 \notin \sigma(B)$ the operator $I + A^{-1}T$ has bounded inverse. Therefore

$$u = \lambda(I + A^{-1}T)^{-1}A^{-1}u + (I + A^{-1}T)^{-1}A^{-1}f. \tag{8.286}$$

The operator $(I + A^{-1}T)^{-1}A^{-1}$ is compact being a product of a bounded and compact operator. Equations (8.285) and (8.286) are equivalent. Therefore

$$(B - \lambda I)^{-1} = \left[I - \lambda(I + A^{-1}T)^{-1}A^{-1}\right]^{-1}(I + A^{-1}T)^{-1}A^{-1}. \tag{8.287}$$

It follows from (8.287) that $\lambda \in \sigma(B)$ if and only if $\lambda^{-1} \in \sigma(F)$, $F := (I + A^{-1}T)^{-1}A^{-1}$. Since F is compact, each λ is an isolated eigenvalue of finite algebraic multiplicity and $\sigma(B)$ is discrete. In this part of the argument we did not use the assumption that B is normal.

If B is normal then $|\lambda_n(B)| = s_n(B)$, where $s_n(B)$ are the singular values of B. Since $B = A(I + A^{-1}T)$ and $A^{-1}T$ is compact, since $s_n(B) = s_n^{-1}(B^{-1})$, and since A^{-1} is compact, we can apply Theorem 8.2 and get

$$\lim_{n\to\infty} \frac{s_n(B^{-1})}{s_n(A^{-1})} = \lim_{n\to\infty} \frac{s_n(A)}{s_n(B)} = 1. \tag{8.288}$$

Since $A > 0$, we have $s_n(A) = \lambda_n(A)$. Therefore the desired result (8.284) will be proved if we prove that

$$\lim_{n\to\infty} \frac{|\lambda_n(B)|}{\lambda_n(B)} = 1. \tag{8.289}$$

Let us prove (8.289). Let

$$A\phi_j + T\phi_j = \lambda_j\phi_j. \tag{8.290}$$

Since B is normal we can assume that

$$(\phi_j, \phi_m) = \delta_{jm}. \tag{8.291}$$

Rewrite (8.290) as

$$\phi_j + A^{-1}T\phi_j = \lambda_j A^{-1}\phi_j. \tag{8.292}$$

Multiply (8.292) by ϕ_j to get

$$1 + (A^{-1}T\phi_j, \phi_j) = \lambda_j \quad (A^{-1}\phi_j, \phi_j). \tag{8.293}$$

Since $A^{-1}T$ is compact and $\phi_j \rightharpoonup 0$ as $j \to \infty$, we have:

$$(A^{-1}T\phi_j, \phi_j) \to 0 \quad \text{as} \quad j \to \infty. \tag{8.294}$$

Note that $(A^{-1}\phi_j, \phi_j) > 0$. Therefore it follows from (8.293) that

$$\frac{Im\lambda_j}{Re\lambda_j} = \frac{Im(A^{-1}T\phi_j, \phi_j)}{1 + Re(A^{-1}T\phi_j, \phi_j)} \to 0 \quad \text{as} \quad j \to \infty. \tag{8.295}$$

This implies (8.289). Theorem 8.6 is proved. □

8.3.2.7 *Asymptotics of eigenvalues (continuation)*

In this section we continue to study perturbations preserving asymptotics of the spectrum of linear operators. Let us give a criterion for compactness of the resolvent $(A - \lambda I)^{-1} := R(\lambda)$ for $\lambda \notin \sigma(A)$, where A is a closed densely defined linear operator in H.

Theorem 8.7 *The operator $(A-\lambda I)^{-1}$, $\lambda \notin \sigma(A)$ is compact if and only if the operator $(I + A^*A)^{-1}$ is compact.*

Proof. *Sufficiency.* Suppose $(I + A^*A)^{-1}$ is compact and $\lambda \notin \sigma(A)$. Let

$$\| g_n \| \leq c, \quad (A - \lambda I)^{-1} g_n = f_n. \tag{8.296}$$

Then

$$\| f_n \| \leq c \quad \text{and} \quad \| Af_n \| \leq c, \tag{8.297}$$

where c denotes various positive constants. Therefore

$$\left\| (I + A^*A)^{1/2} f_n \right\|^2 = \| f_n \|^2 + \| Af_n \|^2 \leq c. \tag{8.298}$$

The operators $(I + A^*A)^{-1}$ and $(I + A^*A)^{-1/2}$ are selfadjoint positive operators. They are simultaneously compact or non-compact. Therefore if $(I + A^*A)^{-1}$ is compact then $(I + A^*A)^{-1/2}$ is compact and (8.298) implies that the sequence $\{f_n\}$ is relatively compact. Therefore the operator $(A - \lambda I)^{-1}$, $\lambda \notin \sigma(A)$, maps any bounded sequence g_n into a relatively compact sequence f_n. This means that $(A - \lambda I)^{-1}$ is compact.

Necessity. Assume that $(A - \lambda I)^{-1}$ is compact and $\| h_n \| \leq c$. Then the sequence $(A - \lambda I)^{-1} h_n$ is relatively compact. We wish to prove that the sequence $q_n := (I + A^*A)^{-1} h_n$ is relatively compact. The sequence $(I + A^*A) q_n = h_n$ is bounded. Thus

$$((I + A^*A)q_n, q_n) = \| q_n \|^2 + \| Aq_n \|^2 \leq c. \tag{8.299}$$

Define $p_n := (A - \lambda I)q_n$, $q_n = (A - \lambda I)^{-1} p_n$. We have

$$\| p_n \| \leq \| Aq_n \| + |\lambda| \, \| q_n \| \leq c \tag{8.300}$$

where c denotes various constants. From (8.300) and compactness of $(A - \lambda I)^{-1}$ it follows that the sequence $q_n = (A - \lambda I)^{-1} p_n$ is relatively compact. Theorem 8.7 is proved. □

Remark 8.5 *Let T be a linear operator in H, $D(A) \subset D(T)$ and let $0 \notin \sigma(A)$.*

Definition 8.4 If for any sequence f_n such that

$$\| f_n \| + \| Af_n \| \leq c$$

the sequence Tf_n is relatively compact then T is called A-compact.

In other words, T is A-compact if it is a compact operator from the space G_A into H. The space G_A is the closure of $D(A)$ in the graph norm $\| f \|_{G_A} := \| f \| + \| Af \|$. If A is closed, which we assume, then $D(A) = G_A$ is a Banach space if it is equipped with the graph norm.

Proposition 8.2 *The operator T is A-compact if and only if the operator TA^{-1} is compact in H.*

Proof. Suppose T is A-compact. Let $\| f_n \| \leq c$, and define $g_n = A^{-1} f_n$. Then $\| g_n \| + \| Ag_n \| \leq c$. Therefore the sequence Tg_n is relatively compact. This means that the sequence $TA^{-1} f_n$ is relatively compact. Therefore TA^{-1} is compact in H. Conversely, suppose TA^{-1} is compact in H and $\| f_n \| + \| Af_n \| \leq c$. Then the sequence $Tf_n = TA^{-1}Af_n$ is relatively compact. Proposition 8.2 is proved. □

8.3.2.8 *Asymptotics of s-values*

In this section we prove

Theorem 8.8 *Let A be a closed linear operator in H. Suppose that $\sigma(A)$, the spectrum of A, is discrete and $0 \notin \sigma(A)$. Let T be a linear operator, $D(A) \subset D(T)$, $B = A + T$, $D(B) = D(A)$.*

If the operator TA^{-1} is compact then B is closed. If, in addition, A^{-1} is compact and, for some number $k \notin \sigma(A)$, the operator $B + kI$ is injective, then $\sigma(B)$ is discrete and

$$\lim_{n\to\infty} \frac{s_n(B)}{s_n(A)} = 1 \quad as \quad n \to \infty. \tag{8.301}$$

The following lemma is often useful.

Lemma 8.20 *Suppose that $\{f_n\} \in H$ is a bounded sequence which does not contain a convergent subsequence. Then there is a sequence $\{\psi_m\} = \{f_{n_{m+1}} - f_{n_m}\}$ such that*

$$\psi_m \rightharpoonup 0 \quad as \quad m \to \infty \tag{8.302}$$

and $\{\psi_m\}$ does not contain a convergent subsequence.

Proof. Since $\{f_n\}$ is bounded we can assume that it converges weakly:

$$f_n \rightharpoonup f \tag{8.303}$$

(passing to a subsequence and using the well known fact that bounded sets in a Hilbert space are relatively weakly compact). Since $\{f_n\}$ does not contain a convergent subsequence, one can find a subsequence such that

$$\| f_{n_m} - f_{n_k} \| \geq \epsilon > 0 \quad \text{for all} \quad m \neq k. \tag{8.304}$$

If

$$\psi_m := f_{n_{m+1}} - f_{n_m} \tag{8.305}$$

then (8.303) implies (8.302), and the sequence $\{\psi_m\}$ does not contain a convergent subsequence because

$$\| \psi_m \| \geq \epsilon > 0, \tag{8.306}$$

and if there would be a convergent subsequence ψ_{m_j} it would have to converge to zero since its weak limit is zero. Lemma 8.20 is proved. □

This lemma can be found in [Glazman (1965), §5] where it is used in the proof of the following result: if A is a closed linear operator in H and K is a compact operator then $\sigma_c(A+K) = \sigma_c(A)$, where $\sigma_c(A)$ is continuous spectrum of A that is the set of points λ such that there exists a bounded sequence $\psi_m \in D(A)$ which does not contain a convergent subsequence and which has the property $\| A\psi_m - \lambda\psi_m \| \to 0$ as $m \to \infty$.

Proof of Theorem 8.8 (1) Let us first prove that B is closed. Assume that

$$f_n \to f, \quad Bf_n = Af_n + Tf_n \to g, \tag{8.307}$$

and $f_n \subset D(B) = D(A)$. Suppose we have the estimate

$$\| Af_n \| \leq c. \tag{8.308}$$

Then the sequence $Tf_n = TA^{-1}Af_n$ contains a convergent subsequence since TA^{-1} is compact. This and the second equation (8.307) imply that the sequence $\{Af_n\}$ contains a convergent subsequence which we denote again Af_n. Since A is closed by the assumption, we conclude that $f \in D(A) = D(B)$ and

$$Af + Tf = g$$

where we took into account that

$$\lim_{n\to\infty} Tf_n = \lim_{n\to\infty} TA^{-1}Af_n = TA^{-1}Af = Tf.$$

Thus, the operator B is closed provided that (8.308) holds. Let us prove inequality (8.308). Suppose

$$\| Af_n \| \to \infty, \quad n \to \infty. \tag{8.309}$$

Define

$$g_n := \frac{f_n}{\| Af_n \|}, \quad \| g_n \| \to 0, \quad \| Ag_n \| = 1. \tag{8.310}$$

Equation (8.307) implies

$$Ag_n + Tg_n \to 0, \quad n \to \infty. \tag{8.311}$$

As above, compactness of the operator TA^{-1} and the last equation (8.310) imply that one can assume that the subsequence Tg_{n_m}, which we denote again Tg_n, converges in H. This and equation (8.311) imply that Ag_n converges to an element h:

$$Ag_n \to h. \tag{8.312}$$

Since A is closed and $g_n \to 0$, one concludes that $h = 0$. This is a contradiction:

$$1 = \lim_{n\to\infty} \| Ag_n \| = \| h \| = 0.$$

This contradiction proves estimate (8.308). We have proved that the operator B is closed.

(2) Let us prove that $\sigma(B)$ is discrete. We have

$$(B - \lambda I)^{-1} = (A + T - \lambda I)^{-1} = (A + kI)^{-1}(I + Q - \mu S)^{-1}, \tag{8.313}$$

where

$$Q := TS, \quad S = (A + kI)^{-1} \tag{8.314}$$

$$\mu = \lambda + k, \quad k \notin \sigma(A). \tag{8.315}$$

The operators S and Q are compact. If $B + kI$ is injective then $I + Q$ is injective. Since Q is compact this implies that $I + Q$ is an isomorphism of H onto H. Therefore

$$(I + Q - \mu S)^{-1} = (I + Q)^{-1}(I - \mu K)^{-1} \tag{8.316}$$

where

$$K := S(I + Q)^{-1} \quad \text{is compact.} \tag{8.317}$$

Therefore the set μ for which the operator $B - \lambda I$ is not invertible is a discrete set, namely the set of the characteristic values of the compact operator K. Recall that μ_j is a characteristic value of K if

$$\phi_j = \mu_j K \phi_j, \quad \phi_j \neq 0. \tag{8.318}$$

Thus the set $\{\mu_j\}$ has the only possible limit point at infinity. Each μ_j is an isolated eigenvalue of k of finite algebraic multiplicity and therefore $\lambda_j = \mu_j - k$ is an isolated eigenvalue of B of finite algebraic multiplicity. Finally, the corresponding to λ_j projection operator (8.261) is finite dimensional, so that λ_j is a normal eigenvalue. We have proved that $\sigma(B)$ is discrete.

(3) Let us prove the last statement of the theorem, i.e. formula (8.301). We have

$$s_n(B) = s_n^{-1}(B^{-1}) = s_n^{-1}\left\{A^{-1}(I + TA^{-1})^{-1}\right\}. \tag{8.319}$$

We can assume without loss of generality that $k = 0$. In this case the operator $I + TA^{-1}$ is invertible and since TA^{-1} is compact one can write $(I + TA^{-1})^{-1} = I + S$, where S is a compact operator. The operator A^{-1} is compact by the assumption. We can apply now Theorem 8.3 and obtain

$$\lim_{n\to\infty} \frac{s_n\left\{A^{-1}(I + S)\right\}}{s_n(A^{-1})} = 1. \tag{8.320}$$

This is equivalent to the desired result (8.301). □

8.3.2.9 *Asymptotics of the spectrum for quadratic forms*

In this section we study perturbations preserving spectral asymptotics for quadratic forms. As a motivation to this study let us consider the following classical problem.

Let $D \subset R^r$ be a bounded domain with a smooth boundary Γ. Consider the problems

$$(-\Delta + 1)u_j = \lambda_j u_j \quad \text{in} \quad D, \quad u_N = 0 \quad \text{on} \quad \Gamma \tag{8.321}$$

$$(-\Delta + 1)u_j = \mu_j u_j \quad \text{in} \quad D, \quad u_N + \sigma u = 0 \quad \text{on} \quad \Gamma \tag{8.322}$$

where $\sigma = \sigma(s) \in C^1(\Gamma)$ and N is the outer normal to Γ.

The question is: how does one see that

$$\lim_{n\to\infty} \frac{\mu_n}{\lambda_n} = 1. \tag{8.323}$$

The usual argument uses relatively complicated variational estimates. The eigenvalues λ_n are minimums of the ratio of the quadratic forms

$$\frac{\int_D [|\nabla u|^2 + |u|^2]dx}{\int_D |u|^2 dx} = \min, \quad u \in H^1(D) \tag{8.324}$$

while μ_n are minimums of the ratio

$$\frac{\int_D [|\nabla u|^2 + |u|^2]dx + \int_\Gamma \sigma |u|^2 ds}{\int_D |u|^2 dx} = \min, \quad u \in H^1(D). \tag{8.325}$$

The desired conclusion (8.323) follows immediately from the abstract result we will prove and from the fact that the quadratic form $\int_\Gamma \sigma|u|^2 ds$ is compact with respect to the quadratic form $\int_D [|\nabla u|^2 + |u|^2]dx$.

Let $A[u,v]$ and $T[u,v]$ be bounded from below quadratic forms in a Hilbert space, $T[u,u] \geq 0$ and $A[u,u] > m \parallel u \parallel^2$, $m > 0$. Assume that $D[A] \subset D[T]$, where $D[A]$ is the domain of definition of the form A, and that the form A is closed and densely defined in H. The form A is called closed if $D[A]$ is closed in the norm

$$\parallel u \parallel_A := \{A[u,u]\}^{1/2}. \tag{8.326}$$

If $A[u,u]$ is not positive definite but bounded from below: $A[u,u] \geq -m \parallel u \parallel^2$, $\forall u \in D[A]$, then the norm $\parallel u \parallel_A$ is defined by

$$\parallel u \parallel_A = \{A[u,u] + (m+1)(u,u)\}^{1/2}. \tag{8.327}$$

The following proposition is well-known (see e.g. [Kato (1995)]).

Proposition 8.3 *Every closed bounded from below quadratic form $A[u,v]$ is generated by a uniquely defined selfadjoint operator A.*

This means that

$$A[u,v] = (Au,v) \quad \forall u \in D(A), \quad v \in D[A]$$

and $D(A) \subset D[A] \subset H$ is dense in $D[A]$ in the norm (8.327). The spectrum of the closed bounded from below quadratic form is the spectrum of the corresponding selfadjoint operator A.

Definition 8.5 A quadratic form T is called A-compact if from any sequence f_n such that $\| f_n \| \leq c$ one can select a subsequence f_{n_k} such that $T[f_{n_k} - f_{n_m}, f_{n_k} - f_{n_m}] \to 0$ as $m, k \to \infty$.

Theorem 8.9 *If $A[u,u]$ is a closed positive definite quadratic form in H with discrete spectrum $\lambda_n(A)$, and $T[u,u]$ is a positive A-compact quadratic form, $D(A) \subset D(T)$, then the form $B[u,u] := A[u,u] + T[u,u]$, $D[B] = D[A]$, is closed, its spectrum is discrete and*

$$\lim_{n\to\infty} \frac{\lambda_n(B)}{\lambda_n(A)} = 1. \tag{8.328}$$

The conclusions of the theorem remain valid if $T[u,u]$ is not positive but $|T[u,u]| \leq T_1[u,u]$ and T_1 is A-compact.

We need a couple of lemmas for the proof.

Lemma 8.21 *Under the assumptions of Theorem 8.9 the quadratic form $T[u,u] > 0$ can be represented as*

$$T[u,v] = [Tu,v] \tag{8.329}$$

where $[u,v]$ is the inner product in $H_A := D[A]$ and $T > 0$ is a compact selfadjoint operator in H_A.

Proof. Consider the quadratic form $T[u,v]$ in the Hilbert space H_A. Since $T[u,u]$ is A-compact, it is bounded in H_A. If $T[u,v]$ is not closed in H_A consider its closure and denote it again by $T[u,v]$. By Proposition 8.2 there exists a selfadjoint in H_A operator $T > 0$ such that (8.329) holds. Let us prove that T is compact in H_A. Suppose $\| u_n \|_A \leq c$. Since $T[u,u]$ is A-compact there exists a subsequence, which we denote u_n again, such that

$$T[u_n - u_m, \quad u_n - u_m] \to 0, \quad n, m \to \infty.$$

Thus

$$[T(u_n - u_m), u_n - u_m] \to 0, \quad n, m \to \infty. \tag{8.330}$$

Since $T > 0$ is selfadjoint, $T^{1/2}$ is well defined and (8.330) can be written as

$$\left\| T^{1/2}(u_n - u_m) \right\|_A \to 0, \quad n, m \to \infty. \tag{8.331}$$

This implies that $T^{1/2}$ is compact in H_A. Therefore T is compact. Lemma 8.21 is proved. □

Lemma 8.22 *Under the assumptions of Theorem 8.9 one has $H_B = H_A$.*

Proof. It is sufficient to prove that

$$T[u, u] \leq \epsilon A[u, u] + c(\epsilon) \parallel u \parallel^2 \quad \forall \epsilon > 0. \tag{8.332}$$

If (8.332) holds then

$$\begin{aligned}(1 - \epsilon)A[u, u] - c(\epsilon) \parallel u \parallel^2 \leq B[u, u] &\leq (1 + \epsilon)A[u, u] + c(\epsilon) \parallel u \parallel^2 \\ &\leq c_2(\epsilon)A[u, u]\end{aligned}$$

so that the norm $\left\{B[u, u] + c(\epsilon) \parallel u \parallel^2\right\}^{1/2}$ is equivalent to the norm $\parallel u \parallel_A$. This means that $H_B = H_A$. The proof of (8.332) is the same as the proof of Lemma 8.19 used in the proof of Theorem 8.5. Lemma 8.22 is proved. □

Proof of Theorem 8.9 We need only prove formula (8.328) and the fact that B has a discrete spectrum. The other conclusions of Theorem 8.9 have been proved in Lemmas 8.21 and 8.22. Since the form $B[u, u]$ is bounded from below in H we may assume that it is positive definite. If not we choose a constant m such that $B_m[u, u] := B[u, u] + m(u, u)$ is positive definite. Since $\lambda_n(B_m) = \lambda_n(B) + m$ and since $\lambda_n(A) \to +\infty$, the equation

$$\lim_{n\to\infty} \frac{\lambda_n(B_m)}{\lambda_n(A)} = 1, \quad n \to \infty$$

is equivalent to (8.328).

Note first that the spectrum of the form $B[u, u]$ is discrete. Indeed, the following known proposition (Rellich's lemma) implies this. □

Proposition 8.4 *Let $B[u, u]$ be a positive definite closed quadratic form in H. The spectrum of B is discrete if and only if the imbedding operator $i : H_B \to H$ is compact.*

For the convenience of the reader we prove Proposition 8.4 after we finish the proof of Theorem 8.9.

Returning to the proof of Theorem 8.9 we note that A has a discrete spectrum by the assumption. Therefore $i : H_A \to H$ is compact. Since $H_A = H_B$ the imbedding $i : H_B \to H$ is compact. By Proposition 8.4 this implies that the spectrum of B is discrete.

To prove formula (8.328) we use the minimax principle:

$$\begin{aligned}\lambda_{n+1}(B) &= \sup_{L_n} \inf_{u \perp L_n} \frac{B[u,u]}{(u,u)} \\ &\geq \inf_{f \perp \mathcal{L}_n(A)} \frac{A[u,u]}{(u,u)} \left(1 + \frac{T[u,u]}{A[u,u]}\right) \\ &\geq \lambda_{n+1}(A) \left(1 - \sup_{f \perp \mathcal{L}_n(A)} \frac{(Tu,u)}{\| u \|_A^2}\right) \\ &= \lambda_{n+1}(A)(1 - \gamma_n), \quad \gamma_n \to \infty. \end{aligned} \tag{8.333}$$

Here we used Theorem 8.2 and denoted by $\mathcal{L}_n(A)$ the linear span of the first n eigenvectors of the operator A generated by the quadratic form $A[u,u]$. Interchanging A and B we get

$$\lambda_{n+1}(A) \geq \lambda_{n+1}(B)(1 - \delta_n), \quad \delta_n \to 0, \quad n \to \infty. \tag{8.334}$$

From (8.333) and (8.334) formula (8.328) follows.

The last statement of Theorem 8.9 follows from the following proposition 8.4.

Proposition 8.5 *If* $|T[u,u]| \leq T_1[u,u]$ *and* T_1 *is* A*-compact then the operator* $A^{-1}T$ *is compact in* H_A.

We will prove this proposition after the proof of Proposition 8.4.

Proposition 8.5 granted, the proof of the last statement of Theorem 8.9 is quite similar to the given above and is left to the reader. Theorem 8.9 is proved. □

Proof of Proposition 8.4 Assume that the spectrum of $B[u,u]$ is discrete. Then the corresponding selfadjoint operator B has only isolated eigenvalues $0 < m \leq \lambda_n(B) \to +\infty$. Therefore the operator B^{-1} is compact in H. This implies that $B^{-1/2}$ is compact in H. Assume that $\| u_n \|_B \leq c$, that is, $\| B^{1/2}u_n \| \leq c$. Then the sequence $u_n = B^{-1/2}B^{1/2}u_n$ contains a convergent in H subsequence. Thus, the imbedding $i : H_B \to H$ is compact.

Conversely, suppose $i : H_B \to H$ is compact. Then any sequence u_n such that $\| u_n \|_B = \| B^{1/2}u_n \| < c$ contains a convergent subsequence. This means that $B^{-1/2}$ is compact. Since $B^{-1/2}$ is selfadjoint it follows that B^{-1} is compact. Since $B \geq m > 0$ this implies that the spectrum of B is discrete. Proposition 8.4 is proved. □

Proof of Proposition 8.5 Denote $Q := A^{-1}T$, $Q_1 = A^{-1}T_1$. Then

$$|[Qu, u]| \leq [Q_1 u, u]. \tag{8.335}$$

where the brackets denote the inner product in H_A. The operator Q_1 is nonnegative and compact in H_A. Indeed,

$$[Q_1 u, u] = (T_1 u, u) \geq 0$$

so $Q_1 \geq 0$ in H_A. Suppose $\| u_n \|_A \leq c$. Then $T_1 u_n$ contains a Cauchy sequence in H_{T_1}, that is

$$(T_1 u_{n_m} - T_1 u_{n_k}, u_{n_m} - u_{n_k}) \to 0, \quad m, k \to \infty. \tag{8.336}$$

Thus

$$[Q_1 (u_{n_m} - u_{n_k}), u_{n_m} - u_{n_k}] \to 0, \quad m, k \to \infty. \tag{8.337}$$

Since $Q_1 \geq 0$ equation (8.337) implies that $Q_1^{1/2}$ is compact in H_A. Therefore Q_1 is compact in H_A. Conversely, if Q_1 is compact in H_A then $T_1 \geq 0$ is A-compact. Indeed, if $\| u_n \|_A \leq c$ then $T_1 u_n = A Q_1 u_n$ so that there is a subsequence u_{n_m} such that

$$(T_1 (u_{n_m} - u_{n_k}), u_{n_m} - u_{n_k}) = [Q_1 (u_{n_m} - u_{n_k}), u_{n_m} - u_{n_k}] \to 0$$

as $m, k \to \infty$, because Q_1 is compact in H_A.

So we have proved that

$$T_1 \geq 0 \quad \textit{is A compact if and only if} \quad A^{-1}T_1 \quad \textit{is compact in } H_A. \tag{8.338}$$

Let us prove that if Q_1 is compact in H_A. This is the conclusion of Proposition 8.5. It is sufficient to prove that if $f_n \rightharpoonup 0$ and $g_n \rightharpoonup 0$ in H_A then

$$[Qf_n, g_n] \to 0, \quad n \to \infty. \tag{8.339}$$

Indeed, if

$$\{f_n \rightharpoonup 0 \quad \text{and} \quad g_n \rightharpoonup 0\} \Rightarrow [Qf_n, g_n] \to 0 \tag{8.340}$$

then

$$\{f_n \rightharpoonup f \quad \text{and} \quad g_n \rightharpoonup g\} \Rightarrow [Qf_n, g_n] \to [Qf, g], \tag{8.341}$$

that is, Q is compact. To check that (8.340) implies (8.341) one writes

$$\begin{aligned}[Qf_n, g_n] &= [Q(f_n - f), g_n] + [Qf, g_n] \\ &= [Q(f_n - f), g_n - g] + [Q(f_n - f), g] + [Qf, g_n - g] + [Qf, g] \\ &= [Q(f_n - f), (g_n - g)] + [f_n - f, Q^* g] + [Qf, g_n - g] \\ &\qquad + \quad [Qf, g].\end{aligned} \tag{8.342}$$

It follows from (8.342) that (8.340) implies (8.341).

Let us check that (8.335) implies (8.340). One uses the well known polarization identity

$$\begin{aligned}[Qf, g] = \frac{1}{4}\{[(f+g), f+g] - [(f-g), f-g] - i\,[Q(f+ig), f+ig] \\ + \quad i\,[Q(f-ig), f-ig]\}.\end{aligned} \tag{8.343}$$

It is clear from (8.343) and (8.335) that

$$\begin{aligned}|[Qf_n, g_n]| \le \frac{1}{4}\{&|[Q_1(f_n + g_n), f_n + g_n]| + |[Q_1(f_n - g_n), f_n - g_n]| \\ &+ |[Q_1(f_n + ig_n), f_n + ig_n]| \\ &+ |[Q_1(f_n - ig_n), f_n - ig_n]|\} \to 0, \quad n \to \infty.\end{aligned} \tag{8.344}$$

The last conclusion follows from the assumed compactness of Q_1 and the fact that if $f_n \rightharpoonup 0$ and $g_n \rightharpoonup 0$ then any linear combination $c_1 f_n + c_2 g_n$ converges weakly to zero. Proposition 8.5 is proved. □

Example 8.2 It is now easy to see that (164) holds. Indeed, the imbedding $i : H^1(D) \to L^2(\Gamma, |\sigma|)$ is compact. Therefore the quadratic form $\int_\Gamma \sigma |u|^2 ds$ is A-compact, where $A[u, u] = \int_D \left(|\nabla u|^2 + |u|^2\right) dx$. From this and Theorem 8.9 the formula (8.323) follows.

8.3.2.10 *Proof of Theorem 2.3*

In this section we prove Theorem 2.3.

First let us note that if

$$R(x, y) = \int_{-\infty}^{\infty} \omega(\lambda) \Phi(x, y, \lambda) d\rho(\lambda) \tag{8.345}$$

where $\omega(\lambda) \in C(R^1)$, $\omega(\infty) = 0$ then the operator $R : L^2(D) \to L^2(D)$, where $D \subset R^r$ is a bounded domain, and

$$Rh := \int_D R(x,y)h(y)dy, \tag{8.346}$$

is compact. This is proved in Section 4.2 (cf. the argument after formula (4.71)). If $\omega(\lambda) \geq 0$ then $R = R^* \geq 0$. Suppose that

$$\omega_1(\lambda) = \omega(\lambda)\left[1 + \phi(\lambda)\right], \quad \phi(\pm\infty) = 0 \tag{8.347}$$

where $\phi(\lambda) \in C(R^1)$, $1 + \phi(\lambda) > 0$. Then the corresponding operator R_1 can be written as

$$R_1 = R(I + Q), \tag{8.348}$$

where Q is a compact operator with the kernel

$$Q(x,y) = \int_{-\infty}^{\infty} \phi(\lambda)\Phi(x,y,\lambda)d\rho(\lambda). \tag{8.349}$$

By Theorem 8.3 one has

$$\lim_{n\to\infty} \frac{s_n(R_1)}{s_n(R)} = 1. \tag{8.350}$$

Note that the operator $I + Q$ is injective since $1 + \phi(\lambda > 0$. Since $R_1 \geq 0$ and $R \geq 0$ one has $s_n(R_1) = \lambda_n(R_1)$, $s_n(R) = \lambda_n(R)$. Therefore (191) can be written as

$$\lim_{n\to\infty} \frac{\lambda_n(R_1)}{\lambda_n(R)} = 1. \tag{8.351}$$

Therefore it is sufficient to prove formula (2.31) for $\omega(\lambda) = (1 + \lambda^2)^{-a/2}$.

Secondly, let us note that if one defines

$$N(\lambda) := \sum_{\lambda_n \leq \lambda} 1, \tag{8.352}$$

then formulas

$$\lambda_n = cn^p\left[1 + o(1)\right] \quad \text{as} \quad n \to +\infty, \quad c = \text{const} > 0, \quad p > 0, \tag{8.353}$$

and

$$N(\lambda) = c^{-1/p}\lambda^{1/p}\left[1 + o(1)\right], \quad \lambda \to +\infty, \quad p > 0, \tag{8.354}$$

are equivalent. This follows from the fact that the function $N(\lambda)$ is the inverse function for $\lambda(N) := \lambda_N$ in the sense that $N(\lambda_N) = N$ and $\lambda(N(\lambda)) = \lambda_N$. Therefore if one knows the asymptotics of $N(\lambda)$ then one knows the asymptotics of λ_n and vice versa. In [Ramm (1975), p. 339] it is proved that if $o(1)$ in (8.353) is $0(n^{-p_1})$, $p_1 > 0$, then $o(1)$ in (8.354) is $0(\lambda^{-p_1/p})$.

Thirdly, let us recall a well known fact that an elliptic operator $\mathcal{L}$ of order s with smooth coefficients and regular boundary conditions in a bounded domain $D \subset R^r$ with a smooth boundary (these are the regularity assumptions) has a discrete spectrum and

$$N(\lambda, \mathcal{L}) = \gamma\lambda^{r/s}[1 + o(1)], \quad \lambda \to +\infty, \tag{8.355}$$

where $N(\lambda, \mathcal{L})$ is defined by (8.352) with $\lambda_n = \lambda_n(\mathcal{L})$, and $\gamma = \text{const} > 0$ is defined by formula (2.32). By formula (8.353) one obtains

$$\lambda_n(\mathcal{L}) = \gamma^{-s/r}n^{s/r}[1 + o(1)], \quad n \to +\infty. \tag{8.356}$$

The operator R is a rational function of $\mathcal{L}$ so that by the spectral mapping theorem one obtains

$$\lambda_n(R) = \lambda_n^{-a}(\mathcal{L})[1 + o(1)] = \gamma^{as/r}n^{-as/r}[1 + o(1)], \quad n \to \infty. \tag{8.357}$$

This is formula (2.31).

For the function $\omega(\lambda) = (1 + \lambda^2)^{-a/2}$ and even a a proof of the formula (2.31) is given in [Ramm (1980), p. 62].

This proof goes as follows. The problem

$$R\phi_n := \int_D R(x, y)\phi_n(y)dy = \lambda_n\phi_n(x), \quad x \in D \tag{8.358}$$

is equivalent to the problem

$$\int_{R^r} R(x, y)\phi_n(y)dy = \begin{cases} \lambda_n\phi_n(x), & x \in D \\ u_n(x), & x \in \Omega, \end{cases} \tag{8.359}$$

where

$$\phi_n(x) := 0 \quad \text{in} \quad \Omega \tag{8.360}$$

$$Q(\mathcal{L})u_n = 0 \quad \text{in} \quad \Omega \tag{8.361}$$

$$u_n(\infty) = 0, \quad \partial_N^j u_n = \lambda_n \partial_N^j \phi_n \quad \text{on} \quad \Gamma, \quad 0 \leq j \leq \frac{as}{2} - 1, \tag{8.362}$$

where $Q(\lambda) := (1 + \lambda^2)^{a/2}$. The equivalence means that every solution to (8.358) generates the solution to (8.359)- (8.362) and vice versa. The problem (8.359)- (8.362) can be written as

$$\lambda_n Q(\mathcal{L})\phi_n = \chi_D(x)\phi_n(x) \quad \text{in} \quad R^r, \tag{8.363}$$

where

$$\chi_D(x) = \begin{cases} 1, & x \in \overline{D} \\ 0, & x \in \Omega. \end{cases}$$

This problem has been studied [Tulovskii (1979)] and formula (8.357) has been established.

For the general case of $\omega(\lambda) = (1 + \lambda^2)^{-a/2}$, $a > 0$ one can use the results from the spectral theory of elliptic pseudo-differential operators. Under suitable regularity assumptions the following formula for the number $N(\lambda) := \{\#\lambda_n : \lambda_n \leq \lambda\}$ of eigenvalues of such an operator R in a bounded domain $D \subset R^r$ is valid:

$$N(\lambda) = (2\pi)^{-r} \text{ meas } \{(x, \xi) \in D \times R^r : r(x, \xi) < \lambda\} \left[1 + o(1)\right], \quad \lambda \to +\infty. \tag{8.364}$$

Here meas is the Lebesgue measure, and $r(x, \xi)$ is the symbol of the pseudo-differential operator R. This means that

$$Rh := (2\pi)^{-r} \iint \exp\{i(x - y) \cdot \xi\} r(x, \xi) h(y) dy d\xi, \quad \int = \int_{R^r}. \tag{8.365}$$

The symbol of the elliptic operator (2.5) is $\sum_{|j| \leq s} a_j(x)(i\xi)^j$. Only the principal symbol, that is $i^s \sum_{|j|=s} a_j(x)\xi^j$, defines the main term of the asymptotics of $N(\lambda)$. Since $s = \text{ord}\mathcal{L}$ is even one chooses $\mathcal{L}$ so that $\mathcal{L}_0(x, \xi) := \sum_{|j|=s} a_j(x)\xi^j > 0$ for $|\xi| \neq 0$. For example, one chooses $\mathcal{L} = -\Delta$ rather than Δ. In this case

$$(2\pi)^r \text{ meas } \{(x, \xi) \in D \times R^r : \mathcal{L}_0(x, \xi) < \lambda\} = \lambda^{r/s}(2\pi)^{-r} \int_D \eta dx = \gamma \lambda^{r/s}, \tag{8.366}$$

where η is given by (2.33) for the operator $\mathcal{L}_0$ in the selfadjoint form.

The asymptotic behavior of the function $N(\lambda)$ has been studied extensively for wide classes of differential and pseudo-differential operators (see, e.g., [Levitan (1971); Hörmander (1983-85); Safarov and Vassiliev (1997); Shubin (1986)] and references therein).

8.3.3 *Trace class and Hilbert-Schmidt operators*

In this section we summarize for convenience of the reader some rsults on Hilbert-Schmidt and trace-class operators. This material can be used for references.

One writes $A \in \sigma_p$, $1 \leq p < \infty$ if

$$\sum_{j=1}^{\infty} s_j^p(A) < \infty.$$

8.3.3.1 *Trace class operators*

The operators $A \in \sigma_1$ are called trace class (or nuclear) operators. The operators $A \in \sigma_2$ are called Hilbert-Schmidt (HS) operators. The class σ_∞ denotes the class of compact operators. If $A \in \sigma_p$ then $A \in \sigma_q$ with $q \geq p$. We summarize some basic known results about trace class and HS operators.

Lemma 8.23 *If and only if $A \in \sigma_1$ the sum $TrA := \sum_{j=1}^{\infty}(A\phi_j, \phi_j)$ is finite for any orthonormal basis of H and does not depend on the choice of the basis. In fact $TrA = \sum_{j=1}^{\nu(A)} \lambda_j(A)$, where $\nu(A)$ is the sum of algebraic multiplicities of the eigenvalues of A.*

The following properties of the trace are useful.

Lemma 8.24 *If $A_j \in \sigma_1$, $j = 1, 2$, then*

1) $Tr(c_1A_1 + c_2A_2) = c_1TrA_1 + c_2TrA_2$, $c_j = const$, $j = 1, 2$,
2) $TrA^* = \overline{(TrA)}$, *the bar stands for complex conjugate,*
3) $Tr(A_1A_2) = Tr(A_2A_1)$,
4) $Tr(B^{-1}AB) = TrA$ *where B is a linear isomorphism of H onto H,*
5) $Tr(A_1A_2) \geq 0$ *if* $A_1 \geq 0$ *and* $A_2 \geq 0$,
6) $Tr(A_1A_2)^{1/2} \leq \frac{TrA_1 + TrA_2}{2}$ *if* $A_1 \geq 0$ *and* $A_2 \geq 0$,
7) $|TrA| \leq \sum_{j=1}^{\infty} s_j(A) := \| A \|_1$,
8) $\| A \|_1 = \sup \sum_{j=1}^{\infty} |(Af_j, h_j)|$, *where the* sup *is taken over all orthonormal bases $\{f_j\}$ and $\{h_j\}$ of H,*
9) $\| c_1A_1 + c_2A_2 \|_1 \leq |c_1| \| A_1 \|_1 + |c_2| \| A_2 \|_1$,

10) *Assume that* $A_j \in \sigma_1$, $1 \le j < \infty$, $\| A_j \|_1 \le c$, c *does not depend on* j, *and* $A_j \rightharpoonup A$. *Then* $A \in \sigma_1$, *and* $\| A \|_1 \le \sup_j \| A_j \|_1$, *the symbol* $\rightharpoonup$ *denotes weak convergence of operators,*

11) *if* $A \in \sigma_1$ *and* B *is a bounded linear operator, then*

$$\| AB \|_1 \le \| A \|_1 \| B \|,$$
$$\| BA \|_1 \le \| A \|_1 \| B \| .$$

8.3.3.2 *Hilbert-Schmidt operators*

The operators $A \in \sigma_2$ are called Hilbert-Schmidt (HS) operators.

Lemma 8.25 $A \in \sigma_2$ *if and only if the sum* $\| A \|_2^2 := \sum_{j=1}^{\infty} \| A\phi_j \|^2$ *is finite for any orthonormal basis* $\{\phi_j\}$ *of* H. *If* $A \in \sigma_2$ *then*

$$\| A \|_2^2 = \sum_{j=1}^{\infty} s_j^2(A).$$

If $c_j = const$ *and* $A_j \in \sigma_2$ *then*

$$\| c_1 A_1 + c_2 A_2 \| \le |c_1| \| A_1 \|_2 + |c_2| \| A_2 \|_2 .$$

If $A \in \sigma_2$ *and* B *is a bounded linear operator then*

$$\| A \|_2 = \| A^* \|_2,$$
$$\| AB \|_2 \le \| A \|_2 \| B \| . \tag{8.367}$$

Lemma 8.26 $A \in \sigma_2$ *if and only if there exists an orthonormal basis* $\{\phi_j\}$ *of* H *for which*

$$\sum_{j=1}^{\infty} \| A\phi_j \|^2 < \infty. \tag{8.368}$$

In particular, if (8.368) hold for an for an orthonormal basis of H, *then it holds for every orthonormal basis of* H.

Lemma 8.27 $A \in \sigma_1$ *if and only if there exists an orthonormal basis of* H *for which*

$$\sum_{j=1}^{\infty} \| A\phi_j \| < \infty. \tag{8.369}$$

However if (8.369) holds for an orthonormal basis of H *it may not hold for another orthonormal basis of* H.

Example 8.3 Let $H = \ell_2$, $f = c\left(1, \frac{1}{2}, \dots, \frac{1}{n}, \dots\right)$ where $c = \text{const} > 0$ is chosen so that $\| f \| = 1$. Let A be the orthogonal projection on the one-dimensional subspace spanned by f, and let $\{\phi_j\}$, $\phi_j = \delta_{1j}$, be an orthonormal basis of ℓ_2. Then $A\phi_j = \frac{c}{j} f$, $\sum_{j=1}^{\infty} \| A\phi_j \| = \sum_{j=1}^{\infty} \frac{c}{j} = \infty$.

Lemma 8.28 *If and only if $A \in \sigma_1$ it can be represented in the form*

$$A = A_1 A_2 \quad \textit{where} \quad A_j \in \sigma_2, \quad j = 1, 2.$$

Lemma 8.29 *The classes σ_1 and σ_2 are ideals in the algebra $\mathcal{L}(H)$ of all linear bounded operators on H. If H is a separable Hilbert space then the set σ_∞ of all compact operators on H is the only closed proper non-zero ideal in $\mathcal{L}(H)$. The ideal is called proper if it is not $\mathcal{L}(H)$ itself. The closedness is understood as the closedness in the norm of linear operators on H.*

8.3.3.3 *Determinants of operators*

Definition 8.6 If $A \in \sigma_1$ then

$$d(\mu) := \det(I - \mu A) := \prod_{j=1}^{\nu(A)} \left[1 - \mu \lambda_j(A)\right].$$

One has

1) $|d(\mu)| \le \exp\left(|\mu| \, \| A \|_1\right)$.
2) $d(\mu) = \exp\left(-\int_0^\mu Tr\left[A(I - \mu A)^{-1}\right] d\mu\right)$, if the operator $I - \lambda A$, $0 \le \lambda \le \mu$, is invertible.
3) $\det(I - A) = \lim_{n\to\infty} \det\left[\delta_{ij} - (A\phi_i, \phi_j)\right]_{i,j=1,\dots,n}$ where $\{\phi_j\}$ is an arbitrary orthonormal basis of H.
4) $\det(I - AB) = \det(I - BA)$, $AB \in \sigma_1$, $BA \in \sigma_1$, $A \in \sigma_\infty$, $B \in \mathcal{L}(H)$.
5) $\det\left[(I - A)(I - B)\right] = \det\left[(I - B)(I - A)\right]$, $A, B \in \sigma_1$.
6) If $A(z) \in \sigma_1$ is analytic operator function in a domain Δ of the complex plane z then $d(1, z) := \det\left(I - A(z)\right)$ is analytic in Δ; here $d(\mu, z) := \det\left(I - \mu A(z)\right)$.
7) $\frac{d}{dz} Tr\left\{F\left(A(z)\right)\right\} = Tr\left\{F'\left(A(z)\right) \frac{dA(z)}{dz}\right\}$ where $F(\lambda)$ is holomorphic in the domain which contains the spectrum of $A(z)$ for all $z \in \Delta$, and $F(0) = 0$.
8) $\det(I+A) = \exp\left\{Tr \log(I + A)\right\}$, $A \in \sigma_1$ where $\log(I+A)$ can be defined by analytic continuation of $\log(I + zA)$. This function is well defined for $|z| \, \| A \| < 1$.

If $A \in \sigma_2$ then the series $\sum_{j=1}^{\infty} |\lambda_j(A)|$ may diverge and the Definition 8.6 is not applicable. One gives

Definition 8.7

$$d_2(\mu) := \det_2(I - \mu A) := \prod_{j=1}^{\mu(A)} \{[1 - \mu\lambda_j(A)] \exp[\mu\lambda_j(A)]\}.$$

One has:

9) $|d_2(\mu)| \le \exp\left\{\frac{|\mu|^2}{2} Tr(A^*A)\right\}$.

10) $d_2(1) = \lim_{n\to\infty} \det[\delta_{ij} - (A\phi_i, \phi_j)]_{1\le i,j\le n} \exp\left[\sum_{j=1}^{n}(A\phi_j, \phi_j)\right]$ where $\{\phi_j\}$ is an arbitrary orthonormal basis of H.

11) If $A, B \in \sigma_2$ and $I - C = (I - A)(I - B)$ then

$$\det_2(I - C)\exp[Tr(AB)] = \det_2(I - A)\det_2(I - B).$$

If $B \in \sigma_1$ then

12) $$\det_2(I - C) = \det_2(I - A)\det(I - B)\exp\{Tr[(I - A)B]\}.$$

Definition 8.8 If $A \in \sigma_p$ then

13)
$$d_p(\mu) := \det_p(I - \mu A) := \prod_{j=1}^{\infty}\left\{[1 - \mu\lambda_j(A)]\exp\left[\sum_{m=1}^{p-1}\frac{\mu^m}{m}\lambda_j^m(A)\right]\right\}.$$

Carleman's inequality: If $A \in \sigma_2$, λ_j are eigenvalues of A counted according to their multiplicities, $|\lambda_1| \ge |\lambda_2| \ge \cdots$ and $\phi_\lambda(A) := \prod_{j=1}^{\infty}(1 - \lambda_j\lambda^{-1})\exp(\lambda_j\lambda^{-1})$, then

14) $$\| \phi_\lambda(A)(A - \lambda I)^{-1} \| \le |\lambda| \exp\left\{\frac{1}{2}\left(1 + |\lambda|^{-2} \| A \|^2\right)\right\}.$$

8.4 Elements of probability theory

8.4.1 *The probability space and basic definitions*

A *probability space* is a triple $\{\Omega, \mathcal{U}, P\}$ where Ω is a set, $\mathcal{U}$ is a sigma algebra of its subsets, and P is a measure on $\mathcal{U}$, such that $P(\Omega) = 1$, so its a measure space $\{\Omega, \mathcal{U}\}$ equuipped with a normalized countably additive measure

$$P\left(U_{j=1}^{\infty} A_j\right) = \sum_{j=1}^{\infty} P(A_j), \quad A_j \cap A_m = \emptyset \quad \text{for } j \neq m.$$

A random variable ξ is a $\mathcal{U}$-measurable function on Ω, that is, a $\mathcal{U}$-measurable map $\Omega \to R^1$. A random vector ξ is a $\mathcal{U}$-measurable map $\Omega \to R^r$.

A distribution function of ξ is $F(x) = P(\xi < x)$.

It has properties: $F(-\infty = 0$, $F(+\infty) = 1$, F is nondecreasing, $F(x+0) - F(x) = P(\xi = x)$, $F(x-0) = F(x)$.

The probability density $f(x) := F'(x)$ is defined in the classical sense if $F(x)$ is absolutely continuous, so that $F(x) = \int_{-\infty}^{x} f(t)dt$. If ξ is a discrete random variable, that is ξ takes values in a discrete set of points, then its distribution function is $F(x) = \sum_i P_i \theta(x - x_i)$, where

$$\theta(x) := \begin{cases} 1 & x > 0 \\ 0 & x \leq 0 \end{cases}, \quad P_i > 0, \quad \sum_i P_i = 1.$$

The probability density for this distribution is $f(x) = \sum_i P_i \delta(x - x_i)$ where $\delta(x)$ is the delta-function.

The probability density has the properties:

1) $f \geq 0$,

2) $\int_{-\infty}^{\infty} f dt = 1$.

A random vector $\xi = (\xi_1, \ldots, \xi_r)$ has a distribution function

$$F(x_1, \ldots, x_r) := P(\xi_1 < x_1, \ldots, \xi_r < x_r).$$

This function has *characteristic properties*

1) $F(+\infty, \ldots, +\infty) = 1$,
2) $F(x_1, \ldots, x_m = -\infty, \ldots, x_r) = 0$ for any $1 \leq m \leq r$,
3) $F(x_1, \ldots, x_m = +\infty, \ldots, x_r) = F(x_1, \ldots, x_{m-1}, x_{m+1}, \ldots, x_r)$,
4) F is continuous from the left, i.e. $F(x_1, \ldots, x_m - 0, \ldots, x_r) = F(x_1, \ldots, x_m, \ldots, x_r)$ and nondecreasing in each of the variables $x_1, \ldots, x_r$.

The probability density $f(x_1, \ldots, x_r)$ is defined by the formula

$$f(x_1, \ldots, x_r) = \frac{\partial^r F(x_1, \ldots, x_r)}{\partial x_1 \ldots \partial x_r}$$

so that

$$F(x_1,\ldots,x_r) = \int_{-\infty}^{x_1} dx_1 \ldots \int_{-\infty}^{x_r} dx_r f(x_1,\ldots,x_2).$$

This formula holds if the measure defined by the distribution function F is absolutely continuous with respect to the Lebesgue measure in R^r, i.e. if $P(\xi \in \Delta) = 0$ for any $\Delta \subset R^r$ such that $\text{meas}\Delta = 0$ where meas is the Lebesgue measure in R^r.

Example 8.4 A random vector ξ is said to be *uniformly distributed* in a set $\Delta \subset R^r$ if its probability density is

$$f(x) = \begin{cases} 0 & x \notin \Delta \\ \frac{1}{\text{meas}\Delta} & x \in \Delta. \end{cases} \tag{8.370}$$

Example 8.5 A random vector ξ is said to be *Gaussian* (or *normally distributed*, or *normal*) if its probability density is

$$f(x) = (2\pi)^{-r/2}[\det C]^{1/2} \exp\left\{\sum_{i,j=1}^{r} c_{ij}(x_i - m_i)(x_j - m_j)\right\}. \tag{8.371}$$

Here $C = (c_{ij})$ is a positive definite matrix, $M[\xi] = \overline{\xi} = \mathbf{m} = (m_1,\ldots,m_r)$, the matrix C^{-1} is the *covariance matrix* of ξ:

$$C^{-1} := \left(c_{ij}^{(-1)}\right) = \overline{(\xi_i - m_i)(\xi_j - m_j)}, \quad \overline{\xi}_i = m_i, \tag{8.372}$$

where all the quantities are real-valued, the bar denotes the mean value, defined as follows:

$$\overline{\xi}_i := \int x_i dF, \quad \int := \int_{R^r}.$$

If $g(x_1,\ldots,x_r)$ is a measurable function defined on R^r, $g : R^r \to R^1$, then $\eta = g(\xi_1,\ldots,\xi_r)$ is a random variable with the distribution function:

$$F_\eta(x) := P(\eta < x) = \int_{g(x_1,\ldots,x_r)<\eta} dF(x_1,\ldots,x_r)$$

where $F(x_1,\ldots,x_r)$ is the distribution function for the vector $(\xi_1,\ldots,\xi_r)$. One has

$$\overline{g(\xi_1,\ldots,\xi_r)} = \int g(x_1,\ldots,x_r)dF(x_1,\ldots,x_r).$$

In particular the *variance* of a random variable is defined as

$$\mathcal{D}[\xi] := \overline{(\xi - \overline{\xi})^2} = \int_{-\infty}^{\infty} (x-m)^2 dF(x)$$

where

$$\overline{\xi} = m = \int_{-\infty}^{\infty} x dF(x).$$

Let us define *conditional probabilities.* If A and B are random events then

$$P(A \mid B) := \frac{P(AB)}{P(B)}, \tag{8.373}$$

where AB is the event $A \cap B$ and $P(A \mid B)$ is called the conditional probability of A under the assumption that B occured. The *conditional mean value* (expectation) of a random variable $\xi = f(u)$ under the assumption that B occured is defined by

$$M(\xi \mid B) = P^{-1}(B) \int_B f(u) P(du). \tag{8.374}$$

In particular, if

$$\xi = \begin{cases} 1 & \text{if } A \text{ occurs} \\ 0 & \text{otherwise} \end{cases}$$

then (8.374) reduces to (8.373).

If $A = U_{j=1}^n E_j$, $E_j \cap E_{j'} = \emptyset$, $j \neq j'$, then

$$P(A) = \sum_{j=1}^{n} P(A \mid E_j) P(E_j) \tag{8.375}$$

and

$$P(E_j \mid A) = P(A \mid E_j) P(E_j) P^{-1}(A). \tag{8.376}$$

This is the *Bayes formula.* The *conditional distribution function* of a random variable ξ with respect to an event A is defined as

$$F_\xi(x \mid A) := \frac{P(\{\xi < x\} \cap A)}{P(A)}. \tag{8.377}$$

One has

$$M(\xi \mid A) = \int_{-\infty}^{\infty} x dF_\xi (x \mid A). \tag{8.378}$$

The *characteristic function* of a random variable ξ with the distribution function $F(x)$ is defined by

$$\phi(t) := M[\exp(it\xi)] = \int_{-\infty}^{\infty} \exp(itx) dF(x). \tag{8.379}$$

It has the properties

1) $\phi(0) = 1$, $|\phi(t)| \leq 1$, $-\infty < t < \infty$, $\phi(-t) = \phi^*(t)$
2) $\phi(t)$ is uniformly continuous on R^1
3) $\phi(t)$ is positive definite in the sense

$$\sum_{j,m=1}^{n} \phi(t_j - t_m) z_j z_m^* \geq 0 \tag{8.380}$$

for any complex numbers z_j and real numbers t_j, $1 \leq j \leq n$.

Theorem 8.10 **(Bochner-Khintchine)** *A function $\phi(t)$ is a characteristic function if and only if it has properties 1)-3).*

One has

$$\lim_{T\to\infty} \frac{1}{2T} \int_{-T}^{T} \phi(t) \exp(-itx) dt = 0$$

if $F(x)$ is continuous at the point x. More generally

$$\lim_{T\to\infty} \frac{1}{2T} \int_{-T}^{T} \phi(t) \exp(-itx) dt = F(x+0) - F(x).$$

If conditions 2) and 3) hold but condition 1) does not hold then one still has the formula

$$\phi(t) = \int_{-\infty}^{\infty} \exp(itx) dF(x) \tag{8.381}$$

where $F(x)$ is a monotone nondecreasing function, but the condition $F(+\infty) = 1$ does not hold and is replaced by $F(+\infty) < \infty$.

Let us define the notion of *random function*. Let $\{\Omega, \mathcal{U}, P\}$ be the probability space and $\xi(\mathbf{t}, \omega)$, $\omega \in \Omega$, be a family of random variables depending on a parameter $\mathbf{t} \in D \subset R^r$. The space X to which the variable $1/2\xi(\mathbf{t}, \omega)$ belongs for a fixed $\omega \in \Omega$ is called the phase space of the random function $\xi(x,\omega)$. If $r=1$ the random function is called a random process, if $r>1$ it is called a random field. Usually one writes $\xi(\mathbf{t})$ in place of $\xi(\mathbf{t},\omega)$. If $X=R^m$ then $\xi(\mathbf{t})$ is a vector random field, if $m=1$ then $\xi(\mathbf{t})$ is a scalar random field. One assumes that X is a measurable space $(X,\mathcal{B})$ where $\mathcal{B}$ is the Borel sigma-algebra generated by all open sets of X.

If one takes n points $\mathbf{x}_1,\dots,\mathbf{x}_n\in D\subset R^r$ then one obtains a random vector
$\{\xi(\mathbf{t}_1,\omega),\dots,\xi(\mathbf{t},\omega)\}$. Let $F(\mathbf{t}_1,\dots,\mathbf{t}_n)$ be the distribution function for this random vector. For various choices of the points $\mathbf{t}_j$ one obtains various distribution functions. The collection of these functions is *consistent* in the following sense:

1) $F_{\xi_1,\dots,\xi_n}(\mathbf{x}_1,\mathbf{x}_2,\dots,\mathbf{x}_{m-1},\mathbf{x}_m\in X,\dots,\mathbf{x}_n)=$
$F(\mathbf{x}_1,\dots,\mathbf{x}_{m-1},\mathbf{x}_{m+1},\dots,\mathbf{x}_n)$
2) $F_{\xi_1,\dots,\xi_n}(\mathbf{x}_1,\dots,\mathbf{x}_n)=F_{\xi_{i_1},\dots,\xi_{i_n}}(\mathbf{x}_{i_1},\dots,\mathbf{x}_{i_n})$
for any permutation $(i_1,\dots,i_n)$ of the set $(1,\dots,n)$.

The following theorem gives conditions for a family $F(\mathbf{x}_1,\dots,\mathbf{x}_n)$ of functions to be the family of finite-dimensional distribution functions corresponding to a random function $\xi(\mathbf{t})$.

Theorem 8.11 **(Kolmogorov)** *If and only if the family of functions F is compatible and each of the functions has the characteristic properties 1)-4) of a distribution function, there exists a random function for which this family is a family of finite-dimensional distribution functions.*

Moment functions of a random function are defined as

$$m_j(t_1,\dots,t_j)=M\left[\xi(t_1)\dots\xi(t_j)\right]=\overline{\xi(t_1)\dots\xi(t_j)}. \tag{8.382}$$

Especially often one uses the *mean values*

$$M\left[\xi(t,\omega)\right]=m(t)=\overline{\xi(t)} \tag{8.383}$$

and *covariance function*

$$R(t,\tau):=\left[\xi(t)-m(t)\right]^*\left[\xi(\tau)-m(\tau)\right]. \tag{8.384}$$

The characteristic property of the class of covariance functions is positive definiteness in the sense

$$\sum_{i,j=1}^{n}R(t_i,t_j)z_i^*z_j\geq 0 \tag{8.385}$$

for any choice of real t_i and complex z_i. The star stands for complex conjugate.

8.4.2 Hilbert space theory

Let us assume that a family of random variables with finite second moments is equipped with the inner product defined as

$$(\xi, \eta) := \overline{\xi \eta^*} \tag{8.386}$$

and the norm is defined as

$$\| \xi \| = (\xi, \xi)^{1/2}. \tag{8.387}$$

Then the random variables belong to the space $L^2 = L^2(\Omega, \mathcal{U}, P)$. Convergence in this space is defined as convergence in the norm (8.387).

If $\xi(t)$ is a random function then its *correlation function* is defined as

$$B(t, \tau) := \overline{\xi^*(t)\xi(\tau)}. \tag{8.388}$$

If $\overline{\xi(t)} = 0$ then $B(t, \tau) = R(t, \tau)$, where $R(t, \tau)$ is the covariance function (8.384). The *characteristic property of the class of correlation functions is positive definiteness in the sense (8.385).*

A random function $\xi(t) \in L^2(\Omega, \mathcal{U}, P)$ is called *continuous* (in L^2 sense) at a point t_0 if

$$\| \xi(t) - \xi(t_0) \| \to 0 \quad \text{as} \quad \rho(t, t_0) \to 0, \tag{8.389}$$

where $\rho(t, t_0)$ is the distance between 1/2 and t_0.

Lemma 8.30 *For (8.389) to hold it is necessary and sufficient that $B(t, \tau)$ be continuous at the point (t_0, t_0).*

A random function $\xi(t) \in L^2$ is called *differentiable* (in L^2 sense) at a point t_0 if there exists in L^2 the limit

$$\xi'(t_0) := \text{l.i.m.}_{\epsilon \to 0} \epsilon^{-1} \left[\xi(t_0 + \epsilon) - \xi(t_0) \right], \tag{8.390}$$

where l.i.m. = limit in mean stands for the limit in L^2.

Lemma 8.31 *For (8.390) to hold it is necessary and sufficient that*

$\frac{\partial^2 B(t_0,t_0)}{\partial t \partial \tau}$ exists, that is

$$\frac{\partial^2 B(t_0,t_0)}{\partial t \partial \tau} := \lim_{\substack{\epsilon_1 \to 0 \\ \epsilon_2 \to 0}} \frac{1}{\epsilon_1 \epsilon_2} [B(t_0+\epsilon_1, t_0+\epsilon_2) - B(t_0, t_0+\epsilon_2) - B(t_0+\epsilon_1, t_0) + B(t_0,t_0)]. \quad (8.391)$$

If $\frac{\partial^2 B(t,\tau)}{\partial t \partial \tau}$ exists then

$$\overline{\xi'^*(t)\xi'(\tau)} = \frac{\partial^2 B(t,\tau)}{\partial t \partial \tau}$$

$$\overline{\xi'^*(t)\xi(\tau)} = \frac{\partial B(t,\tau)}{\partial t}. \quad (8.392)$$

Let $\xi(x)$, $x \in D \subset R^r$ be a random function and $\mu(x)$ be a finite measure. *The Lebesgue integral* of $\xi(x)$ is defined as

$$\int_D \xi(x) d\mu(x) = \text{l.i.m.}_{n\to\infty} \int_D \xi_n(x) d\mu(x) \quad (8.393)$$

where $\xi_n(x) \leq \xi_{n+1}(x)$, $\xi_n(x) \in L^2$ and $\xi_n(x) \underset{P}{\to} \xi(x)$, that is

$$\lim_{n\to\infty} P(|\xi(x) - \xi_n(x)| > \epsilon) = 0, \quad \forall x \in D \quad (8.394)$$

for every $\epsilon > 0$.

If $\mu(D) < \infty$ and

$$\int_D B(x,x) d\mu(x) < \infty \quad (8.395)$$

then

$$\overline{\int_D |\xi(x)|^2 d\mu(x)} = \int_D B(x,x) d\mu(x). \quad (8.396)$$

Assume that $\xi(x) \in L^2$ and $B(x,y)$ is continuous in $D \times D$, where $D \subset R^r$ is a finite domain. Then, by Mercer's theorem (see p. 61) one has

$$B(x,y) = \sum_{j=1}^{\infty} \lambda_j \phi_j(x) \phi_j^*(y) \quad (8.397)$$

where

$$B\phi_j = \lambda_j \phi_j, \quad \lambda_1 \geq \lambda_2 \geq \cdots > 0 \quad (8.398)$$

$$(\phi_j, \phi_m) = \int_D \phi_j \phi_m^* dx = \delta_{jm}, \tag{8.399}$$

$$B\phi := \int_D B(x,y)\phi(y)dy. \tag{8.400}$$

Put

$$\xi_n := \int_D \xi(x)\phi_n(x)dx. \tag{8.401}$$

Then

$$\overline{\xi_n^* \xi_m} = \int_D \int_D B(x,y)\phi_m(y)\phi_n^*(x)dydx = \lambda_m \delta_{nm}, \tag{8.402}$$

and

$$\overline{\xi^*(x)\xi_n} = \int_D B(x,y)\phi_n(y)dy = \lambda_n \phi_n(x). \tag{8.403}$$

Lemma 8.32 *The series*

$$\xi(x) = \sum_{j=1}^{\infty} \xi_j \phi_j^*(x) \tag{8.404}$$

converges in L^2 for every $x \in D$ if the function $B(x,y)$ is continuous in $D \times D$.

Remark 8.6 *If one defines $B_1(t,\tau)$ by the formula*

$$B_1(t,\tau) := \overline{\xi(t)\xi^*(\tau)} \tag{8.405}$$

so that $B_1 = B^*$, then formulas (8.397)-(8.400) hold for B_1, in formula (8.401) in place of ϕ_n one puts ϕ_n^* and in formula (8.404) one puts ϕ_j in place of ϕ_j^*.

Let us define the notion of a *stochastic* or *random measure.*

Consider a random function $\zeta(x) \in L^2$, $x \in D$. Let $\mathcal{B}$ be a sigma-algebra of Borel subsets of D. Suppose that to any $\Delta \subset \mathcal{B}$ there corresponds a random variable $\mu(\Delta)$ with the properties

1) $\mu(\Delta) \in L^2$, $\mu(\emptyset) = 0$
2) $\mu(\Delta_1 \cup \Delta_2) = \mu(\Delta_1) + \mu(\Delta_2)$ if $\Delta_1 \cap \Delta_2 = \emptyset$
3) $\overline{\mu(\Delta_1)\mu^*(\Delta_2)} = m(\Delta_1 \cap \Delta_2)$,

where $m(\Delta)$ is a certain deterministic function on $\mathcal{B}$. Note that

$$m(\Delta) = \overline{|\mu(\Delta)|^2} \geq 0$$
$$\text{and } m(\Delta_1 \cup \Delta_2) = \overline{|\mu(\Delta_1) + \mu(\Delta_2)|^2} = m(\Delta_1) + m(\Delta_2)$$

provided that $\Delta_1 \cap \Delta_2 = \emptyset$, so that $m(\Delta)$ has some of the basic properties of a measure. It is called the *structural function* of $\mu(\Delta)$ and $\mu(\Delta)$ is called an *elementary orthogonal stochastic measure.*

Assume that $m(\Delta)$ is *semiadditive* in the following sense: for any $\Delta \in \mathcal{B}$ the inclusion $\Delta \subset \bigcup_{j=1}^{\infty} \Delta_j$, $\Delta_j \in \mathcal{B}$, implies

$$m(\Delta) \leq \sum_{j=1}^{\infty} m(\Delta_j). \tag{8.406}$$

Then $m(\Delta)$ can be extended to a Borel measure on $\mathcal{B}$.

Let $f(x) \in L^2\ (D, \mathcal{B}, m)$. Define *the stochastic integral* as the following limit

$$\int_D f(x)\zeta(dx) = \text{l.i.m.} \int_D f_n(x)\zeta(dx). \tag{8.407}$$

where $f_n(x)$ is a sequence of *simple functions* such that

$$\| f - f_n \|_{L^2(D,m)} := \left\{ \int_D |f - f_n|^2 m(dx) \right\}^{1/2} \to 0 \qquad n \to \infty. \tag{8.408}$$

A *simple function* is a function of the type

$$f_n(x) = \sum_{j=1}^{n} c_j \chi_{A_j} \tag{8.409}$$

where $c_j = \text{const}$, χ_{A_j} is the characteristic function of the set $A_j \subset \mathcal{B}$.

Lemma 8.33 *If $f_i \in L^2(D, m)$, $i = 1, 2$, and $c_i = const$, then*

$$\int_D (c_1 f_1 + c_2 f_2)\zeta(dx) = c_1 \int_D f_1 \zeta(dx) + c_2 \int_D f_2 \zeta(dx) \tag{8.410}$$

and

$$\overline{\int_D f_1(x)\zeta(dx) \int_D f_2^*(x)\zeta^*(dx)} = \int_D f_1(x) f_2^*(x) m(dx). \tag{8.411}$$

Using the notion of the stochastic integral one can construct an *integral representation of random functions.*

Suppose that a random function $\xi(x)$, $x \in D$, has the covariance function of the form

$$B(x,y) = \int_\Lambda g^*(x,\lambda)g(y,\lambda)m(d\lambda) \tag{8.412}$$

where $m(d\lambda)$ is a Borel measure on the set Λ, $g(x,\lambda) \in L^2(\Lambda, m(d\lambda))$ $\forall x \in D$, and the set of functions $\{g(x,\lambda), x \in D\}$ is complete in $L^2(\Lambda, m(d\lambda))$.

Lemma 8.34 *Under the above assumptions there exists an orthogonal stochastic measure $\zeta(d\lambda)$ such that*

$$\xi(x) = \int_\Lambda g(x,\lambda)\zeta(d\lambda). \tag{8.413}$$

Equation (8.413) holds with probability one, and $m(\Delta) := \int_\Delta m(d\lambda)$ is the structural function corresponding to $\zeta(d\lambda)$. There is an isometric isomorphism between $L^2(\Lambda, m(d\lambda))$ and L^2_ξ, where L^2_ξ is the closure of the set of random variables of the form $\sum_{j=1}^n c_j\zeta(\Delta_j)$, $\Delta_j \in \Lambda$, in the norm (8.387). This isomorphism is established by the correspondence

$$\xi(x) \leftrightarrow g(x,\lambda), \quad \zeta(\Delta) \leftrightarrow \chi_\Delta(\lambda). \tag{8.414}$$

If $h_i(\lambda) \in L^2(\Lambda, m(d\lambda))$, $i = 1, 2$ then

$$(h_1, h_2)_{L^2(\Lambda,m)} := \int_\Lambda h_1 h_2^* m(d\lambda) = \overline{\xi_1\xi_2^*}, \tag{8.415}$$

where

$$\xi_i = \int_\Lambda h_i(\lambda)\zeta(d\lambda). \tag{8.416}$$

This theory extends to the case of random vector-functions.

8.4.3 *Estimation in Hilbert space $L^2(\Omega, \mathcal{U}, P)$*

Assume that L^2_ξ is the subspace of $L^2(\Omega, \mathcal{U}, P)$, a random variable $\eta \in L^2(\Omega, \mathcal{U}, P)$ and we want to give the best estimate of η by an element of L^2_ξ, that is to find $\eta_0 \in L^2_\xi$ such that

$$\delta := \| \eta - \eta_0 \| = \inf_{\phi \in L^2_\xi} \| \eta - \phi \|, \tag{8.417}$$

where the norm is defined by (8.387).

The element $\eta_0 \in L^2_\xi$ does exist, is unique, and is the projection of η onto the subspace L^2_ξ in the Hilbert space $L^2(\Omega, \mathcal{U}, P)$.

The error of the estimate, the quantity δ defined by (8.417), can be calculated analytically in some cases. For example, if $(\xi_1, \dots, \xi_n)$ is a finite set of random variables then

$$\eta_0 = \frac{1}{\Gamma} \begin{vmatrix} (\xi_1, \xi_1) & \dots & (\xi_1, \xi_n) & \xi_1 \\ & & & \vdots \\ (\xi_n, \xi_1) & \dots & (\xi_n, \xi_n) & \xi_n \\ (\eta, \xi_1) & \dots & (\eta, \xi_n) & 0 \end{vmatrix}, \tag{8.418}$$

where $\Gamma = \Gamma(\xi_1, \dots, \xi_n)$ is the Gramian of $(\xi_1, \dots, \xi_n)$:

$$\Gamma := \begin{vmatrix} (\xi_1, \xi_1) & \dots & (\xi_1, \xi_n) \\ \dots & \dots & \dots \\ (\xi_n, \xi_1) & & (\xi_n, \xi_n) \end{vmatrix}. \tag{8.419}$$

One has

$$\delta^2 = \frac{\Gamma(\xi_1, \dots, \xi_n, \eta)}{\Gamma(\xi_1, \dots, \xi_n)}. \tag{8.420}$$

The optimal estimate $\eta_0 \in L^2_\xi$ satisfies *the orthogonality equation*

$$(\eta_0 - \eta, \xi(x)) = 0, \quad \forall x \in D \tag{8.421}$$

which means geometrically that $\eta - \eta_0$ is orthogonal to L^2_ξ.

Equation (1.6) for the optimal filter is a particular case of (8.421): if, in the notation of Chapter 1,

$$\eta_0(z) = \int_D h(z,y)\mathcal{U}(y)dy, \quad \eta = s(z), \tag{8.422}$$

then (8.421) becomes

$$\int_D h(z,y)\overline{\mathcal{U}(y)\mathcal{U}^*(x)}dy = \overline{s(z)\mathcal{U}^*(x)},$$

which, according to (1.3), can be written as

$$\int_D h(z,y)R(x,y)dy = f(x,z). \tag{8.423}$$

This is equation (1.6).

8.4.4 *Homogeneous and isotropic random fields*

If $\xi(x)$ is a random field and

$$\overline{\xi(x)} = 0, \quad \overline{\xi^*(x)\xi(y)} = R(x-y) \tag{8.424}$$

then $\xi(x)$ is called *a (wide-sense) homogeneous random field.* It is called a *homogeneous random field* if for any n and any $x_1, \dots, x_n, x$ the distribution function of n random variables $\xi(x_1 + x), \dots, \xi(x_n + x)$ does not depend on x. Here $x \in R^r$ or, if $x \in D \subset R^r$, then one assumes that $x, y \in D$ implies $x + y \in D$. The function $R(x)$ is positive definite in the sense (8.380). Therefore, by Bochner-Khintchine theorem, there exists a monotone nondecreasing function $F(x)$, $F(+\infty < \infty$, such that

$$R(x) = \int \exp(ix \cdot y) dF(y), \quad \int = \int_{R^r}, \tag{8.425}$$

$$x \cdot y = \sum_{j=1}^{r} x_j y_j. \tag{8.426}$$

One often writes $dF(y) = F(dy)$ to emphasize that F determines a measure on R^r. Monotonicity in the case $r > 1$ is understood as monotonicity in each of the variables.

If $r > 1$ then a positive definite function $R(x)$ is the Fourier transform of a positive finite measure on R^r. This measure is given by the function $F(x)$ which satisfies characteristic properties 2)-4) of a distribution function. It follows from (8.425) that

$$0 < R(0) = F(R^r) < \infty \tag{8.427}$$

$$|R(x)| \leq R(0) \tag{8.428}$$

$$R(-x) = R^*(x). \tag{8.429}$$

A homogeneous random field is called *isotropic* if

$$\overline{\xi^*(x)\xi(y)} = \overline{\xi^*(gx)\xi(gy)} \tag{8.430}$$

for all $x, y \in R^r$ and all $g \in SO(n)$, where $SO(n)$ is the group of rotations of R^r around the origin. Equation (8.430) for homogeneous random fields is equivalent to

$$R(x) = R(gx) \quad \forall g \in SO(n). \tag{8.431}$$

This means that

$$R(x) = R(|x|) \tag{8.432}$$

where $|x| = (x_1^2 + \cdots + x_r^2)^{1/2}$ is the length of the vector x. This and formula (8.425) imply that $dF(y) = d\phi(|y|)$. If

$$\int |R(x)|dx < \infty \tag{8.433}$$

then $dF = f(y)dy$ and $f(y)$ is continuous:

$$f(y) = (2\pi)^{-r} \int \exp(-ix \cdot y)R(x)dx. \tag{8.434}$$

If

$$\int |R|^2 dx < \infty \tag{8.435}$$

then $dF = f(y)dy$, $f(y) \in L^2(R^r)$, and formula (8.434) holds in L^2-sense.

It is known that

$$\int_{|x|=\rho} \exp(ix \cdot y)ds = \left(\frac{2\pi\rho}{|y|}\right)^{r/2} |y| J_{(r-2)/2}(\rho|y|) \tag{8.436}$$

where $J_n(t)$ is the Bessel function and ds is the element of the surface area of the sphere $|x| = \rho$ in R^r. Using formula (8.436) one obtains

Lemma 8.35 *Assume that $R(\rho)$ is continuous function. This function is a correlation function of a homogeneous isotropic random field in R^r if and only if it is of the form:*

$$R(\rho) = 2^{(r-2)/2}\Gamma\left(\frac{r}{2}\right) \int_0^\infty \frac{J_{(n-2)/2}(\lambda\rho)}{(\lambda\rho)^{(r-2)/2}} dg(\lambda) \tag{8.437}$$

where $g(\lambda)$ is a monotone nondecreasing bounded function, $g(+\infty) < \infty$, and $\Gamma(z)$ is the Gamma-function.

If $r = 2$ formula (8.437) becomes

$$R(\rho) = \int_0^\infty J_0(\lambda\rho)dg(\lambda), \tag{8.438}$$

for $n = 3$ one gets

$$R(\rho) = 2\int_0^\infty \frac{\sin(\lambda\rho)}{\lambda\rho} dg(\lambda). \tag{8.439}$$

From formula (8.425) and Lemma 8.34 it follows that a homogeneous random field admits the spectral representation of the form

$$\xi(x) = \int \exp(ix \cdot y)\zeta(dy), \quad \int = \int_{R^r}, \tag{8.440}$$

where $\zeta(dy)$ is an orthogonal stochastic measure on R^r.

If the random field is homogeneous isotropic and continuous in the L^2-sense then

$$\xi(x) = c_r \sum_{m=0}^{\infty} \sum_{j=1}^{h(m,r)} S_{m,j}(\theta) \int_0^\infty \frac{J_{m+(r-2)/2}(\lambda|x|)}{(\lambda|x|)^{(r-2)/2}} \zeta_{mj}(d\lambda). \tag{8.441}$$

Here $c_r = \text{const}$, $S_{m,j}(\theta)$ is the orthonormalized in $L^2(S^{r-1})$ system of the spherical harmonics, S^{r-1} is the unit sphere in R^r, $\theta \in S^{r-1}$,

$$h(m,r) = (2m + r - 2)\frac{(m + r - 3)!}{(r - 2)!m!},$$

$r \geq 2$ is the number of linearly independent spherical harmonics corresponding to the fixed m. For example, if $r = 3$ then $h(m,3) = 2m+1$. The stochastic orthogonal measures $\zeta_{mj}(d\lambda)$ have the properties

$$\overline{\zeta_{mj}(d\lambda)} = 0, \tag{8.442}$$

$$\overline{\zeta_{mj}(\Delta_1)\zeta^*_{pq}(\Delta_2)} = \delta_{mp}\delta_{jq}m(\Delta_1 \cap \Delta_2) \tag{8.443}$$

where Δ_1 and Δ_2 are arbitrary Borel sets in the interval $(0,\infty)$, and $m(\Delta)$ is a finite measure on $(0,\infty)$.

If $\xi(x)$ is a homogeneous random field with correlation function (8.425), then one can aply a differential operator $Q(-i\partial)$, $\partial = (\partial_1, \dots, \partial_r)$, $\partial_j = \frac{\partial}{\partial x_j}$, to $\xi(x)$ in L^2 sense if and only if

$$\int |Q(y)|^2 F(dy) < \infty. \tag{8.444}$$

If condition (8.444) holds then $Q(-i\partial)\xi(x)$ is a homogeneous random field and its correlation function is $Q^*(-i\partial)Q(-i\partial)R(x)$ and the corresponding *spectral density* is $|Q(y)|^2 f(y)$, where $F(dy) = f(y)dy$. By the spectral density of the homogeneous random field with correlation function (8.425) one means the function $f(y)$ defined by $F(dy) = f(y)dy$ in the case when $F(dy)$ is absolutely continuous with respect to Lebesgue's measure, so that $f(y) \in L^1(R^r)$.

8.4.5 *Estimation of parameters*

Let ξ be a random variable with the distribution function $F(x,\theta)$ which depends on a parameter θ. The problem is to estimate the unknown θ given n sample values of ξ to which $F(x,\theta)$ belongs. The estimated value of θ let us denote $\hat{\theta} = \hat{\theta}(x_1,\dots,x_n)$ where x_j, $1 \le j \le n$, are observed values of ξ. What are the properties one seeks in an estimate? If $\rho(\theta,\hat{\theta})$ is the risk function which measures the distance of the estimate $\hat{\theta}$ from the true value of the parameter θ, then the estimate is good if

$$\overline{\rho(\theta,\hat{\theta})} \le \overline{\rho(\theta,\hat{\theta}_1)}, \tag{8.445}$$

where $\hat{\theta}_1$ is any other estimate, and

$$\overline{\rho(\theta,\hat{\theta})} = \int \rho\left(\theta,\hat{\theta}(x_1,\dots,x_n)\right) dF(x_1,\theta)\dots dF(x_n,\theta) \tag{8.446}$$

and $F(x,\theta)$ is the distribution function of ξ for a fixed θ.

A *minimax estimate* is the one for which

$$\sup_{\theta} \overline{\rho(\theta,\hat{\theta})} = \min \tag{8.447}$$

where θ runs through the set Θ in which it takes values.

A *Bayes estimate* is the one for which

$$\int_{\Theta} \overline{\rho(\theta,\hat{\theta})} d\mu(\theta) = \min \tag{8.448}$$

where $\mu(\theta)$ is an *a priori* given distribution function on the set Θ. This means that one prescribes a priori more weight to some distribution of θ.

An *unbiased* estimate is the one for which

$$\overline{\hat{\theta}} = \theta. \tag{8.449}$$

An *efficient* estimate is the one for which

$$\overline{|\theta - \hat{\theta}|^2} \le \overline{|\theta - \hat{\theta}_1|^2} \quad \text{for any} \quad \hat{\theta}_1. \tag{8.450}$$

The Cramer-Rao inequality gives a lower bound for the variance of the estimate:

$$\sigma_\theta^2(\hat{\theta}) := \overline{|\hat{\theta} - \theta|^2} \ge \frac{1}{nI(\theta)}, \tag{8.451}$$

where one assumes that (78) holds,

$$I(\theta) := \int \left| \frac{\partial \log p(x,\theta)}{\partial \theta} \right|^2 p(x,\theta)\nu(dx), \tag{8.452}$$

and one assumes that dF has a density $p(x,\theta)$ with respect to a *σ-finite measure* $\nu(dx)$

$$dF = p(x,\theta)\nu(dx) \tag{8.453}$$

and $p(x,\theta)$ is differentiable in θ.

A mesure ν on a set E in a measure space is called *σ-finite* if E is a countable union of sets E_j with $\nu(E_j) < \infty$.

In particular, if *measure* ν *is concentrated* on the discrete finite set of points $y_1, \dots, y_n$, then

$$I(\theta) = \sum_{j=1}^{n} \left| \frac{\partial \log p(y_j,\theta)}{\partial \theta} \right|^2 p(y_j,\theta). \tag{8.454}$$

The quantity $I(\theta)$ is called the information quantity. *A measure* ν *is called concentrated on a set* $A \subset E$ *if* $\nu(B) = \nu(B \cap A)$ *for every* $B \subset E$, *that is, if* $\nu(B) = 0$ *whenever* $B \cap A = \emptyset$.

Sometimes an estimate $\hat{\theta}$ is called *efficient* if the equality sign holds in (8.451). An estimate $\hat{\theta}$ is called *sufficient* if $dF(x,\theta) = p(x,\theta)\nu(dx)$ and

$$p(x_1,\theta)\dots p(x_n,\theta) = g_\theta(\hat{\theta})h(x_1,\dots,x_n)$$

where g_θ and h are nonnegative functions, h does not depend on θ and g_θ depends on $x_1,\dots,x_n$ only through $\hat{\theta} = \hat{\theta}(x_1,\dots,x_n)$. Suppose that the volume of the sample grows, i.e. $n \to \infty$. An estimate $\hat{\theta}(x_1,\dots,x_n) := \hat{\theta}_n$ is called *consistent* if

$$\lim_{n\to\infty} P\left(|\hat{\theta}_n - \theta| > \epsilon\right) = 0 \quad \text{for every} \quad \epsilon > 0. \tag{8.455}$$

There are many methods for constructing estimates.

Maximum-likelihood estimates of θ is the estimate obtained from the equations

$$\frac{\partial \log L(\theta, x_1, \dots, x_n)}{\partial \theta_j} = 0, \quad 1 \le j \le m. \tag{8.456}$$

Here θ is a vector parameter $\theta = (\theta_1,\dots,\theta_m)$, the function $L(\theta,x_1,\dots,x_n)$ is called the *likelihood function* and is defined by

$$L(\theta,x_1,\dots,x_n) := \Pi_{i=1}^{n} p(x_i,\theta), \tag{8.457}$$

where $p(x, \theta)$ is the density of dF defined by $dF = p(x, \theta)dx$.

Cramer proved that if:

1) $\frac{\partial^k \log p(x,\theta)}{\partial \theta^k}$, $k \le 3$, exists for all $\theta \in \Theta$ and almost all $x \in R^1$
2) $\left|\frac{\partial^k p(x,\theta)}{\partial \theta^k}\right| \le g_k(x)$ where $g_k(x) \in L^1(R^1)$, $k = 1, 2$, and $\sup_{\theta \in \Theta} \int_{-\infty}^{\infty} g_3(x) p(x, \theta) dx < \infty$
3) $I(\theta)$ is positive and finite for every $\theta \in \Theta$, where $I(\theta)$ is defined by (8.452) with $\nu(dx) = dx$,

then equation (8.456) has a solution $\hat{\theta}(x_1, \dots, x_n)$ which is a consistent, asymptotically efficient and asymptotically Gaussian estimate of θ.

Here asymptotic efficiency is understood in the sense that inequality (8.451) becomes an equality asymptotically as $n \to \infty$. More precisely, define

$$\text{eff}(\hat{\theta}) := \left[nI(\theta)\sigma_\theta^2(\hat{\theta}) \right]^{-1}. \tag{8.458}$$

Then the estimate $\hat{\theta}$ is asymptotically efficient if

$$\lim_{n \to \infty} \text{eff}(\hat{\theta}_n) = 1. \tag{8.459}$$

The estimate $\hat{\theta}_n$ is asymptotically Gaussian in the sense that

$$I(\theta) n^{1/2} \left[\hat{\theta}(x_1, \dots, x_n) - \theta \right] \sim N(0, 1) \quad \text{as} \quad n \to \infty \tag{8.460}$$

where $N(0, 1)$ is the Gaussian distribution with zero mean value and variance one, and we assumed for simplicity that θ is a scalar parameter.

We do not discuss other methods for constructing estimates (such as the method of moments, the minimum of χ^2 method, intervals of confidency, the Bayes estimates etc.).

8.4.6 *Discrimination between hypotheses*

One observes n values $x_1, \dots, x_n$, of a random quantity ξ, and assumes that there are two hypotheses H_0 and H_1 about ξ. If H_0 occurs then the probability density of the observed values is $f_n(x_1, \dots, x_n \mid H_0)$, otherwise it is $f_n(x_1, \dots, x_n \mid H_1)$. Given the observed values $x_1, \dots, x_n$ one has to decide whether H_0 or H_1 occured. Let us denote γ_i, $i = 0, 1$, the decision that H_i occured. The decision γ_0 is taken if $(x_1, \dots, x_n) \in D_0$, where D_0 is a certain domain in R^n. The choice of such a domain is the choice of the decision rule. If $(x_1, \dots, x_n) \notin D_0$ then the decision γ_1 is taken. The error α_{10} of the first kind is defined as

$$\alpha_{10} = P(\gamma_1 \mid H_0) = \int_{D_1} f(x \mid H_0) dx, \quad x = (x_1, \dots, x_n) \tag{8.461}$$

where $D_1 = R^n \setminus D_0$. Thus α_{10} is the probability to take the decision that H_1 occured when in fact H_0 occured. The error of the second kind is

$$\alpha_{01} = P(\gamma_0 \mid H_1) = \int_{D_0} f(x \mid H_1) dx, \quad x = (x_1, \dots, x_n). \tag{8.462}$$

The conditional probabilities to take the right decisions are

$$P(\gamma_0 \mid H_0) = 1 - \alpha_{10}, \quad P(\gamma_1 \mid H_1) = 1 - \alpha_{01}. \tag{8.463}$$

One cannot decrease both α_{10} and α_{01} without limit: if α_{10} decreases then D_1 decreases, therefore D_0 increases and α_{01} increases. The problem is to choose an optimal in some sense decision rule.

Let us describe some approaches to this problem. The *Neyman-Pearson* approach gives the decision rule which minimizes α_{01} under the condition that $\alpha_{10} \le \alpha$, where α is a fixed *confidence level.*

Let us define the *likelihood ratio*

$$\ell(x) := \frac{f(x \mid H_1)}{f(x \mid H_0)}, \quad x = (x_1, \dots, x_n), \tag{8.464}$$

and the threshold $c > 0$ which is given by the equation

$$P\left\{\ell(x) \ge c \mid H_0\right\} = \alpha. \tag{8.465}$$

The *Neyman-Pearson decision rule* is

$$\text{if } \ell(x) < c \text{ then } H_0 \text{ occured, otherwise } H_1 \text{ occured.} \tag{8.466}$$

If $\phi(t)$ is a monotone increasing function then $\ell(x) < c$ if and only if $\phi(\ell(x)) < \phi(c)$. In particular, the rule

$$\text{if } \log \ell(x) < \log c \text{ then } H_0 \text{ occured, otherwise } H_1 \text{ occured} \tag{8.467}$$

is equivalent to (8.466).

Assume that the a priori probability p_0 of H_0 is known, so that the a priori probability of H_1 is $1 - p_0$.

Then the *maximum a posteriori probability decision rule* is:

$$\text{if } \ell(x) < \frac{p_0}{1 - p_0} \text{ then } H_0 \text{ occured, otherwise } H_1 \text{ occured.} \tag{8.468}$$

If no a priori information about p_0 is known, then one can use the *maximum likelihood decision rule*

$$\text{if } \ell(x) < 1 \text{ then } H_0 \text{ occured, otherwise } H_1 \text{ occured.} \tag{8.469}$$

All these rules are threshold rules with various thresholds, and other decision rules are discussed in the litrature.

8.4.7 *Generalized random fields*

A *generalized random field* $\xi(x)$, $x \in R^r$ is defined as follows. Suppose that $\{\phi_1(x), \dots, \phi_m(x)\}$ is a set of $C_0^\infty(R^r)$ functions, and to each such set there corresponds a random vector $\{\xi(\phi_1), \dots, \xi(\phi_m)\}$, such that the distribution functions for all such vectors for all m and all choices of $\{\phi_1, \dots, \phi_m\}$ are consistent. Then one says that a generalized random field ξ is defined.

The theories of generalized random functions of one and several variables are similar. A linear combination $\sum_{j=1}^N c_j \xi_j(x)$ of generalized random functions is defined by the formula

$$\sum_{j=1}^N c_j \xi_j(x)(\phi_k) = \sum_{j=1}^N c_j \xi_j(\phi_k).$$

Similarly, if $\psi \in C^\infty(R^r)$, then $\psi\xi(\phi) := \xi(\psi\phi)$, $\xi(x+h)(\phi) := \xi\left(\phi(x-h)\right)$. If $\overline{\xi(\phi)} := m(\phi)$ is a continuous linear functional on C_0^∞, then $m := \overline{\xi}$ is called the mean value of ξ,

$$m(\phi) = \int x dF, \quad \text{where} F(x) = P\left\{\xi(\phi) < x\right\}. \tag{8.470}$$

Recall that $m(\phi)$ is called continuous in C_0^∞ if $\phi_n(x) \to \phi(x)$ implies $m(\phi_n) \to m(\phi)$, where $\phi_n \to \phi$ means that all ϕ_n and ϕ vanish outside of a fixed compact set E of R^r and $\max_{x\in E} |\phi_n^{(j)} - \phi^{(j)}| \to 0$ as $n \to \infty$ for all multiindices j.

The correlation functional of a generalized random field is defined as

$$B(\phi, \psi) = \overline{\xi^*(\phi)\xi(\psi)}. \tag{8.471}$$

The covariance functional is defined as

$$R(\phi, \psi) = \overline{\left[\xi^*(\phi) - m^*(\phi)\right]\left[\xi(\psi) - m(\psi)\right]} = B(\phi, \psi) - m(\phi)m(\psi). \tag{8.472}$$

Both functionals are nonnegative definite:

$$B(\phi, \phi) \geq 0, \quad R(\phi, \phi) \geq 0.$$

If the random vectors $\{\xi(\phi_1), \dots, \xi(\phi_m)\}$ for all m are Gaussian, then ξ is called a generalized random Gaussian field. If $B(\phi, \psi)$ and $m(\phi)$ are continuous in $C_0^\infty(R^r)$ bilinear and, respectively, linear functionals and $R(\phi, \phi) \geq 0$ then there is a generalized random Gaussian field for which $B(\phi, \psi)$ is the correlation functional and $m(\phi)$ is the mean value functional.

A generalized random field is called homogeneous (stationary) if the random vectors
$\{\xi(\phi_1(x+h)), \dots, \xi(\phi_m(x+h))\}$ and $\{\xi(\phi_1(x)), \dots, \xi(\phi_m(x))\}$
have the same distribution function for any $h \in R^r$. The mean value functional for a homogeneous generalized random field is

$$m(\phi) = \text{const} \int \phi dx$$

and its correlation functional is

$$B(\phi, \psi) = \int \tilde{\phi}(\lambda) \tilde{\psi}^*(\lambda) \mu(d\lambda)$$

where $\tilde{\phi}(\lambda)$ is the Fourier transform of $\phi(x)$, and $\mu(d\lambda)$ is a positive measure on R^r satisfies for some $p \geq 0$ the condition

$$\int_{R^r} \left(1 + |\lambda|^2\right)^{-p} \mu(d\lambda) < \infty.$$

The measure μ is called the *spectral measure* of ξ. One can introduce the spectral representation of the generalized random field similar to (8.440). An important example of a Gaussian generalized random field is the Brownian motion.

8.4.8 *Kalman filters*

Let us start with the basic equation for the optimal Wiener filter for the filtering problem

$$\sigma^2 h(t, \tau) + \int_{t_0}^{t} h(t, \tau') R_s(\tau, \tau') d\tau' = f(\tau, t), \quad t_0 < \tau < t, \tag{8.473}$$

where $\mathcal{U} = s + n$ is the observed signal, s and n are uncorrelated,

$$\begin{aligned} &\overline{n^*(t) n(\tau)} = \sigma^2 \delta(t - \tau), \quad && R_s(\tau, t) := \overline{s^*(\tau) s(t)}, \\ &\overline{s(t)} = \overline{n(t)} = 0, && f(\tau, t) := \overline{\mathcal{U}^*(\tau) s(t)}. \end{aligned} \tag{8.474}$$

The optimal estimate of s is

$$\hat{s}(t) = \int_{t_0}^{t} h(t,\tau)\mathcal{U}(\tau)d\tau. \tag{8.475}$$

The error of this estimate

$$\tilde{s}(t) := s(t) - \hat{s}(t) \tag{8.476}$$

and

$$\mathcal{D}\left[\tilde{s}(t)\right] = R_s(t,t) - \int_{t_0}^{t} h^*(t,\tau)f(\tau,t)d\tau. \tag{8.477}$$

Let us assume that $s(t)$ satisfies the following differential equation

$$\dot{s}(t) = A(t)s + w, \tag{8.478}$$

where for simplicity we assume all functions to be scalar functions, w to be white noise, and

$$\overline{w} = 0, \quad \overline{w^*(t)w(\tau)} = Q\delta(t-\tau), \quad Q = \text{const} > 0. \tag{8.479}$$

One could assume that

$$\mathcal{U} = Hs(t) + n, \tag{8.480}$$

where n is white noise, and H is a linear operator, but the argument will be essentially the same, and, for simplicity, we assume (8.479).

Note that

$$f(\tau,t) = \overline{\mathcal{U}^*(\tau)s(t)} = \overline{\left[s^*(\tau) + n^*(\tau)\right]s(t)} = R_s(\tau,t) \tag{8.481}$$

assuming that the noise $n(\tau)$ and the signal $s(t)$ are uncorrelated

$$\overline{n^*(\tau)s(t)} = 0. \tag{8.482}$$

Also

$$R(\tau,t) := \overline{\mathcal{U}^*(\tau)\mathcal{U}(t)} = R_s(\tau,t) + \sigma^2\delta(t-\tau) \tag{8.483}$$

provided that (8.482) holds and

$$\overline{n^*(\tau)n(t)} = \sigma^2\delta(t-\tau), \quad \overline{n} = 0. \tag{8.484}$$

To derive a differential equation for the optimal impulse function, differentiate (8.473) in t using (8.481):

$$\frac{\partial R_s(\tau,t)}{\partial t} = h(t,t)R(\tau,t) + \int_{t_0}^{t} \frac{\partial h(t,\tau')}{\partial t} R(\tau,\tau')d\tau'. \tag{8.485}$$

For $\tau < t$ equation (8.483) becomes

$$R(\tau,t) = R_s(\tau,t), \quad \tau < t. \tag{8.486}$$

This and equation (8.473) yield after multiplication by $h(t,t)$:

$$h(t,t)R(\tau,t) = \int_{t_0}^{t} h(t,t)h(t,\tau')R(\tau,\tau')d\tau'. \tag{8.487}$$

From (8.478) one obtains

$$\frac{\partial}{\partial t} R_s(\tau,t) = AR_s(\tau,t), \tau < t \tag{8.488}$$

where one used the equation

$$\overline{s^*(\tau)w(t)} = 0 \quad \text{for} \quad \tau < t. \tag{8.489}$$

To derive (8.489) note that equation (8.478) implies

$$s(t) = \psi(t,t_0)s(t_0) + \int_{t_0}^{t} \psi(t,\tau)w(\tau)d\tau \tag{8.490}$$

where $\psi(t,\tau)$ is the transition function for the operator $\frac{d}{dt} - A$. Since

$$\overline{w^*(t)w(\tau)} = 0 \quad \text{for} \quad \tau < t, \tag{8.491}$$

it follows from (8.490) that (8.489) holds.

From (8.488), (8.481) and (8.473) one has

$$\frac{\partial}{\partial t} R_s(\tau,t) = A \int_{t_0}^{t} h(t,\tau')R(\tau,\tau')d\tau'. \tag{8.492}$$

From (8.485), (8.492) and (8.487) one obtains

$$0 = \int_{t_0}^{t} \left[-Ah(t,\tau') - h(t,t)h(t,\tau') + \frac{\partial h(t,\tau')}{\partial t} \right] R(\tau,\tau')d\tau' \tag{8.493}$$

for $\tau < t$. Since R is positive definite (see (8.483)), equation (8.493) has only the trivial solution, one gets

$$\frac{\partial h(t,\tau)}{\partial t} = Ah(t,\tau) - h(t,t)h(t,\tau), \quad \tau < t. \tag{8.494}$$

This is a differential equation for the optimal filter $h(t, \tau)$.

Let us find a differential equation for the optimal estimate $\hat{s}(t)$ defined by (8.475):

$$\dot{\hat{s}} = h(t,t)\mathcal{U}(t) + \int_{t_0}^{t} \frac{\partial h(t,\tau)}{\partial t}\mathcal{U}(\tau)d\tau \tag{8.495}$$

where $\dot{f} := \frac{df}{dt}$. From (8.494) and (8.495) one gets

$$\begin{aligned}\dot{\hat{s}} &= h(t,t)\mathcal{U}(t) + \int_{t_0}^{t} \left[Ah(t,\tau) - h(t,t)h(t,\tau)\right]\mathcal{U}(\tau)d\tau \\ &= h(t,t)\mathcal{U}(t) + A\hat{s}(t) - h(t,t)\hat{s}(t) \\ &= A\hat{s}(t) + h(t,t)\left[\mathcal{U}(t) - \hat{s}(t)\right].\end{aligned} \tag{8.496}$$

This is the differential equation for $\hat{s}(t)$. The initial condition is $\hat{s}(t_0) = 0$ according to (8.475). Let us express $h(t,t)$ in terms of the variance of the error (8.373). If this is done then (8.496) can be used for computations. From (8.473) and (8.481) one obtains:

$$R_s(\tau,t) = \sigma^2 h(t,\tau) + \int_{t_0}^{t} R_s(\tau,\tau')h(t,\tau')d\tau'. \tag{8.497}$$

Put $\tau = t$ in (8.497) assume that $h(t,\tau)$ is real-valued and use (8.477) to get

$$h(t,t) = \sigma^{-2}\mathcal{D}\left[\tilde{s}(t)\right]. \tag{8.498}$$

Let us finally derive a differential equation for $h(t,t)$. From (8.476), (8.478) and (8.496) it follows that

$$\begin{aligned}\dot{\tilde{s}} &= As + w - A\hat{s} - h(t,t)\left[s(t) + n(t) - \hat{s}(t)\right] \\ &= \left[A - h(t,t)\right]\tilde{s}(t) + w - h(t,t)n(t).\end{aligned} \tag{8.499}$$

The solution to (8.499) is

$$\tilde{s}(t) = \psi(t,t_0)\tilde{s}(t_0) + \int_{t_0}^{t} \psi(t,\tau)\left[w(\tau) - h(\tau,\tau)n(\tau)\right]d\tau. \tag{8.500}$$

One obtains, using (8.499), that

$$\begin{aligned}\dot{h}(t,t) &= \sigma^{-2}\left[\overline{\dot{\tilde{s}}^*(t)\tilde{s}(t)} + \overline{\tilde{s}^*(t)\dot{\tilde{s}}(t)}\right] \\ &= \sigma^{-2} 2Re\left\{\left[A^*(t) - h(t,t)\right]\sigma^2 h(t,t)\right. \\ &\quad \left. + \overline{w^*(t)\tilde{s}(t)} - h(t,t)\overline{h^*(t)\tilde{s}(t)}\right\} \\ &= \left[A^*(t) + A(t)\right]h(t,t) - 2h^2(t,t) + Q\sigma^{-2} + h^2(t,t) \\ &= \left[A^*(t) + A(t)\right]h(t,t) - h^2(t,t) + Q\sigma^{-2}, \end{aligned} \tag{8.501}$$

where we assumed that w, n and $s(t_0)$ are uncorrelated, took into account that $h(t,t) > 0$ (see (8.498)) and used formula (8.500) to get

$$\overline{w^*(t)\tilde{s}(t)} = \frac{1}{2}Q\psi(t,t) = \frac{Q}{2} \tag{8.502}$$

and

$$\overline{h^*(t)\tilde{s}(t)} = -\frac{h(t,t)}{2}\sigma^2. \tag{8.503}$$

Note that $\psi(t,t) = 1$ and the $\frac{1}{2}$ factor in (8.502) and (8.503) appeared since we used the formula $\int_{t_0}^t \delta(t-\tau) f d\tau = \frac{1}{2}f(t)$.

Equation (8.501) is the Riccati equation for $h(t,t)$. Equations (8.494), (8.496) and (8.501) define Kalman's filter. This filter consists in computing the optimal estimate (8.475) by solving the differential equation (8.496) in which $h(t,t)$ is obtained by solving the Riccati equation (8.501). The initial data for equation (8.501) is

$$h(t_0,t_0) = \sigma^2 \mathcal{D}\left[\tilde{s}(t_0)\right] = \sigma^{-2}\mathcal{D}\left[s(t)\right] = \sigma^{-2}R_s(t_0,t_0). \tag{8.504}$$

Here we used equation (8.476) and took into account that $\hat{s}(t_0) = 0$.

The ideas of the derivation of Kalman's filter are the same for random vector-functions. In this case $A(t)$ is a matrix. For random fields there is no similar theory due to the fact that there is no causality in the space variables in contrast with the time variable.

Appendix A

Analytical Solution of the Basic Integral Equation for a Class of One-Dimensional Problems

In this Section we develop the theory for a class of random processes, because this theory is analogous to the estimation theory for random fields, developed in the next Section.

Let

$$(\star)Rh = f, \quad 0 \le x \le L, \qquad Rh = \int_0^L R(x,y)h(y)\,dy$$

where the kernel $R(x,y)$ satisfies the equation $QR = P\delta(x-y)$. Here Q and P are formal differential operators of order n and $m < n$, respectively, n and m are nonnegative even integers, $n > 0$, $m \ge 0$, $Qu := q_n(x)u^{(n)} + \sum_{j=0}^{n-1} q_j(x)u^{(j)}$, $Ph := h^{(m)} + \sum_{j=0}^{m-1} p_j(x)h^{(j)}$, $q_n(x) \ge c > 0$, the coefficients $q_j(x)$ and $p_j(x)$ are smooth functions defined on $\mathbb{R}$, $\delta(x)$ is the delta-function, $f \in H^\alpha(0,L)$, $\alpha := \frac{n-m}{2}$, H^α is the Sobolev space.

An algorithm for finding analytically the unique solution $h \in \dot{H}^{-\alpha}(0,L)$ to $(\star)$ of minimal order of singularity is given. Here $\dot{H}^{-\alpha}(0,L)$ is the dual space to $H^\alpha(0,L)$ with respect to the inner product of $L^2(0,L)$.

Under suitable assumptions it is proved that $R : \dot{H}^{-\alpha}(0,L) \to H^\alpha(0,L)$ is an isomorphism.

Equation $(\star)$ is the basic equation of random processes estimation theory. Some of the results are generalized to the case of multidimensional equation $(\star)$, in which case this is the basic equation of random fields estimation theory. The presentation in Appendix A follows the paper [Ramm (2003)].

A.1 Introduction

In Chapter 2 estimation theory for random fields and processes is constructed. The estimation problem for a random process is as follows. Let $u(x) = s(x) + n(x)$ be a random process observed on the interval $(0, L)$, $s(x)$ is a useful signal and $n(x)$ is noise. Without loss of generality we assume that $\overline{s(x)} = \overline{n(x)} = 0$, where the overbar stands for the mean value, $\overline{u^*(x)u(y)} := R(x, y)$, $R(x, y) = \overline{R(y, x)}$, $\overline{u^*(x)s(y)} := f(x, y)$, and the star here stands for complex conjugate. The covariance functions $R(x, y)$ and $f(x, y)$ are assumed known. One wants to estimate $s(x)$ optimally in the sense of minimum of the variance of the estimation error. More precisely, one seeks a linear estimate

$$Lu = \int_0^L h(x, y)u(y)\, dy, \tag{A.1}$$

such that

$$\overline{|(Lu)(x) - s(x)|^2} = \min. \tag{A.2}$$

This is a filtering problem. Similarly one can formulate the problem of optimal estimation of $(As)(x)$, where A is a known operator acting on $s(x)$. If $A = I$, where I is the identity operator, then one has the filtering problem, if A is the differentiation operator, then one has the problem of optimal estimation of the derivative of s, if $As = s(x + x_0)$, then one has an extrapolation problem, etc. The kernel $h(x, y)$ is, in general, a distribution. As in Chapter 1, one derives a necessary condition for h to satisfy (A.2):

$$\int_0^L R(x, y)h(y, z)\, dy = f(x, z), \quad 0 \le x, z \le L. \tag{A.3}$$

Since z enters as a parameter in (A.3), the basic equation of estimation theory is:

$$Rh := \int_0^L R(x, y)h(y)\, dy = f(x), \quad 0 \le x \le L. \tag{A.4}$$

The operator in $L^2(0, L)$ defined by (A.4) is symmetric. In Chapter 1 it is assumed that the kernel

$$R(x, y) = \int_{-\infty}^{\infty} \frac{P(\lambda)}{Q(\lambda)} \Phi(x, y, \lambda)\, d\rho(\lambda), \tag{A.5}$$

where $P(\lambda)$ and $Q(\lambda)$ are positive polynomials, $\Phi(x, y, \lambda)$ and $d\rho(\lambda)$ are spectral kernel and, respectively, spectral measure of a selfadjoint ordinary differential operator ℓ in $L^2(\mathbb{R})$, $\deg Q(\lambda) = q$, $\deg P(\lambda) = p < n$, $p \geq 0$, $\text{ord}\ell := \sigma > 0$, $x, y \in \mathbb{R}^r$, $r \geq 1$, and ℓ is a selfadjoint elliptic operator in $L^2(\mathbb{R}^r)$.

It is proved in Chapter 4 that the operator $R : \dot{H}^{-\alpha}(0, L) \to H^{\alpha}(0, L)$, $\alpha := \frac{n-m}{2}\sigma$ is an isomorphism. By $H^{\alpha}(0, L)$ the Sobolev space $W^{\alpha,2}(0, L)$ is denoted, and $\dot{H}^{-\alpha}(0, L)$ is the dual space to $H^{\alpha}(0, L)$ with respect to $L^2(0, L) := H^0(0, L)$ inner product. Namely, $\dot{H}^{-\alpha}(0, L)$ is the space of distributions h which are linear bounded functionals on $H^{\alpha}(0, L)$. The norm of $h \in \dot{H}^{-\alpha}(0, L)$ is given by the formula

$$\|h\|_{\dot{H}^{-\alpha}(0,L)} = \sup_{g \in H^{\alpha}(0,L)} \frac{|(h, g)|}{\|g\|_{H^{\alpha}(0,L)}}, \tag{A.6}$$

where (h, g) is the $L^2(0, L)$ inner product if h and g belong to $L^2(0, L)$.

One can also define $\dot{H}^{-\alpha}(0, L)$ as the subset of the elements of $H^{-\alpha}(\mathbb{R})$ with support in $[0, L]$.

We generalize the class of kernels $R(x, y)$ defined in (A.5): we do not use the spectral theory, do not assume ℓ to be selfadjoint, and do not assume that the operators Q and P commute.

We assume that

$$QR = P\delta(x - y), \tag{A.7}$$

where Q and P are formal differential operators of orders n and m respectively, $n > m \geq 0$, n and m are even integers, $\delta(x)$ is the delta-function,

$$Qu := \sum_{j=0}^{n} q_j(x) u^{(j)}, \quad q_n(x) \geq c > 0, \quad Ph := h^{(m)} + \sum_{j=0}^{m-1} p_j(x) h^{(j)}, \tag{A.8}$$

q_j and p_j are smooth functions defined on $\mathbb{R}$. We also assume that the equation $Qu = 0$ has $\frac{n}{2}$ linearly independent solutions $u_j^- \in L^2(-\infty, 0)$ and $\frac{n}{2}$ linearly independent solutions $u_j^+ \in L^2(0, \infty)$. In particular, this implies that if $Qh = 0$, $h \in H^{\alpha}(\mathbb{R})$, $\alpha > 0$, then $h = 0$, and the same conclusion holds for $h \in H^{\beta}(\mathbb{R})$ for any fixed real number β, including negative β, because any solution to the equation $Qh = 0$ is smooth: it is a linear combination of n linearly independent solution to this equation, each of which is smooth and none belongs to $L^2(\mathbb{R})$.

Let us assume that $R(x, y)$ is a selfadjoint kernel such that

$$c_1\|\varphi\|_-^2 \leq (R\varphi, \varphi) \leq c_2\|\varphi\|_-^2, \quad c_1 = \text{const} > 0, \quad \forall\varphi \in C_0^\infty(\mathbb{R}), \qquad \text{(A.9)}$$

where $(\cdot, \cdot)$ is the $L^2(\mathbb{R})$ inner product, $\|\varphi\|_- := \|\varphi\|_{H^{-\alpha}(\mathbb{R})} := \|\varphi\|_{-\alpha}$, $\alpha := \frac{n-m}{2}$, $\|\varphi\|_\beta := \|\varphi\|_{H^\beta(\mathbb{R})}$, and we use below the notation $\|\varphi\|_+ := \|\varphi\|_{H^\alpha(0,L)} := \|\varphi\|_{H^+}$. The spaces $H^\alpha(0, L)$ and $\dot{H}^{-\alpha}(0, L)$ are dual of each other with respect to the $L^2(0, L)$ inner product, as was mentioned above. If $\varphi \in \dot{H}^{-\alpha}(0, L)$, then $\varphi \in H^{-\alpha}(\mathbb{R})$, and the inequality (A.9) holds for such φ. By this reason we also use (for example, in the proof of Theorem A.1 below) the notation H^- for the space $\dot{H}^{-\alpha}(0, L)$.

Assumption (A.9) holds, for example, for the equation

$$Rh = \int_{-1}^{1} \exp(-|x - y|)h(y)dy = f(x), \quad -1 \leq x \leq 1.$$

Its solution of minimal order of singularity is

$$h(x) = (-f'' + f)/2 + \delta(x+1)[-f'(-1) + f(-1)]/2 + \delta(x-1)[f'(1) + f(1)]/2.$$

One can see that the solution is a distribution with support at the boundary of the domain D if the following inequalities (A.10) and (A.11) hold:

$$c_3\|\varphi\|_{-\alpha+n} \leq \|Q^*\varphi\|_{-\alpha} \leq c_4\|\varphi\|_{-\alpha+n}, \ c_3, \ c_4 = \text{const} > 0, \ \forall\varphi \in C_0^\infty(\mathbb{R}), \qquad \text{(A.10)}$$

$$c_5\|\varphi\|^2_{\frac{n+m}{2}} \leq (PQ^*\varphi, \varphi) \leq c_6\|\varphi\|^2_{\frac{n+m}{2}}, \quad \forall\varphi \in C_0^\infty(\mathbb{R}), \qquad \text{(A.11)}$$

where Q^* is a formally adjoint to Q differential expression, and c_5 and c_6 are positive constants independent of $\varphi \in C_0^\infty(\mathbb{R})$. The right inequality (A.11) is obvious because $\text{ord}PQ^* = n+m$, and the right inequality (A.10) is obvious because $\text{ord}Q^* = n$.

Let us formulate our basic results.

Theorem A.1 *If (A.9) holds, then the operator R, defined in (A.5), is an isomorphism of $\dot{H}^{-\alpha}(0, L)$ onto $H^\alpha(0, L)$, $\alpha = \frac{n-m}{2}$.*

Theorem A.2 *If (A.7), (A.10) and (A.11) hold, then (A.9) holds and $R : \dot{H}^{-\alpha}(0, L) \to H^\alpha(0, L)$ is an isomorphism.*

Theorem A.3 *If (A.7), (A.10) and (A.11) hold, and $f \in H^n(0, L)$, then the solution to (A.4) in $\dot{H}^{-\alpha}(0, L)$ does exist, is unique, and can be* *calculated analytically by the following formula:*

$$h = \int_0^x G(x,y)Qf\,dy + \sum_{j=0}^{n-\alpha-1} \left[a_j^-(-1)^j G_y^{(j)}(x,0) + a_j^+(-1)^j G_y^{(j)}(x,L)\right], \tag{A.12}$$

where $a_j^\pm$ *are some constants and* $G(x,y)$ *is the unique solution to the problem*

$$PG = \delta(x-y), \quad G(x,y) = 0 \text{ for } x < y. \tag{A.13}$$

The constants $a_j^\pm$ *are uniquely determined from the condition* $h(x) = 0$ *for* $x > L$.

Remark A.1 *The solution* $h \in \dot{H}^{-\alpha}(0,L)$ *is the solution to equation (A.4) of minimal order of singularity.*

Remark A.2 *If* $P = 1$ *in (A.7) then the solution* h *to (A.4) of minimal order of singularity,* $h \in \dot{H}^{-\frac{n}{2}}(0,L)$, *can be calculated by the formula* $h = QF$, *where* F *is given by (A.22) (see below) and* u_+ *and* u_- *are the unique solutions of the problems* $Qu_+ = 0$ *if* $x > L$, $u_+^{(j)}(L) = f^{(j)}(L)$, $0 \le j \le \frac{n}{2} - 1$, $u_+(\infty) = 0$, *and* $Qu_- = 0$ *if* $x < 0$, $u_-^{(j)}(0) = f^{(j)}(0)$, $0 \le j \le \frac{n}{2} - 1$, $u_-(-\infty) = 0$.

A.2 Proofs

Proof of Theorem A.1. The set $C_0^\infty(0,L)$ is dense in $\dot{H}^{-\alpha}(0,L)$ (in the norm of $H^{-\alpha}(\mathbb{R})$). Using the right inequality (A.9), one gets:

$$\|R\|_{H^- \to H^+} = \sup_{h \in H^-} \frac{(Rh,h)}{\|h\|_-^2} \le c_2, \tag{A.14}$$

by the symmetry of R in $L^2(0,L)$. This implies $\|R\|_{H^- \to H^+} \le c_2$. Using the left inequality (A.9), one gets: $c_1\|h\|_-^2 \le \|Rh\|_+ \|h\|_-$, so

$$c_1\|h\|_- \le \|Rh\|_+. \tag{A.15}$$

Therefore

$$\|R^{-1}\|_{H^+ \to H^-} \le \frac{1}{c_1}. \tag{A.16}$$

Consequently, the range $Ran(R)$ of R is a closed subspace of H^+. In fact, $Ran(R) = H^+$. Indeed, if $Ran(R) \ne H^+$, then there exists a $g \in H^-$ such that $0 = (R\psi, g)\ \forall\psi \in H^-$. Taking $\psi = g$ and using the left inequality (A.9) one gets $\|g\|_- = 0$, so $g = 0$. Thus $Ran(R) = H^+$.

Theorem A.1 is proved. □

Proof of Theorem A.2. From (A.7) and (A.8) it follows that the kernel $R(x, y)$ defines a pseudodifferential operator of order $-2\alpha = m - n$. In particular, this implies the right inequality (A.9). In this argument inequalities (A.10) and (A.11) were not used.

Let us prove that (A.10) and (A.11) imply the left inequality (A.9).

One has

$$\|Q^*\varphi\|_{-\alpha} \le C\|\varphi\|_{n-\alpha}, \quad \forall\varphi \in C_0^\infty(\mathbb{R}), \tag{A.17}$$

because ord$Q^* = n$. Inequality (A.10) reads:

$$c_3\|\varphi\|_{-\alpha+n} \le \|Q^*\varphi\|_{-\alpha} \le c_4\|\varphi\|_{-\alpha+n}, \quad \forall\varphi \in C_0^\infty(\mathbb{R}), \tag{A.18}$$

where c_3 and c_4 are positive constants. If (A.18) holds, then $Q^* : H^{-\alpha+n}(\mathbb{R}) \to H^{-\alpha}(\mathbb{R})$ is an isomorphism of $H^{-\alpha+n}(\mathbb{R})$ onto $H^{-\alpha}(\mathbb{R})$ provided that $N(Q) := \{w : Qw = 0,\ w \in H^\alpha(\mathbb{R})\} = \{0\}$. Indeed, if the range of Q^* is not all of $H^{-\alpha}(\mathbb{R})$, then there exists an $w \ne 0$, $w \in H^\alpha(\mathbb{R})$ such that $(Q^*\varphi, w) = 0\ \ \forall\varphi \in C_0^\infty(\mathbb{R})$, so $Qw = 0$. If $Qw = 0$ and $w \in H^\alpha(\mathbb{R})$, then, as was mentioned below formula (A.8), it follows that $w = 0$. This proves that $Ran(Q^*) = H^{-\alpha}(\mathbb{R})$.

Inequality (A.11) is necessary for the left inequality (A.9) to hold. Indeed, let $\psi = Q^*\varphi$, $\varphi \in C_0^\infty(\mathbb{R})$, then (A.9) implies

$$c_5\|\varphi\|^2_{-\alpha+n} \le c\|Q^*\varphi\|^2_{-\alpha} \le (RQ^*\varphi, Q^*\varphi) = (QRQ^*\varphi, \varphi) = (PQ^*\varphi, \varphi), \tag{A.19}$$

where $c > 0$ here (and elsewhere in this paper) stands for various estimation constants. Because $-\alpha + n = \frac{n+m}{2}$, inequality (A.19) is the left inequality (A.11). The right inequality (A.11) is obvious because the order of the operator PQ^* equals to $n + m$.

Let us prove now that inequalities (A.11) and (A.10) are sufficient for the left inequality (A.9) to hold.

Using the right inequality (A.10) and the left inequality (A.11), one gets:

$$c\|\psi\|^2_{-\alpha} \le c_5\|\varphi\|^2_{\frac{n+m}{2}} \le (PQ^*\varphi, \varphi) = (R\psi, \psi), \quad \psi = Q^*\varphi, \quad \forall\varphi \in C_0^\infty(\mathbb{R}). \tag{A.20}$$

Let us prove that the set $\{\psi\} = \{Q^*\varphi\}_{\forall\varphi\in C_0^\infty(\mathbb{R})}$ is dense in $\dot{H}^{-\alpha}(0,L)$. Assume the contrary. Then there is an $h \in \dot{H}^{-\alpha}(0,L)$, $h \neq 0$, such that $(Q^*\varphi, h) = 0$ for all $\varphi \in C_0^\infty(\mathbb{R})$. Thus, $(\varphi, Qh) = 0$ for all $\varphi \in C_0^\infty(\mathbb{R})$. Therefore $Qh = 0$, and, by the argument given below formula (A.8), it follows that $h = 0$. This contradiction proves that the set $\{Q^*\varphi\}_{\forall\varphi\in C_0^\infty(\mathbb{R})}$ is dense in $\dot{H}^{-\alpha}(0,L)$.

Consequently, (A.20) implies the left inequality (A.9). The right inequality (A.9) is an immediate consequence of the observation we made earlier: (A.7) and (A.8) imply that R is a pseudodifferential operator of order $-2\alpha = -(n+m)$.

Theorem A.2 is proved. □

Proof of Theorem A.3. Equations (A.4) and (A.7) imply

$$Ph = g := QF. \tag{A.21}$$

Here

$$F := \begin{cases} u_-, & x < 0, \\ f, & 0 \le x \le L, \\ u_+, & x > L, \end{cases} \tag{A.22}$$

where

$$Qu_- = 0, \quad x < 0, \tag{A.23}$$

$$Qu_+ = 0, \quad x > L, \tag{A.24}$$

and u_- and u_+ are chosen so that $F \in H^\alpha(\mathbb{R})$. This choice is equivalent to the conditions:

$$u_-^{(j)}(0) = f^{(j)}(0), \quad 0 \le j \le \alpha - 1, \tag{A.25}$$

$$u_+^{(j)}(L) = f^{(j)}(L), \quad 0 \le j \le \alpha - 1. \tag{A.26}$$

If $F \in H^\alpha(\mathbb{R})$, then $g := QF \in H^{\alpha-n}(\mathbb{R}) = H^{-\frac{n+m}{2}}(\mathbb{R})$, and, by (A.22), one gets:

$$g = Qf + \sum_{j=0}^{n-\alpha-1} \left[a_j^- \delta^{(j)}(x) + a_j^+ \delta^{(j)}(x-L) \right], \tag{A.27}$$

where $a_j^\pm$ are some constants. There are $n - \alpha = \frac{n+m}{2}$ constants a_j^+ and the same number of constants a_j^-.

Let $G(x, y)$ be the fundamental solution of the equation

$$PG = \delta(x - y) \text{ in } \mathbb{R}, \tag{A.28}$$

which vanishes for $x < y$:

$$G(x, y) = 0 \text{ for } x < y. \tag{A.29}$$

Claim. *Such $G(x, y)$ exists and is unique. It solves the following Cauchy problem:*

$$PG = 0, \quad x > y, \quad G_x^{(j)}(x, y)\Big|_{x=y+0} = \delta_{j,m-1}, \quad 0 \le j \le m - 1, \tag{A.30}$$

satisfies condition (A.29), and can be written as

$$G(x, y) = \sum_{j=1}^{m} c_j(y)\varphi_j(x), \quad x > y, \tag{A.31}$$

where $\varphi_j(x)$, $1 \le j \le m$, is a linearly independent system of solutions to the equation:

$$P\varphi = 0. \tag{A.32}$$

Proof of the claim. The coefficients $c_j(y)$ are defined by conditions (A.30):

$$\sum_{j=1}^{m} c_j(y)\varphi_j^{(k)}(y) = \delta_{k,m-1}, \quad 0 \le k \le m - 1. \tag{A.33}$$

The determinant of linear system (A.33) is the Wronskian $W(\varphi_1, \ldots, \varphi_m) \neq 0$, so that $c_j(y)$ are uniquely determined from (A.33).

The fact that the solution to (A.30), which satisfies (A.29), equals to the solution to (A.28) – (A.29) follows from the uniqueness of the solution to (A.28) – (A.29) and (A.30) – (A.29), and from the observation that the solution to (A.28) – (A.29) solves (A.30) – (A.29). The uniqueness of the solution to (A.30) – (A.29) is a well-known result.

Let us prove uniqueness of the solution to (A.28) – (A.29). If there were two solutions, G_1 and G_2, to (A.28) – (A.29), then their difference $G := G_1 - G_2$, would solve the problem:

$$PG = 0 \text{ in } \mathbb{R}, \quad G = 0 \text{ for } x < y. \tag{A.34}$$

By the uniqueness of the solution to the Cauchy problem, it follows that $G \equiv 0$. Note that this conclusion holds in the space of distributions as well, because equation (A.34) has only the classical solutions, as follows from the ellipticity of P. Thus the claim is proved. □

From (A.21) and (A.27) – (A.29) one gets:

$$\begin{aligned} h &= \int_0^x G(x,y)Qf\,dy + \int_0^x G(x,y) \sum_{j=0}^{n-\alpha-1} \left[a_j^- \delta^{(j)}(y) + a_j^+ \delta^{(j)}(y-L)\right] dy \\ &= \int_0^x G(x,y)Qf\,dy + \sum_{j=0}^{n-\alpha-1} (-1)^j \left[G_y^{(j)}(x,y)\Big|_{y=0} a_j^- + G_y^{(j)}(x,y)\Big|_{y=L} a_j^+ \right] \\ &:= \int_0^x G(x,y)Qf\,dy + H(x). \end{aligned} \tag{A.35}$$

It follows from (A.35) that $h \in H^{-\alpha}(\mathbb{R})$ and

$$h = 0 \text{ for } x < 0, \tag{A.36}$$

that is, $(h, \varphi) = 0 \ \forall \varphi \in C_0^\infty(\mathbb{R})$ such that $\operatorname{supp}\varphi \subset (-\infty, 0)$. In order to guarantee that $h \in \dot{H}^{-\alpha}(0,L)$ one has to satisfy the condition

$$h = 0 \text{ for } x > L. \tag{A.37}$$

Conditions (A.36) and (A.37) together are equivalent to $\operatorname{supp} h \subset [0, L]$. Note that although $Qf \in \dot{H}^{-\frac{n+m}{2}}(0,L)$, so that Qf is a distribution, the integral $\int_0^x G(x,y)Qf\,dy = \int_{-\infty}^{\infty} G(x,y)Qf\,dy$ is well defined as the unique solution to the problem $Pw = Qf$, $w = 0$ for $x < 0$.

Let us prove that conditions (A.36) and (A.37) determine the constants $a_j^{\pm}$, $0 \le j \le \frac{n+m}{2} - 1$, uniquely.

If this is proved, then Theorem A.3 is proved, and formula (A.35) gives an analytical solution to equation (A.4) in $\dot{H}^{-\alpha}(0,L)$ provided that an algorithm for finding $a_j^{\pm}$ is given. Indeed, an algorithm for finding $G(x,y)$ consists of solving (A.29) – (A.30). Solving (A.29) – (A.30) is accomplished analytically by solving the linear algebraic system (A.33) and then using formula (A.31). We assume that m linearly independent solutions $\varphi_j(x)$ to (A.32) are known.

Let us derive an algorithm for calculation of the constants $a_j^{\pm}$, $0 \leq j \leq \frac{n+m}{2} - 1$, from conditions (A.36) – (A.37).

Because of (A.29), condition (A.36) is satisfied automatically by h defined in (A.35).

To satisfy (A.37) it is necessary and sufficient to have

$$\int_0^L G(x,y)Qf\,dy + H(x) \equiv 0 \text{ for } x > L. \tag{A.38}$$

By (A.31), and because the system $\{\varphi_j\}_{1\leq j\leq m}$ is linearly independent, equation (A.38) is equivalent to the following set of equations:

$$\int_0^L c_k(y)Qf\,dy + \sum_{j=0}^{\frac{n+m}{2}-1} (-1)^j \left[c_k^{(j)}(0)a_j^- + c_k^{(j)}(L)a_j^+\right] = 0, \quad 1 \leq k \leq m. \tag{A.39}$$

Let us check that there are exactly m independent constants $a_j^{\pm}$ and that all the constants $a_j^{\pm}$ are uniquely determined by linear system (A.39).

If there are m independent constants $a_j^{\pm}$ and other constants can be linearly represented through these, then linear algebraic system (A.39) is uniquely solvable for these constants provided that the corresponding homogeneous system has only the trivial solution. If $f = 0$, then $h = 0$, as follows from Theorem 1.1, and $g = 0$ in (A.27). Therefore $a_j^{\pm} = 0$ $\forall j$, and system (A.39) determines the constants $a_j^{\pm}$ $\forall j$ uniquely.

Finally, let us prove that there are exactly m independent constants $a_j^{\pm}$. Indeed, in formula (A.21) there are $\frac{n}{2}$ linearly independent solutions $u_j^- \in L^2(-\infty, 0)$, so

$$u_- = \sum_{j=1}^{\frac{n}{2}} b_j^- u_j^-, \tag{A.40}$$

and, similarly, u_+ in (A.21) is of the form

$$u_+ = \sum_{j=1}^{\frac{n}{2}} b_j u_j^+, \tag{A.41}$$

where $u_j^+ \in L^2(0,\infty)$. Condition $F \in H^\alpha(\mathbb{R})$ implies

$$\sum_{j=1}^{\frac{n}{2}} b_j^-(u_j^-)^{(k)} = f^{(k)} \text{ at } x=0, \quad 0 \le k \le \alpha - 1 = \frac{n-m}{2} - 1, \tag{A.42}$$

and

$$\sum_{j=1}^{\frac{n}{2}} b_j^+(u_j^+)^{(k)} = f^{(k)} \text{ at } x=L, \quad 0 \le k \le \frac{n-m}{2} - 1. \tag{A.43}$$

Equations (A.42) and (A.43) imply that there are $\frac{n}{2} - \frac{n-m}{2} = \frac{m}{2}$ independent constants b_j^- and $\frac{m}{2}$ independent constants b_j^+, and the remaining $n-m$ constants b_j^- and b_j^+ can be represented through these m constants by solving linear systems (A.42) and (A.43) with respect to, say, first $\frac{n-m}{2}$ constants, for example, for system (A.42), for the constants b_j^-, $1 \le j \le \frac{n-m}{2}$. This can be done uniquely because the matrices of the linear systems (A.42) and (A.43) are nonsingular: they are Wronskians of linearly independent solutions $\{u_j^-\}_{1 \le j \le \frac{n-m}{2}}$ and $\{u_j^+\}_{1 \le j \le \frac{n-m}{2}}$.

The constants $a_j^\pm$ can be expressed in terms of $b_j^\pm$ and f by linear relations. Thus, there are exactly m independent constants $a_j^\pm$. This completes the proof of Theorem A.3. □

Remark A.3 *In Chapter 5 a theory of singular perturbations for the equations of the form*

$$\varepsilon h_\varepsilon + Rh_\varepsilon = f \tag{A.44}$$

is developed for a class of integral operators with a convolution kernels $R(x,y) = R(x-y)$. This theory can be generalized to the class of kernels $R(x,y)$ studied in the present paper. The basic interesting problem is: for any $\varepsilon > 0$ equation (A.44) has a unique solution $h_\varepsilon \in L^2(0,L)$; how can one find the asymptotic behavior of h_ε as $\varepsilon \to 0$? The limit h of h_ε as $\varepsilon \to 0$ should solve equation $Rh = f$ and, in general, h is a distribution, $h \in \dot{H}^{-\alpha}(0,L)$. The theory presented in Chapter 5 allows one to solve the above problem for the class of kernels studied here.

Remark A.4 *Theorems A.1 and A.2 and their proofs remain valid in the case when equation (A.4) is replaced by the*

$$Rh := \int_D R(x,y)h(y)dy = f, \quad x \in \overline{D}. \tag{A.45}$$

Here $D \subset \mathbb{R}^r$, $r > 1$, is a bounded domain with a smooth boundary S, $\overline{D}$ is the closure of D, $R(x, y)$ solves (A.7), where P and Q are uniformly elliptic differential operators with smooth coefficients, $ordP = m \geq 0$, $ordQ = n > m$, equation $Qh = 0$ has only the trivial solution in $H^{\beta}(\mathbb{R}^r)$ for any fixed real number β. Under the above assumptions, one can prove that the operator defined by the kernel $R(x, y)$ is a pseudodifferential elliptic operator of order -2α, where $\alpha := \frac{n-m}{2}$. We do not assume that P and/or Q are selfadjoint or that P and Q commute. An analog of Remark 2.1 holds for the multidimensional equation (A.44) as well. Equation (A.45) is the basic integral equation of random fields estimation theory.

Appendix B

Integral Operators Basic in Random Fields Estimation Theory

B.1 Introduction

Integral equations theory is well developed starting from the beginning of the last century. Of special interest are the classes of integral equations which can be solved in closed form or reduced to some boundary-value problems for differential equations. There are relatively few such classes of integral equations. They include equations with convolution kernels with domain of integration which is the whole space. These equations can be solved by applying the Fourier transform. The other class of integral equations solvable in closed form is the Wiener-Hopf equations. Yet another class consists of one-dimensional equations with special kernels (singular integral equations which are reducible to Riemann-Hilbert problems for analytic functions, equations with logarithmic kernels, etc). (See e.g. [Zabreiko *et al.* (1968)], [Gakhov (1966)].) In Chapter 5 a new class of multidimensional integral equations is introduced. Equations of this class are solvable in closed form or reducible to a boundary-value problem for elliptic equations. This class consists of equations (B.3) (see below), whose kernels $R(x, y)$ are kernels of positive rational functions of an arbitrary selfadjoint elliptic operator in $L^2(\mathbb{R}^n)$, where $n \geq 1$. In Appendix A this theory is generalized to the class of kernels $R(x, y)$ which solve problem $QR = P\delta(x-y)$, where $\delta(x)$ is the delta-function, Q and P are elliptic differential operators, and $x \in \mathbb{R}^1$. Ellipticity in this case means that the coefficient in front of the senior derivative does not vanish. In Appendix A integral equations (B.3) with the kernels of the above class are solved in closed form by reducing them to a boundary-value problem for ODE. Our aim is to generalize the approach proposed in Appendix A to the multidimensional equations (B.3) whose kernel solves equation $QR = P\delta(x - y)$ in $\mathbb{R}^n$, where $n > 1$. This is not only of theoretical interest, but also of great practical interest, because, as shown in Chapter 1, equations (B.3) are basic equations of random fields estimation theory. Thus, solving such equations with larger class of kernels amounts to solving estimation problems for larger class of random fields. The kernel $R(x,y)$ is the covariance function of a random field. The class of kernels R, which solve equation $QR = P\delta(x-y)$ in $\mathbb{R}^n$, contains the class of kernels introduced and studied in Chapters 1-4.

Our theory is not only basic in random fields estimation theory, but can be considered as a contribution to the general theory of integral equations. Any new class of integral equations, which can be solved analytically or reduced to some boundary-value problems is certainly of interest, and potentially can be used in many applied areas.

For convenience of the reader, the notations and auxiliary material are put in Section B.4. This appendix follows closely the paper [Kozhevnikov and Ramm (2005)].

Let P be a differential operator in $\mathbb{R}^n$ of order μ,

$$P := P(x,D) := \sum_{|\alpha|\le\mu} a_\alpha(x) D^\alpha,$$

where $a_\alpha(x) \in C^\infty(\mathbb{R}^n)$.

The polynomials

$$p(x,\xi) := \sum_{|\alpha|\le\mu} a_\alpha(x)\xi^\alpha \quad \text{and} \quad p_0(x,\xi) := \sum_{|\alpha|=\mu} a_\alpha(x)\xi^\alpha$$

are called respectively symbol and principal symbol of P.

Suppose that the symbol $p(x,\xi)$ belongs to the class $SG^{(\mu,0)}(\mathbb{R}^n)$ consisting of all C^∞ functions $p(x,\xi)$ on $\mathbb{R}^n\times\mathbb{R}^n$, such that for any multiindices α,β there exists a constant $C_{\alpha,\beta}$ such that

$$\left|D_x^\alpha D_\xi^\beta p(x,\xi)\right| \le C_{\alpha,\beta}\langle\xi\rangle^{\mu-|\beta|}\langle x\rangle^{-|\alpha|}, \quad x,\xi\in\mathbb{R}^n, \quad \langle\xi\rangle := (1+|\xi|^2)^{1/2} \tag{B.1}$$

It is known (cf. [Wloka *et al.* (1995), Prop. 7.2]) that the map $P(x,D) : \mathcal{S}(\mathbb{R}^n) \to \mathcal{S}(\mathbb{R}^n)$ is continuous, where $\mathcal{S}(\mathbb{R}^n)$ is the Schwartz space of smooth rapidly decaying functions. Let $H^s(\mathbb{R}^n)$ $(s\in\mathbb{R})$ be the usual Sobolev space. It is known that the operator $P(x,D)$ acts naturally on the Sobolev spaces, that is, the operator $P(x,D)$ is (cf. [Wloka *et al.* (1995), Sec. 7.6]) a bounded operator: $H^s(\mathbb{R}^n)\to H^{s-\mu}(\mathbb{R}^n)$ for all $s\in\mathbb{R}$. The operator $P(x,D)$ is called elliptic, if $p_0(x,\xi)\ne 0$ for any $x\in\mathbb{R}^n$, $\xi\in\mathbb{R}^n\setminus\{0\}$.

Let $P(x, D)$ and $Q(x, D)$ be both elliptic differential operators of even orders μ and ν respectively, $0 \leq \mu < \nu$, with symbols satisfying (B.1) (for $Q(x, D)$ we replace p and μ in (B.1) respectively by q and ν). The case $\mu \geq \nu$ is a simpler case which leads to an elliptic operator perturbed by a compact integral operator in a bounded domain.

We assume also that $P(x, D)$ and $Q(x, D)$ are invertible operators, that is, there exist the inverse bounded operators $P^{-1}(x, D) : H^{s-\mu}(\mathbb{R}^n) \rightarrow H^s(\mathbb{R}^n)$ and $Q^{-1}(x, D) : H^{s-\nu}(\mathbb{R}^n) \rightarrow H^s(\mathbb{R}^n)$ for all $s \in \mathbb{R}$.

Let $R := Q^{-1}(x, D)\, P(x, D)$. The invertibility of $P(x, D)$ and $Q(x, D)$ imply that R is an invertible pseudodifferential operator of negative order $\mu - \nu$ acting from $H^s(\mathbb{R}^n)$ onto $H^{s+\nu-\mu}(\mathbb{R}^n)$ $(s \in \mathbb{R})$.

Since P and Q are elliptic, their orders μ and ν are even for $n > 2$. If $n = 2$, we assume that μ and ν are even numbers. Therefore, the number $a := (\nu - \mu)/2 > 0$ is an integer.

Let Ω denote a bounded connected open set in $\mathbb{R}^n$ with a smooth boundary $\partial\Omega$ (C^∞-class surface) and $\overline{\Omega}$ its closure in $L^2(\Omega)$, $\overline{\Omega} = \Omega \cup \partial\Omega$. The smoothness restriction on the domain can be weakened, but we do not go into detail.

The restriction R_Ω of the operator R to the domain $\Omega \subset \mathbb{R}^n$ is defined as

$$R_\Omega := r_\Omega R e_{\Omega_-}, \tag{B.2}$$

where e_{Ω_-} is the extension by zero to $\Omega_- := \mathbb{R}^n \backslash \overline{\Omega}$ and r_Ω is the restriction to Ω.

It is known (cf. [Grubb (1990), Th. 3.11, p. 312]) that the operator R_Ω defines a continuous mapping

$$R_\Omega : H^s(\Omega) \rightarrow H^{s+\nu-\mu}(\Omega) \qquad (s > -1/2),$$

where $H^s(\Omega)$ is the space of restrictions of elements of $H^s(\mathbb{R}^n)$ to Ω with the usual infimum norm (see Section B.4).

The pseudodifferential operator R of negative order $\mu - \nu$ and its restriction R_Ω can be represented as integral operators with kernel $R(x, y)$:

$$Rh = \int_{\mathbb{R}^n} R(x, y)\, h(y)\, dy, \quad R_\Omega h = \int_\Omega R(x, y)\, h(y)\, dy \;\; (x \in \Omega),$$

where $R(x, y) \in C^\infty(\mathbb{R}^n \times \mathbb{R}^n \setminus Diag)$, $Diag$ is the diagonal in $\mathbb{R}^n \times \mathbb{R}^n$, Moreover, $R(x, y)$ has a weak singularity:

$$|R(x, y)| \leq C\, |x - y|^{-\sigma} \qquad n + \mu - \nu \leq \sigma < n.$$

For $n+\mu-\nu<0$, $R(x,y)$ is continuous.

Let $\gamma := n+\mu-\nu$ and $r_{xy} := |x-y| \to 0$. Then $R(x,y) = O(r_{xy}^{-\gamma})$ if n is odd or if n is even and $\nu < n$, and $R(x,y) = O(r_{xy}^{-\gamma} \log r_{xy})$ if n is even and $\nu > n$.

In Chapter 1, the equation

$$R_\Omega h = f \in H^a(\Omega), \quad h \in H_0^{-a}(\Omega), \quad a = \frac{\nu-\mu}{2}, \tag{B.3}$$

is derived as a necessary and sufficient condition for the optimal estimate of random fields by the criterion of minimum of variance of the error of the estimate. The kernel $R(x,y)$ is a known covariance function, and $h(x,y)$ is the distributional kernel of the operator of optimal filter. The kernel $h(x,y)$ should be of minimal order of singularity, because only in this case this kernel solves the estimation problem: the variance of the error of the estimate is infinite for the solutions to equation (B.3), which do not have minimal order of singularity. In Chapters 1-4, equation (B.3) was studied under the assumption that P and Q are polynomial functions of a selfadjoint elliptic operator defined in the whole space. In Appendix A some generalizations of this theory are given. In particular, the operators P and Q are not necessarily selfadjoint and commuting.

In this appendix an extension to multidimensional integral equations of some results from Appendix A is given.

We want to prove that, under some natural assumptions, the operator R_Ω is an isomorphism of the space $H_0^{-a}(\Omega)$ onto $H^a(\Omega)$, where $a = (\nu-\mu)/2 > 0$, and $H_0^s(\Omega)$, $s \in \mathbb{R}$, denotes the subspace of $H^s(\mathbb{R}^n)$ that consists of the elements supported in $\overline{\Omega}$.

To prove the isomorphism property, we reduce the integral equation (B.3) to an equivalent elliptic exterior boundary-value problem. Since we look for a solution u belonging to the space $H^a(\Omega_-) = H^{(\nu-\mu)/2}(\Omega_-)$, and the differential operator Q is of order ν, then Qu should belong to some Sobolev space of negative order. This means that we need results on the solvability of equation (B.3) in Sobolev spaces of negative order. Such spaces as well as solvability in them of elliptic differential boundary value problems in *bounded* domains have been investigated in [Roitberg (1996)] and later in [Kozlov *et al.* (1997)]. The case of pseudodifferential boundary value problems has been studied in [Kozhevnikov (2001)]. In [Erkip and Schrohe (1992)] and in [Schrohe (1999)] the solvability of elliptic differential and pseudodifferential boundary value problems for unbounded manifolds, and in particular for exterior domains, has been established.

These solvability results have been obtained in weighted Sobolev spaces of positive order s. To obtain the isomorphism property, we need similar solvability results for exterior domain in the weighted Sobolev spaces of *negative* order. One can find in Section B.4 the definition of these spaces (cf. [Roitberg (1996)]).

B.2 Reduction of the basic integral equation to a boundary-value problem

In Theorem B.1 the differentiation along the normal to the boundary $D^j_{\mathbf{n}}$ is used. This operator is defined in Section B.4.

Theorem B.1 *Integral equation (B.3) is equivalent to the following system (B.4), (B.5), (B.6):*

$$\begin{cases} Qu = 0 & in\ \Omega_- \\ D^j_{\mathbf{n}} u = D^j_{\mathbf{n}} f & on\ \partial\Omega, \quad 0 \le j \le a-1, \end{cases} \tag{B.4}$$

$$Ph = QF, \qquad h \in H_0^{-a}(\Omega), \tag{B.5}$$

where $u \in H^a(\Omega_-)$ *is an extension of* f:

$$F \in H^a(\mathbb{R}^n), \quad F := \begin{cases} f \in H^a(\Omega) & in\ \Omega, \\ u \in H^a(\Omega_-) & in\ \Omega_-. \end{cases} \tag{B.6}$$

Proof. Let $h \in H_0^{-a}(\Omega)$ solve equation (B.3), $R_\Omega h = f \in H^a(\Omega)$. Let us define $F := Q^{-1}Ph$. Since $h \in H_0^{-a}(\Omega)$, it follows that $Ph \in H^{-a-\mu}(\mathbb{R}^n)$ and $F = Q^{-1}Ph \in H^{-a+\nu-\mu}(\mathbb{R}^n) = H^a(\mathbb{R}^n)$. We have $f = R_\Omega h = r_\Omega Q^{-1}Ph = r_\Omega F$, so F is an extension of f. Therefore, F can be represented in the form (B.6). Furthermore, since $F = Q^{-1}Ph$, then $Ph = QF$, that is, h solves (B.5). Since $h \in H_0^{-a}(\Omega)$, then $QF = Ph \in H_0^{-a-\nu}(\Omega)$. It follows, that $Qu = 0$ in Ω_-. Since $F \in H^a(\mathbb{R}^n)$, we get $D^j_{\mathbf{n}} u = D^j_{\mathbf{n}} f$ on $\partial\Omega_-$, $0 \le j \le a-1$. This means that $u \in H^a(\Omega_-)$ solves the boundary-value problem (B.4). Thus, it is proved that any solution to (B.3) solves problem (B.4), (B.5).

Conversely, let a pair $(u,h) \in H^a(\Omega_-) \times H_0^{-a}(\Omega)$ solve system (B.4), (B.5), (B.6). Since $Ph = QF$, then $Rh = Q^{-1}Ph = F$. It follows from (B.6) that $R_\Omega h = Rh|_\Omega = F|_\Omega = f$, i.e. h solves (B.3). □

Remark B.1 *If* $\mu > 0$, *the boundary value problem (B.4) is underdetermined because* Q *is an elliptic operator of order* ν *which needs* $\nu/2$ *boundary* *conditions, but we have only a ($a < \nu/2$) conditions in (B.4). Therefore, the next step is a transformation of equation (B.5) into $\mu/2$ extra boundary conditions to the boundary value problem (B.4). This will be done in Theorem B.2.*

Let us define $\kappa(\xi',\lambda) := \left(1+|\xi'|^2+\lambda^2\right)^{1/2}$. Choose a function $\rho(\tau) \in \mathcal{S}(\mathbb{R})$ with $\operatorname{supp}\mathcal{F}^{-1}\rho \subset \overline{\mathbb{R}}_-$ and $\rho(0) = 1$. Let $a > 2\sup|\partial_\tau \rho(\tau)|$. Let $\Xi^t_{+,\lambda}$ denote a family ($\lambda \in \mathbb{R}_+$, $t \in \mathbb{Z}$) of order-reducing pseudodifferential operators $\Xi^t_{+,\lambda} := \mathcal{F}^{-1}\chi_+(\xi,\lambda)\mathcal{F}$, where $\chi_+(\xi,\lambda) := \left(\kappa(\xi',\lambda)\rho\left(\frac{\xi_n}{a\kappa(\xi',\lambda)}\right)+i\xi_n\right)^t$ are their symbols. It has been proved in [Grubb (1996), Sec. 2.5] that the operator $\Xi^t_{+,\lambda}$ maps the space $\mathcal{S}_0\left(\overline{\mathbb{R}^n_+}\right) := \left\{u \in \mathcal{S}(\mathbb{R}^n) : \operatorname{supp} u \subset \overline{\mathbb{R}^n_+}\right\}$ onto itself and has the following isomorphism properties for $s \in \mathbb{R}$:

$$\Xi^t_{+,\lambda} : H^s(\mathbb{R}^n) \simeq H^{s-t}(\mathbb{R}^n), \tag{B.7}$$

$$\Xi^t_{+,\lambda} : H^s_0(\mathbb{R}^n_+) \simeq H^{s-t}_0(\mathbb{R}^n_+). \tag{B.8}$$

It is known ([Grubb (1996)], [Schrohe (1999)]) that using $\Xi^t_{+,\lambda}$ and an appropriate partition of unity one can obtain, for sufficiently large λ, the operator Λ^t_+ which is an isomorphism:

$$\Lambda^t_+ : H^s(\mathbb{R}^n) \simeq H^{s-t}(\mathbb{R}^n), \quad \forall s \in \mathbb{R},$$

and

$$\Lambda^t_+ : H^s_0(\Omega) \simeq H^{s-t}_0(\Omega), \quad \forall s \in \mathbb{R}. \tag{B.9}$$

Lemma B.1 *Let $P(x,D)$ be an invertible differential operator of order μ, that is, there exists the inverse operator $P^{-1}(x,D)$ which is bounded: $P^{-1}(x,D) : H^{s-\mu}(\mathbb{R}^n) \to H^s(\mathbb{R}^n)$ for all $s \in \mathbb{R}$. Then a solution h to the equation*

$$P(x,D)h = g, \quad g \in H_0^{-a-\mu}(\Omega)$$

belongs to the space $H_0^{-a}(\Omega)$ if and only if g satisfies the following $\mu/2$ boundary conditions:

$$r_{\partial\Omega} D^j_{\mathbf{n}} \Lambda_+^{-a-\mu/2} P^{-1}(x,D)\, g = 0 \quad (j = 0, ..., \mu/2 - 1).$$

Proof. **Necessity**. Let $h = P^{-1}(x,D)\,g,\ h \in H_0^{-a}(\Omega)$, solve the equation $P(x,D)\,h = g,\quad g \in H_0^{-a-\mu}(\Omega)$. By (B.9), we have $\Lambda_+^{-a-\mu/2}h \in H_0^{\mu/2}(\Omega)$. Therefore, $r_{\partial\Omega}D_{\mathbf{n}}^j\Lambda_+^{-a-\mu/2}h = 0 \quad (j = 0,...,\mu/2-1)$.

Sufficiency. Assume that the equalities $r_{\partial\Omega}D_{\mathbf{n}}^j\Lambda_+^{-a-\mu/2}h = 0 \quad (j = 0,...,\mu/2-1)$ hold. Since $g \in H_0^{-a-\mu}(\Omega) \subset H^{-a-\mu}(\mathbb{R}^n)$, we have $h = P^{-1}(x,D)\,g \in H^{-a}(\mathbb{R}^n)$. Therefore, $\Psi := \Lambda_+^{-a-\mu/2}h \in H^{\mu/2}(\mathbb{R}^n)$. Since $r_{\partial\Omega}D_{\mathbf{n}}^j\Psi = 0 \quad (j = 0,...,\mu/2-1)$, we have $\Psi = \Psi_+ + \Psi_-$, where $\Psi_+ := e_{\Omega_-}r_{\overline{\Omega}}\Psi \in H_0^{\mu/2}(\Omega)$ and $\Psi_- := e_{\Omega}r_{\Omega_-}\Psi \in H_0^{\mu/2}(\Omega_-)$. Since $\Lambda_+^{\nu/2} : H_0^{\mu/2}(\Omega) \simeq H_0^{-a}(\Omega)$, it follows that $\Lambda_+^{\nu/2}\Psi_+ \in H_0^{-a}(\Omega)$. Moreover, $\Lambda_+^{\nu/2}$ is a differential operator with respect to the variable x_n, hence $\mathrm{supp}\Psi_- \subset \overline{\Omega}_-$ implies $\mathrm{supp}\Lambda_+^{\nu/2}\Psi_- \subset \overline{\Omega}_-$. Since P is a differential operator,

$$\mathrm{supp}\left(P\Lambda_+^{\nu/2}\right)\Psi_- \subset \mathrm{supp}\Lambda_+^{\nu/2}\Psi_- \subset \overline{\Omega}_-.$$

On the other hand, we have

$$\Phi := \left(P\Lambda_+^{\nu/2}\right)\Psi = \left(P\Lambda_+^{\nu/2}\right)(\Psi_+ + \Psi_-) = \left(P\Lambda_+^{\nu/2}\right)\Psi_+ + \left(P\Lambda_+^{\nu/2}\right)\Psi_-.$$

For any $\varphi \in C_0^\infty(\Omega_-)$ one has:

$$0 = \langle\Phi,\varphi\rangle = \left\langle\left(P\Lambda_+^{\nu/2}\right)\Psi_+,\varphi\right\rangle + \left\langle\left(P\Lambda_+^{\nu/2}\right)\Psi_-,\varphi\right\rangle = \left\langle\left(P\Lambda_+^{\nu/2}\right)\Psi_-,\varphi\right\rangle$$

Thus, $\mathrm{supp}\left(P\Lambda_+^{\nu/2}\right)\Psi_- \subset \overline{\Omega}$. It follows that $\mathrm{supp}\left(P\Lambda_+^{\nu/2}\right)\Psi_- \subset \partial\Omega$. For any $\Psi_- \in C_0^\infty(\Omega_-)$, we have $\left(P\Lambda_+^{\nu/2}\right)\Psi_- \in C^\infty(\mathbb{R}^n)$ and $\mathrm{supp}\left(P\Lambda_+^{\nu/2}\right)\Psi_- \subset \partial\Omega$. Therefore, $\left(P\Lambda_+^{\nu/2}\right)\Psi_- = 0$. Since P is invertible, $\Lambda_+^{\nu/2}\Psi_- = 0$ for $\Psi_- \in C_0^\infty(\Omega_-)$. Since $C_0^\infty(\Omega_-)$ is dense in $H_0^{\mu/2}(\Omega_-)$, one gets $\Lambda_+^{\nu/2}\Psi_- = 0 \quad$ for $\Psi_- \in H_0^{\mu/2}(\Omega_-)$. It follows that

$$h = \Lambda_+^{\nu/2}\Psi = \Lambda_+^{\nu/2}\Psi_+ + \Lambda_+^{\nu/2}\Psi_- = \Lambda_+^{\nu/2}\Psi_+ \in H_0^{-a}(\Omega).$$

Lemma B.1 is proved. □

Let $F \in C^\infty(\Omega) \cap \mathcal{S}\left(\overline{\Omega}_-\right)$. Assume that F has finite jumps F_k of the normal derivative of order k $(k = 0,1,...)$ on $\partial\Omega$. For $x' \in \partial\Omega$, we will use the following notation:

$$F_0(x') := [F]_{\partial\Omega}(x') := \lim_{\varepsilon\to+0}\left(F(x'+\varepsilon\mathbf{n}) - F(x'-\varepsilon\mathbf{n})\right),$$

$$F_k(x') := \left[D_{\mathbf{n}}^k F\right]_{\partial\Omega}(x').$$

Let $f \in C^\infty\left(\overline{\Omega}\right)$ and $u \in \mathcal{S}\left(\overline{\Omega}_-\right)$, and define $\gamma_k f(x') := r_{\partial\Omega} D_{\mathbf{n}}^k f(x')$, $\gamma_k u(x') := r_{\partial\Omega} D_{\mathbf{n}}^k u(x')$.

Let $\delta_{\partial\Omega}$ denote the Dirac measure supported on $\partial\Omega$, that is, a distribution acting as

$$(\delta_{\partial\Omega}, \varphi) := \int_{\partial\Omega} \overline{\varphi}(x)\, dS, \quad \varphi(x) \in C_0^\infty(\mathbb{R}^n).$$

It is known that for any differential operator Q of order ν there exists a representation $Q = \sum_{j=0}^{\nu} Q_j D_{\mathbf{n}}^j$, where Q_j is a tangential differential operator of order $\nu - j$ (cf Section B.4). We denote by $\{D^\alpha F(x)\}$ the classical derivative at the points where it exists.

The following Lemma B.2 is essentially known but for convenience of the reader a short proof of this lemma is given.

Lemma B.2 *The following equality holds for the distribution QF:*

$$QF = \{QF\} - i \sum_{j=0}^{\nu} Q_j \sum_{k=0}^{j-1} D_{\mathbf{n}}^k \left(F_{j-1-k}\delta_{\partial\Omega}\right). \tag{B.10}$$

Proof. Let $\cos(\mathbf{n}x_j)$ denote cosine of the angle between the exterior unit normal vector $\mathbf{n}$ to the boundary $\partial\Omega$ of Ω and the x_j-axis.

We use the known formulas

$$\int_\Omega \frac{\partial u}{\partial x_j} dx = \int_{\partial\Omega} u(x) \cos(\mathbf{n}x_j)\, d\sigma, \quad u(x) \in C^\infty\left(\overline{\Omega}\right), \quad j = 1, ..., n,$$

$$\int_{\Omega_-} \frac{\partial v}{\partial x_j} dx = -\int_{\partial\Omega} v(x) \cos(\mathbf{n}x_j)\, d\sigma, \quad v(x) \in C_0^\infty\left(\overline{\Omega}_-\right) \quad j = 1, ..., n.$$

where $d\sigma$ is the surface measure on $\partial\Omega$. Applying these formulas to the products $u(x)\varphi(x)$ and $v(x)\varphi(x)$ where $\varphi(x) \in C_0^\infty(\mathbb{R}^n)$, $u(x) \in C^\infty\left(\overline{\Omega}\right)$, $v(x) \in C_0^\infty\left(\overline{\Omega}_-\right)$, we get

$$\int_\Omega \frac{\partial u}{\partial x_j} \varphi(x)\, dx = -\int_\Omega u(x) \frac{\partial \varphi}{\partial x_j} dx + \int_{\partial\Omega} u(x)\varphi(x)\cos(\mathbf{n}x_j)\, d\sigma$$
$$j = 1, ..., n, \tag{B.11}$$

$$\int_{\Omega_-} \frac{\partial v}{\partial x_j} \varphi(x)\, dx = -\int_{\Omega_-} v(x) \frac{\partial \varphi}{\partial x_j} dx - \int_{\partial\Omega} v(x)\, \varphi(x) \cos(\mathbf{n}x_j)\, d\sigma$$

$$j = 1, ..., n. \quad \text{(B.12)}$$

By (B.11), (B.12), we have

$$\left(\frac{\partial F}{\partial x_j}, \varphi\right) = -\left(F, \frac{\partial \varphi}{\partial x_j}\right) = -\int_{\mathbb{R}^n} F(x) \frac{\partial \overline{\varphi}(x)}{\partial x_j} dx$$

$$= \int_{\mathbb{R}^n} \left\{\frac{\partial F(x)}{\partial x_j}\right\} \overline{\varphi}(x)\, dx + \int_{\partial\Omega} [F]_{\partial\Omega}(x) \cos(\mathbf{n}x_j)\, \overline{\varphi}(x)\, dS$$

$$= \left(\left\{\frac{\partial F}{\partial x_j}\right\} + [F]_{\partial\Omega} \cos(\mathbf{n}x_j)\, \delta_{\partial\Omega},\ \varphi(x)\right), \quad \varphi(x) \in C_0^\infty(\mathbb{R}^n).$$

This means,

$$\frac{\partial F}{\partial x_j} = \left\{\frac{\partial F}{\partial x_j}\right\} + [F]_{\partial\Omega} \cos(\mathbf{n}x_j)\, \delta_{\partial\Omega}, \quad j = 1, ..., n.$$

It follows, $D_{\mathbf{n}}F = \{D_{\mathbf{n}}F\} - iF_0\delta_{\partial\Omega}$. Furthermore, using the last formula we have $D_{\mathbf{n}}^2 F = D_{\mathbf{n}}\{D_{\mathbf{n}}F\} - iD_{\mathbf{n}}(F_0\delta_{\partial\Omega}) = \{D_{\mathbf{n}}^2 F\} - iF_1\delta_{\partial\Omega} - iD_{\mathbf{n}}(F_0\delta_{\partial\Omega})$ and so on. By induction one gets:

$$D_{\mathbf{n}}^j F = \{D_{\mathbf{n}}^j F\} - i\sum_{k=0}^{j-1} D_{\mathbf{n}}^k (F_{j-1-k}\delta_{\partial\Omega}) \quad (j = 1, 2, ...).$$

Substituting this formula for $D_{\mathbf{n}}^j F$ into the representation $Q = \sum_{j=0}^{\nu} Q_j D_{\mathbf{n}}^j$, we get (B.10). Lemma B.2 is proved. □

Denoting in the sequel the extensions by zero to $\mathbb{R}^n$ of functions $f(x) \in C^\infty(\overline{\Omega})$, $u(x) \in \mathcal{S}(\overline{\Omega}_-)$, as f^0 and u^0, and using Lemma B.2, we obtain the following formulas:

$$(Qf)^0 = Q(f^0) - i\sum_{j=1}^{\nu} Q_j \sum_{k=0}^{j-1} D_{\mathbf{n}}^k \left(\left(D_{\mathbf{n}}^{j-1-k} f\right)\big|_{\partial\Omega} \delta_{\partial\Omega}\right) \quad \left(f \in C^\infty(\overline{\Omega})\right),$$

$$\text{(B.13)}$$

$$(Qu)^0 = Q\left(u^0\right) + i\sum_{j=1}^{\nu} Q_j \sum_{k=0}^{j-1} D_{\mathbf{n}}^k \left(\left(D_{\mathbf{n}}^{j-1-k}u\right)\big|_{\partial\Omega}\,\delta_{\partial\Omega}\right) \quad \left(u \in \mathcal{S}\left(\overline{\Omega}_-\right)\right), \tag{B.14}$$

where $\left(D_{\mathbf{n}}^j f\right)\big|_{\partial\Omega} := r_{\partial\Omega} D_{\mathbf{n}}^j f$. Using these formulas one can define the action of the operator Q upon the elements of the spaces $\mathfrak{H}^{s,\nu}(\Omega)$ and $\mathfrak{H}^{s,\nu}(\Omega_-)$ $(s \in \mathbb{R})$ (defined in Section B.4) as follows (cf. [Kozlov *et al.* (1997), Sec. 3.2.], [Roitberg (1996), Sec. 2.4]):

$$\left(Q\left(f,\underline{\psi}\right)\right)^0 := Q\left(f^0\right) - i\sum_{j=1}^{\nu} Q_j \sum_{k=0}^{j-1} D_{\mathbf{n}}^k \left(\psi_{j-k}\delta_{\partial\Omega}\right) \quad \left(\left(f,\underline{\psi}\right) \in \mathfrak{H}^{s,\nu}(\Omega)\right), \tag{B.15}$$

$$\left(Q\left(u,\underline{\phi}\right)\right)^0 := Q\left(u^0\right) + i\sum_{j=1}^{\nu} Q_j \sum_{k=0}^{j-1} D_{\mathbf{n}}^k \left(\phi_{j-k}\delta_{\partial\Omega}\right), \quad \left(\left(u,\underline{\phi}\right) \in \mathfrak{H}^{s,\nu}(\Omega_-)\right). \tag{B.16}$$

It is known ([Roitberg (1996)], [Kozlov *et al.* (1997)], [Kozhevnikov (2001)]) that Q, defined respectively in (B.15) and (B.16), is a bounded mapping

$$Q : \mathfrak{H}^{s,\nu}(\Omega) \to \mathcal{H}^{s-\nu}(\Omega) \text{ and } Q : \mathfrak{H}^{s,\nu}(\Omega_-) \to \mathcal{H}^{s-\nu}(\Omega_-).$$

Moreover, Q is respectively the closure of the mapping $f \to Q(x,D)f$ $\left(f \in C^\infty\left(\overline{\Omega}\right)\right)$ or $u \to Q(x,D)u \quad \left(u \in \mathcal{S}\left(\overline{\Omega}_-\right)\right)$ between the corresponding spaces.

Let $W_{m\ell} \quad (m = 1, ..., \mu/2, \quad \ell = a+1, ..., \nu)$ be the operator acting as follows:

$$W_{m\ell}(\phi) := i\gamma_{m-1}\Lambda_+^{-a-\mu/2} P^{-1} \sum_{\ell=a+1}^{\nu} \sum_{j=\ell}^{\nu} Q_j D_{\mathbf{n}}^{j-\ell}\left(\phi\delta_{\partial\Omega}\right), \quad \phi \in C^\infty(\partial\Omega), \tag{B.17}$$

where γ_k is the restriction to $\partial\Omega$ of the $D_{\mathbf{n}}^k$ (cf. Section B.4).

The mapping $W_{m\ell}$ is a pseudodifferential operator of order $m - \mu + \nu/2 - 1 - \ell$. Therefore, for any real s, this mapping is a bounded operator:

$$W_{m\ell} : H^s(\partial\Omega) \to H^{s-m+\mu-\nu/2+1+\ell}(\partial\Omega).$$

For $\left(f,\underline{\psi}\right) \in \mathfrak{H}^{a,\nu}(\Omega)$, one has $g := Q\left(f,\underline{\psi}\right) \in H_0^{a-\nu}(\Omega)$, and we set

$$w_{a+m} := -\gamma_{m-1}\Lambda_+^{-a-\mu/2} P^{-1} g^0 \quad (m = 1, ..., \mu/2), \tag{B.18}$$

where the operator $\gamma_{m-1}\Lambda_+^{-a-\mu/2}P^{-1}(x,D)$ is a trace operator of order $m-1-a-3\mu/2$. It follows that $w_{a+m} \in H^{\mu/2-m+1/2}(\partial\Omega)$.

Theorem B.2 *Integral equation (B.3)*

$$R_\Omega h = f \in H^a(\Omega), \quad h \in H_0^{-a}(\Omega)$$

is equivalent to the following boundary-value problem:

$$\begin{cases} Qu = 0 & \text{in } \Omega_-, \\ D_{\mathbf{n}}^j u = D_{\mathbf{n}}^j f & \text{on } \partial\Omega, \quad 0 \le j \le a-1, \\ \sum\limits_{\ell=a+1}^{\nu} W_{m\ell}(\gamma_{\ell-1}u) = w_{a+m} & \text{on } \partial\Omega, \quad 1 \le m \le \mu/2, \end{cases} \tag{B.19}$$

where the functions u, f and h are related by the formulas

$$h = P^{-1}QF, \quad F \in H^a(\mathbb{R}^n), \quad F := \begin{cases} f \in H^a(\Omega) & \text{in } \Omega, \\ u \in H^a(\Omega_-) & \text{in } \Omega_-. \end{cases}$$

Proof. Our starting point is Theorem B.1. Consider the equation $Ph = QF$, $h \in H_0^{-a}(\Omega)$. Since $F \in H^a(\mathbb{R}^n)$ and $Qu = 0$ in Ω_- by (B.4), then $QF \in H_0^{a-\nu}(\Omega) = H_0^{-a-\mu}(\Omega)$. By Lemma B.1, a solution h to the equation $Ph = QF \in H_0^{-a-\mu}(\Omega)$ belongs to the space $H_0^{-a}(\Omega)$ if and only if QF satisfies the following $\mu/2$ boundary conditions:

$$r_{\partial\Omega} D_{\mathbf{n}}^{m-1}\Lambda_+^{-a-\mu/2}P^{-1}QF = 0, \quad m = 1, ..., \mu/2. \tag{B.20}$$

Since $F = f^0 + u^0$, one has $QF = Q(f^0) + Q(u^0)$. Substituting the last expression into (B.20) we have

$$\gamma_{m-1}\Lambda_+^{-a-\mu/2}P^{-1}Q(u^0) = -\gamma_{m-1}\Lambda_+^{-a-\mu/2}P^{-1}Q(f^0) \quad m = 1, ..., \mu/2.$$

From (B.15) and (B.16), one gets:

$$\begin{aligned} & i\gamma_{m-1}\Lambda_+^{-a-\mu/2}P^{-1}\sum_{j=1}^{\nu} Q_j \sum_{k=0}^{j-1} D_{\mathbf{n}}^k(\phi_{j-k}\delta_{\partial\Omega}) \\ & = \gamma_{m-1}\Lambda_+^{-a-\mu/2}P^{-1}\left(Q\left(f, \underline{\psi}\right)\right)^0 \\ & \quad + i\gamma_{m-1}\Lambda_+^{-a-\mu/2}P^{-1}\sum_{j=1}^{\nu} Q_j \sum_{k=0}^{j-1} D_{\mathbf{n}}^k(\psi_{j-k}\delta_{\partial\Omega}) \end{aligned} \tag{B.21}$$

Since

$$F := \begin{cases} f \in H^a(\Omega) & \text{in } \Omega, \\ u \in H^a(\Omega_-) & \text{in } \Omega_- \end{cases} \quad \text{and } F \in H^a(\mathbb{R}^n),$$

it follows that $\gamma_{j-1}u = \gamma_{j-1}f, \quad j = 1, ..., a$. Therefore, $\phi_j = \gamma_{j-1}u = \gamma_{j-1}f = \psi_j, \quad j = 1, ..., a$.

We identify the space $H^a(\Omega)$ with the subspace of $\mathfrak{H}^{a,(\nu)}(\Omega)$ of all $(f, \underline{\psi}) = (f, \psi_1, ..., \psi_\nu)$ such that $\psi_{a+1} = ... = \psi_\nu = 0$. Let $(f, \underline{\psi})$ belong to this subspace and $(u, \underline{\phi}) = (u, \phi_1, ..., \phi_\nu) \in \mathfrak{H}^{a,(\nu)}(\Omega_-)$. Then we can rewrite (B.21) as

$$\gamma_{m-1}\Lambda_+^{-a-\mu/2}P^{-1}\left(Q\left(f, \underline{\psi}\right)\right)^0 + i\gamma_{m-1}\Lambda_+^{-a-\mu/2}P^{-1}\sum_{j=1}^{\nu} Q_j \sum_{\ell=a+1}^{j} D_{\mathbf{n}}^{j-\ell}\left(\phi_\ell \delta_{\partial\Omega}\right) = 0. \quad \text{(B.22)}$$

Changing the order of the summation

$$\sum_{j=1}^{\nu}\sum_{\ell=a+1}^{j} = \sum_{\ell=a+1}^{\nu}\sum_{j=\ell}^{\nu},$$

we get

$$i\gamma_{m-1}\Lambda_+^{-a-\mu/2}P^{-1}\sum_{\ell=a+1}^{\nu}\sum_{j=\ell}^{\nu} Q_j D_{\mathbf{n}}^{j-\ell}\left(\phi_\ell\delta_{\partial\Omega}\right) = -\gamma_{m-1}\Lambda_+^{-a-\mu/2}P^{-1}\left(Q\left(f, \underline{\psi}\right)\right)^0, \quad \text{(B.23)}$$

where $m = 1, ..., \mu/2$. In view of (B.17) and (B.18), formula (B.23) can be rewritten as $\mu/2$ equations

$$\sum_{\ell=a+1}^{\nu} W_{m\ell}\left(\phi_\ell\right) = w_{a+m} \quad \text{on } \partial\Omega, \quad m = 1, ..., \mu/2.$$

Since $\phi_j = \gamma_{j-1}u$ for $u \in S\left(\overline{\Omega}_-\right)$, we get

$$\sum_{\ell=a+1}^{\nu} W_{m\ell}\left(\gamma_{\ell-1}u\right) = w_{a+m} \quad \text{on } \partial\Omega, \quad m = 1, ..., \mu/2.$$

These equations define $\mu/2$ extra boundary conditions for the boundary-value problem

$$\begin{cases} Qu = 0 & \text{in } \Omega_-, \\ D_{\mathbf{n}}^j u = D_{\mathbf{n}}^j f & \text{on } \partial\Omega, \quad 0 \le j \le a-1. \end{cases}$$

Theorem B.2 is proved. □

B.3 Isomorphism property

We look for a solution $u \in H^a(\Omega_-)$ to the boundary-value problem (B.19). Let us consider the following non-homogeneous boundary-value problem associated with (B.19):

$$\begin{cases} Qu = w & \text{in } \Omega_- \\ \gamma_0 B_j u := \gamma_0 D_{\mathbf{n}}^{j-1} u = w_j & \text{on } \partial\Omega, \quad 1 \leq j \leq a \\ \gamma_0 B_{a+m} u := \sum\limits_{\ell=a+1}^{\nu} W_{m\ell}(u_{\ell-1}) = w_{a+m} & \text{on } \partial\Omega, \quad 1 \leq m \leq \mu/2, \end{cases} \tag{B.24}$$

where w, w_j $j = 1, ..., \nu/2$, are arbitrary elements of the corresponding Sobolev spaces (see below Theorems B.3 and B.4).

For the formulation of the Shapiro-Lopatinskii condition we need some notation.

Let $\varepsilon > 0$ be a sufficiently small number. Denote by U (ε-conic neighborhood) the union of all balls $B(x, \varepsilon\langle x\rangle)$, centered at $x \in \partial\Omega$ with radius $\varepsilon\langle x\rangle$. Let $y = (y', y_n) = (y_1, ..., y_{n-1}, y_n)$ be normal coordinates in an ε-conic neighborhood U of $\partial\Omega$, that is, $\partial\Omega$ may be identified with $\{y_n = 0\}$, y_n is the normal coordinate, and the normal derivative $D_{\mathbf{n}}$ is D_{y_n} near $\partial\Omega$. Each differential operator on $\mathbb{R}^n$ with SG-symbol can be written in U as a differential operator with respect to $D_{y'}$ and D_{y_n}:

$$Q = \sum_{j=0}^{\nu} Q_j(y, D_{y'}) D_{y_n}^j,$$

where $Q_j(y, D_{y'})$ are differential operators with symbols belonging to $SG^{(\nu,0)}(\mathbb{R}^n)$. Let

$$q(y,\xi) = q(y, \xi', \xi_n) = \sum_{j=0}^{\nu} q_j(y, \xi') \xi_n^j$$

be the symbol of Q, where ξ' and ξ_n are cotangent variables associated with y' and y_n.

Assumption 1. *We assume that the operator Q is md-properly elliptic* (cf. [Erkip and Schrohe (1992), Assumption 1, p. 40]), *that is, for all large $|y| + |\xi'|$ the polynomial $q(y, \xi', z)$ with respect to the complex variable z has exactly $\nu/2$ zeros with positive imaginary parts $\tau_1(y', \xi'), ..., \tau_{\nu/2}(y', \xi')$.*

We conclude from Assumption 1 that the polynomial $q(y,\xi',z)$ has no real zeros and has exactly $\nu/2$ zeros with negative imaginary part for all large $|y|+|\xi'|$.

In particular, the Laplacian Δ in the space $\mathbb{R}^n$ $(n \geq 2)$ is elliptic in the usual sense but not *md*- properly elliptic, while the operator $I-\Delta$ is *md*-properly elliptic.

Let

$$\chi(y',\xi') := \left(1+\sum_{i,j=1}^{n-1} \xi_i \left(g(y)^{-1}\right)_{ij} \xi_j\right)^{1/2},$$

where $g=(g_{ij})$ is a Riemannian metric on $\partial\Omega$. We denote

$$q^+(y',\xi',z) := \prod_{j=1}^{\nu/2} \left(z-\chi(y',\xi')^{-1}\tau_j(y',\xi')\right).$$

Consider the operators B_m $(m=1,...,\nu/2)$ from (B.24). Each of them is of the form

$$B_m = \sum_{j=0}^{\nu-1} B_{mj}(y',D_{y'}) D_{y_n}^j$$

in the normal coordinates $y=(y',y_n)=(y_1,...,y_{n-1},y_n)$ in an ε-conic neighborhood of $\partial\Omega$. Here $B_{mj}(y',D_{y'})$ is a pseudodifferential operator of order $\rho_m - j$ $(\rho_m \in \mathbb{N})$ acting on $\partial\Omega$. Let $b_{mj}(y',\xi')$ denote the principal symbol of $B_{mj}(y',D_{y'})$. The operators B_m in the boundary-value problem (B.24) are operators of this type. We set

$$b_m(y',\xi',z) := \sum_{j=0}^{\nu-1} b_{mj}(y',\xi')\chi(y',\xi')^{-\rho_m+j} z^j.$$

Define the following polynomials with respect to z:

$$r_m(y',\xi',z) = \sum_{j=1}^{\nu/2} r_{mj}(y',\xi') z^{j-1}$$

as the residues of $b_m(y',\xi',z)$ modulo $q^+(y',\xi',z)$, i.e. we get $r_{mj}(y',\xi')$ representing $b_m(y',\xi',z)$ in the form

$$b_m(y',\xi',z) = q_m(z)q^+(y',\xi') + \sum_{j=1}^{\nu/2} r_{mj}(y',\xi') z^{j-1},$$

where $q_m(z)$ is a polynomial in z.
Assumption 2. *(Shapiro-Lopatinskii condition) The determinant* $\det(r_{mj}(y', \xi'))$ *is bounded and bounded away from zero, that is, there exist two positive constants* c *and* C *such that*

$$0 < c \leq \det(r_{mj}(y', \xi')) \leq C.$$

Remark B.2 *The following Theorem B.3 has been proved in [Erkip and Schrohe (1992), Th. 3.1] in the more general case of the SG-manifold. The latter includes the exterior of bounded domains which is a particular case of the SG-manifolds. This particular case was chosen for simplicity of the exposition. Moreover, the results in [Erkip and Schrohe (1992), Th. 3.1], [Schrohe (1999)] have been obtained for operators acting in weighted Sobolev spaces. The usual Sobolev spaces in Theorem B.3 are particular cases of the weighted Sobolev spaces with zero order of the weight.*

Theorem B.3 *(cf [Erkip and Schrohe (1992), Th. 3.1], [Schrohe (1999)]). If the differential operator* Q *of even order* ν *satisfies Assumptions 1 and 2, that is* Q *is md-properly elliptic and the Shapiro-Lopatinskii condition holds for the operator* $(Q, \gamma_0 B_1, ..., \gamma_0 B_{\nu/2})$*, then the mapping*

$$(Q, \gamma_0 B_1, ..., \gamma_0 B_{\nu/2}) : H^s(\Omega_-) \to H^{s-\nu}(\Omega_-) \times \prod_{j=1}^{\nu/2} H^{s-\rho_j-1/2}(\partial\Omega), \ s \geq \nu,$$

is a Fredholm operator.

Assumption 3. *The Fredholm operator* $(Q, \gamma_0 B_1, ..., \gamma_0 B_{\nu/2})$ *has the trivial kernel and cokernel.*

For example, if the kernel $R(x, y)$ has the property $(Rh, h) \geq c||h||^2_{H_0^{-a}}$ for all $h \in H_0^{-a}$, where $c = const > 0$ does not depend on h, then the operator in *Assumption 3* is invertible (see Chapter 1).

Corollary B.1 *Under the assumptions of Theorem B.3 and in addition under Assumption 3, for any* $s \in \mathbb{R}$, *there exists a bounded (Poisson) operator*

$$K : \Pi_{j=0}^{m-1} H^{s+2m-j-1/2}(\partial\Omega) \to H^{s+2m}(\Omega_-) \qquad \text{(B.25)}$$

which gives a unique solution $u = K\underline{\chi}$ *to the boundary-value problem*

$$Qu = 0 \ in \ \Omega_-, \quad \gamma_0 B_1 u = \chi_1, ..., \gamma_0 B_{\nu/2} u = \chi_{\nu/2} \qquad \text{(B.26)}$$

with

$$\underline{\chi} = (\chi_1, ..., \chi_{\nu/2}) \in \Pi_{j=0}^{\nu/2-1} H^{s+\nu-j-1/2}(\partial\Omega).$$

More precisely, the operator $u = K\underline{\chi}$ *solves the problem with* $s < 0$ *in the sense that* $u = K\underline{\chi}$ *is the limit in the space* $H^{s+\nu}(\Omega_-)$ *of a sequence* u_n *in* $H^{\nu}(\Omega_-)$ *with*

$$Qu_n = 0, \quad \gamma_0 B_j u_n = \chi_{j,n} \; (j = 1, ..., m), \quad \lim_{n\to\infty} \underline{\chi}_n \to \underline{\chi}$$
$$\text{in } \Pi_{j=0}^{\nu/2-1} H^{s+\nu-j-1/2}(\partial\Omega).$$

Proof. The statement of Corollary is an immediate consequence of Theorem B.3 due to the fact that the solution operator to the boundary-value problem (B.26) with homogeneous equation $Qu = 0$ in Ω_- is a Poisson operator. The latter acts in the full scale of Sobolev spaces [Schrohe (1999)], that is, (B.25) holds for all $s \in \mathbb{R}$. □

Theorem B.4 *Under the assumptions of Theorem B.3 and in addition under Assumption 3, the mapping* R_Ω*, defined in the Introduction, is an isomorphism* : $H_0^{-a}(\Omega) \to H^a(\Omega)$.

Proof. Let us consider the operator $(Q, \gamma_0 B_1, ..., \gamma_0 B_{\nu/2})$ generated by the boundary value problem (B.24). Taking into account that

$$\rho_j = \text{order } B_j = j - 1 \qquad \text{for } j = 1, ..., a,$$
$$\rho_j = \text{order } B_j = j - \mu + \nu/2 - 2 \text{ for } j = a + 1, ..., \nu/2,$$

one concludes by Theorem B.3 that the mapping

$$(u, \underline{\phi}) \mapsto (Q(u, \underline{\phi}), \gamma_0 B_1(u, \underline{\phi}), ..., \gamma_0 B_{\nu/2}(u, \underline{\phi})) = (w, w_1, ..., w_{\nu/2})$$

is a Fredholm operator. It maps the space $H^s(\Omega_-)$ to the space

$$H^{s-\nu}(\Omega_-) \times \prod_{j=1}^{a} H^{s-j+1/2}(\partial\Omega) \times \prod_{j=a+1}^{\nu/2} H^{s-j+\mu-\nu/2+3/2}(\partial\Omega) \quad (s \geq \nu).$$

Assumption 3 implies that this mapping is an isomorphism. By Corollary, the operator K, solving the boundary-value problem

$$Qu = 0, \quad \gamma_0 B_j u = \chi_j \; (j = 1, ..., m),$$

is a Poisson operator

$$K : \Pi_{j=0}^{m-1} H^{s+2m-j-1/2}(\partial\Omega) \to H^{s+2m}(\Omega_-) \quad (s \in \mathbb{R}).$$

Choosing $s = a$ and using Theorem B.2, we conclude, that for any $f \in H^a(\Omega)$ the function u is a unique solution to the boundary-value problem (B.19). Therefore, again by Theorem B.2, the operator R_Ω is an isomorphism of the space $H_0^{-a}(\Omega)$ onto $H^a(\Omega)$. Theorem B.4 is proved.□

Example B.1 Let $P = I$ be the identity operator (its order $\mu = 0$) and $Q = I - \Delta$, $(\nu = \operatorname{ord} Q = 2)$. Then, by Theorem B.4, the corresponding operator R_Ω is an isomorphism: $H_0^{-1}(\Omega) \to H^1(\Omega)$.

Under the assumptions of Theorem B.4 there exists a unique solution to the integral equation (B.3). Let us find this solution.

Examples of analytical formulas for the solution to the integral equation (B.3) can be found in [Ramm (1990)]. The analytical formulas for the solution in the cases when the corresponding boundary-value problems are solvable analytically, can be obtained only for domains Ω of special shape, for example, when Ω is a ball, and for special operators Q and P, for example, for operators with constant coefficients.

We give such a formula for the solution of equation (B.3) assuming $P = I$ and $Q = -\Delta + a^2 I$. Consider the equation

$$R_\Omega h(x) = \int_\Omega \frac{\exp(-a|x-y|)}{4\pi|x-y|} h(y)dy = f(x), \quad x \in \overline{\Omega} \subset \mathbb{R}^3, \quad a > 0, \tag{B.27}$$

with the kernel $R(x,y) := \exp(-a|x-y|)/(4\pi|x-y|)$, $P = I$, and $Q = -\Delta + a^2 I$. By formula (2.24), one obtains a unique solution to the equation (B.27) in $H_0^{-1}(\Omega)$:

$$h(x) = (-\Delta + a^2)f + \left(\frac{\partial f}{\partial \mathbf{n}} - \frac{\partial u}{\partial \mathbf{n}}\right)\delta_{\partial\Omega} \tag{B.28}$$

where u is a unique solution to the exterior Dirichlet boundary-value problem

$$(-\Delta + a^2)u = 0 \quad \text{in} \quad \Omega_-, \quad u|_{\partial\Omega} = f|_{\partial\Omega}. \tag{B.29}$$

For any $\varphi \in C_0^\infty(\mathbb{R}^n)$ one has:

$$((-\Delta + a^2)Rh, \varphi) = (Rh, (-\Delta + a^2)\varphi)$$

$$= \int_\Omega f(-\Delta + a^2)\overline{\varphi}dx + \int_{\Omega_-} u(-\Delta + a^2)\overline{\varphi}dx$$

$$= \int_{\Omega} (-\Delta + a^2) f \overline{\varphi} dx + \int_{\Omega_-} (-\Delta + a^2) u \overline{\varphi} dx$$

$$- \int_{\partial\Omega} (f \partial_{\mathbf{n}} \overline{\varphi} - \partial_{\mathbf{n}} f \overline{\varphi}) \, ds + \int_{\partial\Omega} (u \partial_{\mathbf{n}} \overline{\varphi} - \partial_{\mathbf{n}} u \overline{\varphi}) \, ds$$

$$= \int_{\Omega} (-\Delta + a^2) f \overline{\varphi} dx + \int_{\partial\Omega} (\partial_{\mathbf{n}} f - \partial_{\mathbf{n}} u) \, ds,$$

where the condition $u = f$ on $\partial\Omega$ was used. Thus, we have checked that formula (B.28) gives the unique in $H_0^{-1}(\Omega)$ solution to equation (B.27). This solution has minimal order of singularity.

B.4 Auxiliary material

We denote by $\mathbb{R}$ the set of real numbers, by $\mathbb{C}$ the set of complex numbers. Let $\mathbb{Z} := \{0, \pm 1, \pm 2, ...\}$, $\mathbb{N} := \{0, 1, ...\}$, $\mathbb{N}_+ := \{1, 2, ...\}$, $\mathbb{R}^n := \{x = (x_1, ..., x_n) : x_i \in \mathbb{R},\ i = 1, ..., n\}$.

Let α be a multi-index, $\alpha := (\alpha_1, \ldots, \alpha_n)$, $\alpha_j \in \mathbb{N}$, $|\alpha| := \alpha_1 + \ldots + \alpha_n$, $i := \sqrt{-1}$; $D_j := i^{-1}\partial/\partial x_j$; $D^\alpha := D_1^{\alpha_1} D_2^{\alpha_2} \ldots D_n^{\alpha_n}$.

Let $C^\infty\left(\overline{\Omega}\right)$ be the space of infinitely differentiable up to the boundary functions in $\overline{\Omega}$. Near $\partial\Omega$ there is defined A normal vector field $\mathbf{n}(x) = (n_1(x), ..., n_n(x))$, is defined in a neighborhood of the boundary $\partial\Omega$ as follows: for $x_0 \in \partial\Omega$, $\mathbf{n}(x_0)$ is the unit normal to $\partial\Omega$, pointing into the exterior of Ω. We set

$$\mathbf{n}(x) := \mathbf{n}(x_0) \text{ for } x \text{ of the form } x = x_0 + s\mathbf{n}(x_0) =: \zeta(x_0, s)$$

where $x_0 \in \partial\Omega$, $s \in (-\delta, \delta)$. Here $\delta > 0$ is taken so small that the representation of x in terms of $x_0 \in \partial\Omega$ and $s \in (-\delta, \delta)$ is unique and smooth, that is, ζ is bijective and C^∞ with C^∞ inverse, from $\partial\Omega \times (-\delta, \delta)$ to the set $\zeta(\partial\Omega \times (-\delta, \delta)) \subset \mathbb{R}^n$.

We call differential operators *tangential* when, for $x \in \zeta(\partial\Omega \times (-\delta, \delta))$, they are either of the form

$$Af = \sum_{j=1}^{n} a_j(x) \frac{\partial f}{\partial x_j}(x) + a_0(x) f \quad \text{with} \quad \sum_{j=1}^{n} a_j(x) n_j(x) = 0,$$

or they are products of such operators. The derivative along $\mathbf{n}$ is denoted $\partial_{\mathbf{n}}$:

$$\partial_{\mathbf{n}} f := \sum_{j=1}^{n} n_j(x) \frac{\partial f}{\partial x_j}(x)$$

for $x \in \zeta(\partial\Omega \times (-\delta, \delta))$. Let $D_{\mathbf{n}} := i^{-1}\partial_{\mathbf{n}}$.

Let $\Omega_- := \mathbb{R}^n \setminus \overline{\Omega}$ denote the exterior of the domain Ω, $r_{\partial\Omega}$, r_Ω be respectively the restriction operators to $\partial\Omega$, Ω : $r_{\partial\Omega} f := f|_{\partial\Omega}$, $r_\Omega f := f|_\Omega$

Let $\mathcal{S}(\mathbb{R}^n)$ be the space of rapidly decreasing functions, that is the space of all $u \in C^\infty(\mathbb{R}^n)$ such that

$$\sup_{|\alpha| \le k} \sup_{x \in \mathbb{R}^n} \left| \left(1 + |x|^2\right)^m D^\alpha u(x) \right| < \infty \quad \text{for all } k, m \in \mathbb{N}.$$

Let $\mathcal{S}\left(\overline{\Omega}_-\right)$ be the space of restrictions of the elements $u \in \mathcal{S}(\mathbb{R}^n)$ to $\overline{\Omega}_-$ (this space is equipped with the factor topology).

Let $u \in C^\infty\left(\overline{\Omega}\right)$ and $v \in \mathcal{S}\left(\overline{\Omega}_-\right)$, then we set $\gamma_k u := r_{\partial\Omega} D_{\mathbf{n}}^k u = \left(D_{\mathbf{n}}^k u\right)|_{\partial\Omega}$, $\gamma_k v := r_{\partial\Omega} D_{\mathbf{n}}^k v = \left(D_{\mathbf{n}}^k v\right)|_{\partial\Omega}$.

Let $H^s(\mathbb{R}^n)$ $(s \in \mathbb{R})$ be the usual Sobolev space:

$$H^s(\mathbb{R}^n) := \left\{ f \in \mathcal{S}' \mid \mathcal{F}^{-1}\left(1 + |\xi|^2\right)^{s/2} \mathcal{F} f \in L_2(\mathbb{R}^n) \right\},$$

$$\|f\|_{H^s(\mathbb{R}^n)} := \left\| \mathcal{F}^{-1}\left(1 + |\xi|^2\right)^{s/2} \mathcal{F} f \right\|_{L_2(\mathbb{R}^n)},$$

where $\mathcal{F}$ denotes the Fourier transform $f \mapsto \mathcal{F}_{x \to \xi} f(x) = \int_{\mathbb{R}^n} e^{-ix\xi} f(x) dx$, $\mathcal{F}^{-1}$ its inverse and $\mathcal{S}' = \mathcal{S}'(\mathbb{R}^n)$ denote the space of tempered distributions which is dual to the space $\mathcal{S}(\mathbb{R}^n)$.

Let $H^s(\Omega)$ and $H^s(\Omega_-)$ $(0 \le s \in \mathbb{R})$ be respectively the spaces of restrictions of elements of $H^s(\mathbb{R}^n)$ to Ω and Ω_-. The norms in the spaces $H^s(\Omega)$ and $H^s(\Omega_-)$ are defined by the relations

$$\|f\|_{H^s(\Omega)} := \inf \|g\|_{H^s(\mathbb{R}^n)} \quad (s \ge 0),$$

$$\|f\|_{H^s(\Omega_-)} := \inf \|g\|_{H^s(\mathbb{R}^n)} \quad (s \ge 0),$$

where infimum is taken over all elements $g \in H^s(\mathbb{R}^n)$ which are equal to f in Ω respectively in Ω_-.

By $H_0^s(\Omega)$ $(s \in \mathbb{R})$ and $H_0^s(\Omega_-)$, we denote the closed subspaces of the space $H^s(\mathbb{R}^n)$ which consist of the elements with supports respectively in $\overline{\Omega}$ or in $\overline{\Omega}_-$, that is,

$$H_0^s(\Omega) := \left\{f \in H^s(\mathbb{R}^n) : \operatorname{supp} f \subseteq \overline{\Omega}\right\} \subset H^s(\mathbb{R}^n), \quad s \in \mathbb{R},$$

$$H_0^s(\Omega_-) := \left\{f \in H^s(\mathbb{R}^n) : \operatorname{supp} f \subseteq \overline{\Omega}_-\right\} \subset H^s(\mathbb{R}^n), \quad s \in \mathbb{R}.$$

We define the spaces

$$\mathcal{H}^s(\Omega) := \begin{cases} H^s(\Omega) & s > 0 \\ H_0^s(\Omega) & s \le 0, \end{cases}$$

$$\mathcal{H}^s(\Omega_-) := \begin{cases} H^s(\Omega_-) & s > 0, \\ H_0^s(\Omega_-) & s \le 0. \end{cases}$$

For $s \neq k + 1/2$ $(k = 0, 1, ..., \ell - 1)$, we define the spaces $\mathfrak{H}^{s,\ell}(\Omega)$ and $\mathfrak{H}^{s,\ell}(\Omega_-)$ respectively as the sets of all

$$(u, \underline{\phi}) = (u, \phi_1, ..., \phi_\ell) \text{ and } (v, \underline{\psi}) = (v, \psi_1, ..., \psi_\ell)$$

where $u \in \mathcal{H}^s(\Omega)$, $v \in \mathcal{H}^s(\Omega_-)$, $\underline{\phi} = (\phi_1, ..., \phi_\ell)$ and $\underline{\psi} = (\psi_1, ..., \psi_\ell)$ are vectors in $\prod_{j=1}^{\ell} H^{s-j+1/2}(\partial\Omega)$ satisfying the condition

$$\phi_j = D_{\mathbf{n}}^{j-1} u\big|_{\partial\Omega}, \quad \psi_j = D_{\mathbf{n}}^{j-1} v\big|_{\partial\Omega} \qquad \text{for } j < \min(s, \ell).$$

The norms in $\mathfrak{H}^{s,\ell}(\Omega)$ and $\mathfrak{H}^{s,\ell}(\Omega_-)$ can be defined as

$$\left\|(u, \underline{\phi})\right\|^2_{\mathfrak{H}^{s,\ell}(\Omega)} = \|u\|^2_{\mathcal{H}^s(\Omega)} + \sum_{j=1}^{\ell} \|\phi_j\|^2_{H^{s-j+1/2}(\partial\Omega)},$$

$$\left\|(v, \underline{\psi})\right\|^2_{\mathfrak{H}^{s,\ell}(\Omega)} = \|v\|^2_{\mathcal{H}^s(\Omega)} + \sum_{j=0}^{\ell} \|\psi_j\|^2_{H^{s-j+1/2}(\partial\Omega)}.$$

Since only the components ϕ_j and ψ_j with index $j < s$ can be chosen independently of u, we can identify $\mathfrak{H}^{s,\ell}(\Omega)$ and $\mathfrak{H}^{s,\ell}(\Omega_-)$ with the following spaces.

For $s \neq k + 1/2$ $(k = 0, 1, ..., \ell - 1)$,

$$\mathfrak{H}^{s,\ell}(\Omega) = \begin{cases} \mathcal{H}^s(\Omega), \ \ell = 0, \\ \mathcal{H}^s(\Omega), \ 1 \leq \ell < s + 1/2, \\ \mathcal{H}^s(\Omega) \times \prod\limits_{j=[s+1/2]+1}^{\ell} H^{s-j+1/2}(\partial\Omega), \ 0 < \left[s + \frac{1}{2}\right] < \ell, \\ \mathcal{H}^s(\Omega) \times \prod\limits_{j=1}^{\ell} H^{s-j+1/2}(\partial\Omega), \ s < \frac{1}{2} \end{cases}.$$

$$\mathfrak{H}^{s,\ell}(\Omega_-) = \begin{cases} \mathcal{H}^s(\Omega_-), \ \ell = 0, \\ \mathcal{H}^s(\Omega_-), \ 1 \leq \ell < s + 1/2, \\ \mathcal{H}^s(\Omega_-) \times \prod\limits_{j=[s+1/2]+1}^{\ell} H^{s-j+1/2}(\partial\Omega), \ 0 < \left[s + \frac{1}{2}\right] < \ell, \\ \mathcal{H}^s(\Omega_-) \times \prod\limits_{j=1}^{\ell} H^{s-j+1/2}(\partial\Omega), \ s < \frac{1}{2} \end{cases}$$

Finally, for $s = k + 1/2$ $(k = 0, 1, ..., \ell - 1)$, we define the spaces $\mathfrak{H}^{s,\ell}(\Omega)$, $\mathfrak{H}^{s,\ell}(\Omega_-)$ by the method of complex interpolation.

Let us note that for $s \neq k+1/2$ $(k = 0, 1, ..., \ell - 1)$, the spaces $\mathfrak{H}^{s,\ell}(\Omega)$, $\mathfrak{H}^{s,\ell}(\Omega_-)$ are completion of $C^\infty(\overline{\Omega})$, $\mathcal{S}(\overline{\Omega_-})$ respectively in the norms

$$\|(u, \gamma_0 u, ..., \gamma_{\ell-1} u)\|^2_{\mathfrak{H}^{s,\ell}(\Omega)} = \|u\|^2_{\mathcal{H}^s(\Omega)} + \sum_{j=0}^{\ell-1} \|\gamma_j u\|^2_{H^{s-j-1/2}(\partial\Omega)},$$

$$\|v, \gamma_0 v, ..., \gamma_{\ell-1} v\|^2_{\mathfrak{H}^{s,\ell}(\Omega_-)} = \|v\|^2_{\mathcal{H}^s(\Omega_-)} + \sum_{j=0}^{\ell-1} \|\gamma_j v\|^2_{H^{s-j-1/2}(\partial\Omega)}.$$

Bibliographical Notes

The estimation theory optimal by the criterion of minimum of the error variance has been created by N. Wiener (1942) for stationary random processes and for an infinite interval of the time of observation. A large bibliography one can find in [Kailath (1974)]. The theory has been essentially finished by mid-sixties for a finite interval of the time of observation and for stationary random processes with rational spectral density. Many attempts were made in engineering literature to construct a generalization of the Wiener theory for the case of random fields. The reason is that such a theory is needed in many applications, e.g., TV and optical signal processing, geophysics, underwater acoustics, radiophysics etc. The attempts to give an analytical estimation theory in engineering literature (see [Ekstrom (1982)] and references therein) were based on some type of scanning, and the problem has not been solved as an optimization problem for random fields.

The first analytical theory of random fields estimation and filtering, which is a generalization of Wiener's theory, has been developed in the series of papers [Ramm (1969); Ramm (1969b); Ramm (1969c); Ramm (1970b); Ramm (1970c); Ramm (1970d); Ramm (1971); Ramm (1971c); Ramm (1973c); Ramm (1975); Ramm (1976); Ramm (1978); Ramm (1978b); Ramm (1978c); Ramm (1978d); Ramm (1979); Ramm (1980b); Ramm (1980c); Ramm (1984b); Ramm (1985); Ramm (1987d); Ramm (2002); Ramm (2003); Kozhevnikov and Ramm (2005)] and in [Ramm (1980), Chapter 1]. This theory is presented in Chapters 2-4. Its applications are given in Chapter 7, and its generalizations to a wider class of random fields are given in the Appendices A and B. The material in Chapter 3 is based on the paper [Ramm (1985)]. The material in Section 7.7 is taken from [Ramm (1980)] where a reference to the paper Katznelson, J. and Gould, L., Construction of nonlinear filters and control systems, *Information and Control,*

5, (1962), 108-143, can be found together with a critical remark concerning this paper. In Section 7.2 the paper [Ramm (1973b)] is used; in Section 7.3.4 papers [Ramm (1968); Ramm (1973b); Ramm (1978); Ramm (1981); Ramm (1985b); Ramm (1984); Ramm (1987b); Ramm (1987c)] are used; the stable differentiation formulas (7.66, (7.67) and (7.71) were first given in [Ramm (1968)]; in Section 7.5 the papers [Ramm (1969b); Ramm (1970b); Ramm (1970c); Ramm (1970d)] are used. There is a large literature in which various aspects of the theory presented in Section 7.6 are discussed, see [Fedotov (1982)], [Ivanov *et. al.* (1978)], [Lattes and Lions (1967)], [Lavrentiev and Romanov (1986)], [Morozov (1984)], [Payne (1975)], [Tanana (1981)], [Tikhonov (1977)] and references therein. The presentation in Section 7.6 is self-contained and partly is based on [Ramm (1981)].

The class $\mathcal{R}$ of random fields has been introduced by the author in 1969 [Ramm (1969b); Ramm (1970b); Ramm (1970c); Ramm (1970d)]. It was found (see [Molchan (1975)], [Molchan (1974)]) that the Gaussian random fields have Markov property if and only if they are in the class $\mathcal{R}$ and $P(\lambda) = 1$ (see formula (1.10)).

Chapter 5 contains a singular perturbation theory for the class of integral equations basic in estimation theory. This chapter is based on [Ramm and Shifrin (2005)] (see also [Ramm and Shifrin (1991)], [Ramm and Shifrin (1993)], [Ramm and Shifrin (1995)]).

Random fields have been studied extensively [Adler (1981)], [Gelfand and Vilenkin (1968)], [Koroljuk (1978)], [Pitt (1971)], [Rosanov (1982)], [Vanmarcke (1983)], [Wong (1986)], [Yadrenko (1983)], but there is no intersection between the theory given in this book and the material presented in the literature. In the presentation of the material in Section 8.1.2 the author used the book [Berezanskij (1968)]. Theorem 8.1 in Section 8.1.1 is taken from [Mazja (1986), p. 60], and the method of obtaining the eigenfunction expansion theorem in Section 8.2 is taken from [Berezanskij (1968)]. There is a large literature on the material presented in section 8.2.4. Of course, it is not possible in this book to cover this material in depth (and it was not our goal). Only some facts, useful for a better understanding of the theory presented in this book, are given. For second-order elliptic operators $\mathcal{L}$ a number of stronger conditions sufficient for $\mathcal{L}$ to be selfadjoint or essentially selfadjoint are known (see [Kato (1981)]).

The assumption that $\mathcal{L}$ is selfadjoint is basic for the eigenfunction expansion theory developed in [Berezanskij (1968)]. In some cases an eigenfunction expansion theory sufficient for our purposes can be developed for certain non-selfadjoint operators (see [Ramm (1981b); Ramm (1981c); Ramm (1982); Ramm ([illegible] domain of observation, is finite. For the Schrödinger operator the spect[illegible] and scattering theory in some domains with infinite boundaries is developed in [Ramm (1963b); Ramm (1963); Ramm (1965); Ramm (1968b); Ramm (1969d); Ramm (1970); Ramm (1971b); Ramm (1987); Ramm (1988b)].

The material in Section 8.3.1 is well known. The proofs of the results about s-values can be found in [Gohberg and Krein (1969)].

The material in Section 8.3.2 belongs to the author [Ramm (1980); Ramm (1981b); Ramm (1981c)] and the proofs of all of the results are given in detail. The material in Section 8.3.3 is known and proofs of the results can be found in [Gohberg and Krein (1969)], [Konig (1986)], and [Pietsch (1987)].

In Section 8.4 some reference material in probability theory and statistics is given. One can find much more material in this area in [Koroljuk (1978)]. The purpose of the Chapter 6 is to explain some connection between estimation and scattering theory.

In the presentation of the scattering theory in Section 6.1 the papers [Ramm (1963b); Ramm (1963); Ramm (1965); Ramm (1968b); Ramm (1969d)] are used, where the scattering theory has been developed for the first time in some domains with infinite boundaries. Most of the results in Section 6.1 are well known, except of Theorems 6.1 and 6.2, which are taken from [Ramm (1987e); Ramm (1987f); Ramm (1988); Ramm (1988c); Ramm (1989)]. It is not possible here to give a bibliography on scattering theory. Povsner (1953-1955) and then Ikebe (1960) studied the scattering problem in R^3. Much work was done since then (see [Hörmander (1983-85)] vol. II, IV and references therein). A short and self-contained presentation of the scattering theory given in Section 6.1 may be useful for many readers who would like to get a quick access to basic results and do not worry about extra assumptions on the rate of decay of the potential.

Lemma 6.1 in Section 6.2 is well known, equation (6.13) is derived in [Newton (1982)], our presentation follows partly [Ramm (1987e); Ramm and Weaver (1987)] and Theorem 6.1 is taken from [Ramm (1987e)].

A connection between estimation and scattering theory for one-dimensional problems has been known for quite awhile. In [Levy and Tsitsiklis (1985)] and [Yagle (1988)] some multidimensional problems of estimation theory were discussed. In Section 6.3 some of the ideas from [Yagle (1988)] are used. Our arguments are given in more detail than in [Yagle (1988)].

The estimation problem discussed in [Levy and Tsitsiklis (1985)] and [Yagle (1988)] are the problems in which the noise has a white component and the covariance function has a special structure. In [Levy and Tsitsiklis (1985)] it is assumed that $r = 2$ and $R(x, y) = R(|x - y|)$, that is the random field is isotropic, and in [Yagle (1988)] $r \geq 2$ and $\Delta_x R(x, y) = \Delta_y R(x, y)$. The objective in these papers is to develop a generalization of the Levinson recursion scheme for estimation in one-dimensional case. The arguments in [Levy and Tsitsiklis (1985)] and [Yagle (1988)] are not applicable in the case when the noise is colored, that is, there is no white component in the noise. There was much work done on efficient inversion of Toeplitz's matrices [Friedlander *et. al.* (1979)]. These matrices arise when one discretize equation (3.4). However, as $\epsilon \to 0$, one cannot invert the corresponding Toeplitz matrix since its condition number grows quickly as $\epsilon \to 0$. If $\epsilon = 1$ in (3.4), then there are many efficient ways to solve equation (3.4). It would be of interest to compare the numerical efficiency of various methods. It may be that an iterative method, or a projection method will be more efficient than the discretization method with equidistant nodes used together with the efficient method of inverting the resulting Toeplitz matrix.

The results in Appendix A are taken from [Ramm (2003)] and the results in Appendix B are taken from [Kozhevnikov and Ramm (2005)].

The author has tried to make the material in this book accessible to a large audience. The material from the theory of elliptic pseudodifferential equations (see, e.g., [Hörmander (1983-85)]) was not used. The class $\mathcal{R}$ of kernels is a subset in the set of pseudodifferential operators. The results obtained in this book concerning equations in the class $\mathcal{R}$ are final in the sense that exact description of the range of the operators with kernels in class $\mathcal{R}$ is given and analytical formulas for the solutions of these equations are obtained. The general theory of pseudodifferential operators does not provide analytical formulas for the solutions. It was possible to derive such formulas in this book because of the special structure of the kernels in the class $\mathcal{R}$. In [Eskin (1981), §27] an asymptotic solution is obtained for a class of pseudo-differential equations with a small parameter.

Bibliography

Adler, R. (1981). *The geometry of random fields*, J. Wiley, New York.

Agmon, S. (1982). *Lectures on exponential decay of solutions of second-order elliptic equations*, Princeton Univ. Press, Princeton.

Akhieser, N. (1965). *Lectures on approximation theory*, Nauka, Moscow.

Aronszajn, N. (1950). Theory of reproducing kernels, *Trans. Am. Math. Soc.* 68, pp. 337-404.

Berezanskij, Yu (1968). *Expansions in eigenfunctions of selfadjoint operators*, Amer. Math. Soc., Providence RI.

Deimling, K. (1985). *Nonlinear functional analysis*, Springer Verlag, New York.

Ekstrom, M. (1982). Realizable Wiener filtering in two dimensions, *IEEE Trans. on acoustics, speech and signal processing,* 30, pp. 31-40;

Ekstrom, M and Woods, J. (1976). Two-dimensional spectral factorization with applications in recursive digital filtering, ibid, 2, pp. 115-128.

Erkip, A. and Schrohe, E, (1992). Normal solvability of elliptic boundary-value problems on asymptotically flat manifolds, *J. of Functional Analysis* 109, pp. 22–51.

Eskin, G. (1981). *Boundary value problems for elliptic pseudodifferential equations*, Amer. Math. Soc., Providence, RI.

Fedotov, A. (1982). *Linear ill-posed problems with random errors in the data*, Nauka, Novosibirsk.

Friedlander, B., Morf, Mi, Kailath, T., Ljung, L. (1979). New inversion formulas for matrices classified in terms of their distance from Toeplitz matrices, Linear algebra and its applications, 27, pp. 31-60.

Gakhov, F. (1966). boundary-value problems, Pergamon Press, Oxford.

Glazman, I. (1965). *Direct methods of qualitative spectral analysis of singular differential operators*, Davey, New York.

Gohberg, I. and Krein, M. (1969). *Introduction to the theory of linear nonselfadjoint operators*, AMS, Providence.

Gilbarg, D. and Trudinger, N. (1977). *Elliptic partial differential equations of second order*, Springer Verlag, New York.

Gelfand, I. and Vilenkin, N. (1968). *Generalized functions*, vol. 4., Acad. Press, New York.

Grubb, G. (1990). Pseudo-differential problems in L_p spaces, *Commun. in Partial Differ. Eq.*,15 (3), pp. 289–340.

Grubb, G. (1996). *Functional calculus of pseudodifferential boundary problems*, Birkhäuser, Boston.

Hörmander, L. (1983-85). *The analysis of linear partial differential operators*, vol. I-IV, Springer Verlag, New York.

Ivanov, V., Vasin, V. and Tanana, V (1978). *Theory of linear ill-posed problems and applications*, Moscow, Nauka.

Kato, T. (1995). *Perturbation theory for linear operators*, Springer Verlag, New York.

Kato, T. (1981). *Spectral theory of differential operators*, North Holland, Amsterdam. (Ed. Knowles, I. and Lewis, R.) pp. 253-266.

Kato, T. (1959). Growth properties of solutions of the reduced wave equation,*Comm. Pure Appl. Math*, 12, pp. 403-425.

Kailath, T. (1974). A view of three decades of linear filtering theory, *IEEE Trans. on inform. theory*, IT-20, pp. 145-181.

Kantorovich, L. and Akilov, G. (1980). *Functional analysis*, Pergamon Press, New York.

Klibanov, M. (1985). On uniqueness of the determination of a compactly supported function from the modulus of its Fourier transform, *Dokl. Acad. Sci. USSR*, 32, pp. 668-670.

Konig, H. (1986). *Eigenvalue distribution of compact operators*, Birkhauser, Stuttgart.

Koroljuk, V. ed. (1978). *Reference book in probability and statistics*, Naukova Dumka, Kiev.

Kozhevnikov, A. (2001). Complete scale of isomorphisms for elliptic pseudodifferential boundary-value problems, *J. London Math. Soc.* (2) 64, pp. 409–422.

Kozhevnikov, A. and Ramm, A.G. (2005). Integral operators basic in random fields estimation theory, *Intern. J. Pure and Appl. Math.*, 20, N3, 405-427.

Kozlov, V. Maz'ya, V. and Rossmann, J. (1997). *Elliptic boundary-value problems in domains with point singularities*, AMS, Providence 1997.

Krasnoselskii, M., *et. al.* (1972). *Approximate solution of operator equations*, Walters - Noordhoff, Groningen.

Lattes R. and Lions J. (1967). *Méthode de quasi-reversibilite et applications*, Dunod, Paris.

Lavrentiev, M. and Romanov, V. and Shishatskii, S. (1986). *Ill-posed problems of mathematical physics and analysis*, Amer. Math. Soc., Providence, RI.

Levitan, B. (1971). Asymptotic behavior of spectral function of elliptic equation, *Russ. Math. Survey*, 6, pp. 151-212.

Levy, B. and Tsitsiklis, J. (1985). A fast argorithm for linear estimation of two-dimensional random fields, *IEEE Trans. Inform. Theory*, IT-31, pp. 635-644.

Mazja, V. (1986). *Sobolev spaces*, Springer Verlag, New York.

Molchan, G. (1975). Characterization of Gaussian fields with Markov property, *Sov. Math. Doklady*, 12, pp. 563-567.

Molchan, G. (1974). *L*-Markov Gaussian fields, ibid. 15, pp. 657-662.

Morozov, V. (1984). *Methods for solving incorrectly posed problems*, Springer Verlag, New York.

Naimark, M. (1969). *Linear differential operators*, Nauka, Moscow.

Newton, R. (1982). *Scattering of waves and particles*, Springer Verlag, New York.

Payne, L.E. (1975). *Improperly posed problems*, Part. Dif. Equ., Regional Conf. Appl. Math., Vol. 22, SIAM, Philadelphia.

Pietsch, A. (1987). *Eigenvalues and s-numbers*, Cambridge Univ. Press, Cambridge.

Piterbarg, L. (1981). Investigation of a class of integral equations, *Diff. Uravnenija*, 17, pp. 2278-2279.

Pitt, L. (1971). A Markov property for Gaussian processes with a multidimensional parameter, *Arch. Rat. Mech. Anal.*, 43, pp. 367-391.

Preston, C. (1967). *Random fields*, Lect. notes in math. N34, Springer Verlag, New York.

Ramm, A.G. (1963). Spectral properties of the Schrödinger operator in some domains with infinite boundaries. *Doklady Acad. Sci. USSR*, 152, pp. 282-285.

Ramm, A.G. (1963b). Investigation of the scattering problem in some domains with infinite boundaries I, II, Vestnik 7, pp. 45-66; 19, pp. 67-76.

Ramm, A.G. (1965). Spectral properties of the Schrödinger operator in some infinite domains, Mat. Sbor. 66, pp. 321-343.

Ramm, A.G. (1968). On numerical differentiation. Math., Izvestija vuzov, 11, 1968, 131-135. 40 # 5130.

Ramm, A.G. (1968b). Some theorems on analytic continuation of the Schrödinger operator resolvent kernel in the spectral parameter. *Izv. Ac. Nauk. Arm.* SSR, Mathematics, 3, pp. 443-464.

Ramm, A.G. (1969). Filtering of nonstationary random fields in optical systems. *Opt. and Spectroscopy*, 26, pp. 808-812;

Ramm, A.G. (1969b). Apodization theory. *Optics and Spectroscopy*, 27, (1969), pp. 508-514.

Ramm, A.G. (1969c). Filtering of nonhomogeneous random fields. Ibid. 27, pp. 881-887.

Ramm, A.G. (1969d). Green's function study for differential equation of the second order in domains with infinite boundaries. Diff. eq. 5, pp. 1509-1516.

Ramm, A.G. (1970). Eigenfunction expansion for nonselfadjoint Schrödinger operator. Doklady, 191, pp. 50-53.

Ramm, A.G. (1970b). Apodization theory II. *Opt. and Spectroscopy*, 29, pp. 390-394.

Ramm, A.G. (1970c). Increasing of the resolution ability of the optical instruments by means of apodization. Ibid. 29, pp. 594-599.

Ramm, A.G. (1970d). On resolution ability of optical systems. Ibid., 29, pp. 794-798.

Ramm, A.G. (1971). Filtering and extrapolation of some nonstationary random processes. *Radiotech. i Electron.* 16, pp. 80-87.

Ramm, A.G. (1971b). Eigenfunction expansions for exterior boundary problems. Ibid. 7, pp. 737-742.

Ramm, A.G. (1971c). On multidimensional integral equations with the translation kernel. *Diff. eq.* 7, pp. 2234-2239.

Ramm, A.G. (1972). Simplified optimal differentiators. *Radiotech. i Electron.* 17, pp. 1325-1328.

Ramm, A.G. (1973). On some class of integral equations. Ibid., 9, pp. 931-941.

Ramm, A.G. (1973b). Optimal harmonic synthesis of generalized Fourier series and integrals with randomly perturbed coefficients. Radio technika, 28, pp. 44-49.

Ramm, A.G. (1973c). Discrimination of random fields in noises. *Probl. peredaci informacii*, 9, pp. 22-35. 48 # 13439.

Ramm, A.G. (1975). Approximate solution of some integral equations of the first kind. *Diff. eq.* 11, pp. 582-586. 440-443.

Ramm, A.G. (1976). Investigation of a class of integral equations. *Doklady Acad. Sci. USSR*, 230, pp. 283-286.

Ramm, A.G. (1978). A new class of nonstationary processes and fields and its applications. Proc. 10 all-union sympos. "Methods of representation and analysis of random processes and fields" Leningrad, 3, pp. 40-43.

Ramm, A.G. (1978b). On eigenvalues of some integral equations. *Diff. Equations*, 15, pp. 932-934.

Ramm, A.G. (1978c). Investigation of a class of systems of integral equations. *Proc. Intern. Congr. on appl. math.*, Weimar, DDR, (1978), 345-351.

Ramm, A.G. (1978d). Investigation of some classes of integral equations and their application. In collection "Abel inversion and its generalizations", edited by N. Preobrazhensky, Siberian Dep. of Acad. Sci. USSR, Novosibirsk, pp. 120-179.

Ramm, A.G. (1979). Linear filtering of some vectorial nonstationary random processes, *Math. Nachrichten*, 91, pp. 269-280.

Ramm, A.G. (1980). *Theory and applications of some new classes of integral equations*, Springer Verlag, New York.

Ramm, A.G. (1980b). Investigation of a class of systems of integral equations, *Journ. Math. Anal. Appl.*, 76, pp. 303-308.

Ramm, A.G. (1980c). Analytical results in random fields filtering theory, *Zeitschr. Angew. Math. Mech.*, 60, T 361-T 363.

Ramm, A.G. (1981). Stable solutions of some ill-posed problems. *Math. Meth. in Appl. Sci.* 3, pp. 336-363.

Ramm, A.G. (1981b). Spectral properties of some nonselfadjoint operators, *Bull, Am. Math. Soc.*, 5, N3, pp. 313-315.

Ramm, A.G. (1981c). Spectral properties of some nonselfadjoint operators and some applications in "Spectral theory of differential operators", Math. Studies, North Holland, Amsterdam, ed. I. Knowles and R. Lewis, pp. 349-354.

Ramm, A.G. (1982). Perturbations preserving asymptotics of spectrum with a remainder. *Proc. A.M.S.* 85, N2, pp. 209-212;

Ramm, A.G. (1983). Eigenfunction expansions for some nonselfadjoint operators and the transport equation, *J. Math. Anal. Appl.* 92, pp. 564-580.

Ramm, A.G. (1984). Estimates of the derivatives of random functions. *J. Math. Anal. Appl.* 102, pp. 244-250.

Ramm, A.G. (1984b). Analytic theory of random fields estimation and filtering. Proc. of the intern sympos. on Mathematics in systems theory (Beer Sheva, 1983), Lecture notes in control and inform. sci. N58, Springer Verlag, pp. 764-773.

Ramm, A.G. (1985). Numerical solution of integral equations in a space of distributions. *J. Math. Anal. Appl.* 110, pp. 384-390.

Ramm, A.G. (1985b). Estimates of the derivatives of random functions II. (with T. Miller). J. Math. Anal. Appl. 110, pp. 429-435;

Ramm, A.G. (1986). *Scattering by obstacles*, Reidel, Dordrecht.

Ramm, A.G. (1987). Sufficient conditions for zero not to be an eigenvalue of the Schrödinger operator, *J. Math Phys.*, 28, pp. 1341-1343.

Ramm, A.G. (1987b). Optimal estimation from limited noisy data, *Journ. Math. Anal. Appl.*, 125, pp. 258-266.

Ramm, A.G. (1987c). Signal estimation from incomplete data, *Journ. Math. Anal. Appl.*, 125, pp. 267-271.

Ramm, A.G. (1987d). Analytic and numerical results in random fields estimation theory, *Math. Reports of the Acad. of Sci.*, Canada, 9, pp. 69-74.

Ramm, A.G. (1987e). Characterization of the scattering data in multidimensional inverse scattering problem, in the book: *Inverse Problems: An Interdisciplinary Study.* Acad. Press, New York, pp. 153-167. (Ed. P. Sabatier).

Ramm, A.G. (1987f). Completeness of the products of solutions to PDE and uniqueness theorems in inverse scattering inverse problems, 3, L77-L82.

Ramm, A.G. (1988). Multidimensional inverse problems and completeness of the products of solutions to PDE. *J. Math. Anal. Appl.* 134, 1, pp. 211-253.

Ramm, A.G. (1988b). Conditions for zero not to be an eigenvalue of the Schrödinger operator, *J. Math. Phys.* 29, pp. 1431-1432.

Ramm, A.G. (1988c). Recovery of potential from the fixed energy scattering data. *Inverse Problems*, 4, pp. 877-886;

Ramm, A.G. (1989). Multidimensional inverse scattering problems and completeness of the products of solutions to homogeneous PDE. *Zeitschr. f. angew. Math. u. Mech.*, T305, N4-5, T13-T22.

Ramm, A.G. (1990). *Random fields estimation theory,* Longman Scientific and Wiley, New York, pp. 1-273.

Ramm, A.G. (1990b). Stability of the numerical method for solving the 3D inverse scattering problem with fixed energy data. *Inverse Problems*, 6, L7-L12; J. reine angew. math., 414, (1991), pp. 1-21.

Ramm, A.G. (1990b). Is the Born approximation good for solving the inverse problem when the potential is small? *J. Math. Anal. Appl.* 147, pp. 480-485.

Ramm, A.G. (1991). Symmetry properties of scattering amplitudes and applications to inverse problems. *J. Math. Anal. Appl.* 156, pp. 333-340.

Ramm, A.G. (1992). *Multidimensional inverse scattering problems*, Longman, New York, (Russian edition, Mir, Moscow 1993).

Ramm, A.G. (1996). *Random fields estimation theory*, MIR, Moscow, pp. 1-352.

Ramm, A.G. (2002). Estimation of Random Fields, *Theory of Probability and Math. Statistics*, 66, pp. 95-108.

Ramm, A.G. (2003). Analytical solution of a new class of integral equations. *Differential and Integral Equations,* 16 N2 p. 231-240.

Ramm, A.G. (2003a). On a new notion of regularizer, *J. Phys. A*, 36, p. 2191-2195.

Ramm, A.G. (2004). One dimensional inverse scattering and spectral problems, *Cubo Math. Journ.*, 6, N1, p. 313-426.

Ramm, A.G. (2005). *Inverse problems*, Springer, New York.

Ramm, A.G. and Shifrin, E.I. (1991). Asymptotics of the solution to a singularly perturbed integral equation, Appl. Math. Lett. **4**, pp. 67 - 70.

Ramm, A.G. and Shifrin, E.I. (1993). Asymptotics of the solutions to singularly perturbed integral equations, *Journal of Mathematical Analysis and Applications*, **178**, No 2, pp. 322 - 343.

Ramm, A.G. and Shifrin, E.I. (1995). Asymptotics of the solutions to singularly perturbed multidimensional integral equations, *Journal of Mathematical Analysis and Applications*, **190**, No 3, pp. 667 - 677.

Ramm, A.G and Shifrin, E.I. (2005). Singular pertubation theory for a class of Fredholm integral equations arising in random fields estimation theory, *Journal of integral equations and operator theory.*

Ramm, A.G.and Weaver, O. (1987). A characterization of the scattering data in $3D$ inverse scattering problem. Inverse problems, 3, L49-52.

Ramm, A.G. and Weaver, O. (1989). Necessary and sufficient condition on the fixed energy data for the potential to be spherically symmetric. *Inverse Problems*, 5 pp. 445-447.

Roitberg, Ya. A. (1996). *Elliptic boundary-value problems in the spaces of distributions*, Kluwer, Dordrecht.

Rosanov, Yu (1982). *Markov random fields*, Springer Verlag, New York.

Rudin, W. (1973). *Functional Analysis*, McGraw Hill, New York.

Safarov, Yu. and Vassiliev, D. (1997). *The asymptotic distribution of eigenvalues of partial differential operators*, American Mathematical Society, Providence, RI, 1997.

Saito, Y. (1982). Some properties of the scattering amplitude and the inverse scattering problem, *Osaka J. Math.*, 19, pp. 527-547.

Schrohe, E. (1987). Spaces of weighted symbols and weighted sobolev spaces on manifolds, In: Pseudo-Differential Operators, Cordes, H.O., Gramsch, B., and Widom, H. (eds.) Springer LN Math. 1256, pp. 360-377, Springer-Verlag, Berlin.

Schrohe, E. (1999). Fréchet algebra techniques for boundary value problems on noncompact manifolds, *Math. Nachr.* 199, pp. 145–185.

Shubin, M. (1986). *Pseudodifferential operators and spectral theory*, Springer Verlag, New York.

Skriganov, M. (1978). High-frequency asymptotics of the scattering amplitude, *Sov. Physics, Doklady*, 241, pp. 326-329.

Somersalo, E. *et. al.* (1988). Inverse scattering problem for the Schrödinger equation in three dimensions, IMA preprint 449, pp. 1-7.

Tanana, V. (1981). *Methods for solving operator equations*, Nauka, Moscow.

Tikhonov, A. and Arsenin, V. (1977). *Solutions of ill-posed problems*, Winston, Washington.

Tulovskii, V. (1979). Asymptotic distribution of eigenvalues of differential equations, *Matem. Sborn.*, 89, pp. 191-206.

Vanmarcke, E. (1983) *Random fields: analysis and synthesis*, MIT Press, Cambridge.

Vishik, M.I. and Lusternik, L. (1962). Regular degeneration and boundary layer for linear differential equations with a small parameter, *Amer. Math. Soc. Transl.* **20** pp. 239 - 264.

Wloka, J.T. (1987). *Partial differential equations*, Cambridge University Press.

Wloka, J.T., Rowley, B. and Lawruk, B. (1995). *Boundary value problems for elliptic systems*, Cambridge University Press.

Wong, E. (1986). In search of multiparameter Markov processes, in the book: *Communications and Networks*, ed. I.F. Blake and H.V. Poor, Springer Verlag, New York, pp. 230-243.

Yadrenko, M. (1983). *Spectral theory of random fields*, Optimization Software, New York.

Yagle, A. (1988). Connections between $3D$ inverse scattering and linear least-squares estimation of random fields, *Acta Appl Math.* 13, N3, pp. 267-289.

Yagle, A. (1988). Generalized split Levinson, Schur, and lattice algorithms for $3D$ random fields estimation problem (preprint).

Zabreiko, P., *et. al.* (1968). *Integral equations, Reference Text,* Nauka, Moscow.

Symbols

Index